Treatise on Materials Science and Technology

VOLUME 6

Plastic Deformation of Materials

TREATISE ON MATERIALS SCIENCE AND TECHNOLOGY

VOLUME 6

WITHDRAWN

PLASTIC DEFORMATION OF MATERIALS

EDITED BY

R. J. ARSENAULT

Engineering Materials Group, and
Department of Chemical Engineering
University of Maryland
College Park, Maryland

1975

ACADEMIC PRESS   New York   San Francisco   London

A Subsidiary of Harcourt Brace Jovanovich, Publishers

ACADEMIC PRESS, INC.
111 Fifth Avenue, New York, New York 10003

United Kingdom Edition published by
ACADEMIC PRESS, INC. (LONDON) LTD.
24/28 Oval Road, London NW1

LIBRARY OF CONGRESS CATALOG CARD NUMBER: 77-182672

ISBN 0–12–341806–2

PRINTED IN THE UNITED STATES OF AMERICA

Contents

Low Temperature of Deformation of bcc Metals and Their Solid-Solution Alloys

R. J. Arsenault

Cyclic Deformation of Metals and Alloys

Campbell Laird

High-Temperature Creep

Amiya K. Mukherjee

Review Topics in Superplasticity

Thomas H. Alden

Fatigue Deformation of Polymers

P. Beardmore and S. Rabinowitz

Low Temperature Deformation of Crystalline Nonmetals

R. G. Wolfson

Recovery and Recrystallization during High Temperature Deformation

H. J. McQueen and J. J. Jonas

List of Contributors

Numbers in parentheses indicate the pages on which the authors' contributions begin.

THOMAS H. ALDEN (225), Department of Metallurgy, University of British Columbia, Vancouver, British Columbia, Canada

R. J. ARSENAULT (1), Engineering Materials Group, and Department of Chemical Engineering, University of Maryland, College Park, Maryland

P. BEARDMORE (267), Scientific Research Staff, Ford Motor Company, Dearborn, Michigan

J. J. JONAS (393), Department of Mining and Metallurgical Engineering, McGill University, Montreal, Canada

CAMPBELL LAIRD (101), School of Metallurgy and Materials Science, University of Pennsylvania, Philadelphia, Pennsylvania

H. J. McQUEEN (393), Department of Mechanical Engineering, Concordia University, Montreal, Canada

AMIYA A. MUKHERJEE (163), Department of Mechanical Engineering, University of California, Davis, California

S. RABINOWITZ (267), Scientific Research Staff, Ford Motor Company, Dearborn, Michigan

R. G. WOLFSON (333), Thayer School of Engineering, Dartmouth College, Hanover, New Hampshire

Foreword

The study of the phenomenon of plastic deformation is a very fertile and large branch of materials science. Since ancient times, when the first copper implements were forged, there have been numerous investigations of the plastic deformation characteristics of macroscopic solids. Mechanical properties, i.e., strength, ductility, toughness, are very important characteristics of materials. In addition, plastic deformation is a central means by which millions of tons of materials are shaped into products.

Plastic deformation is clearly a very complex, irreversible process and, therefore, its understanding has evolved slowly. In the late nineteenth century it was first noted that plastic deformation occurred in slip or shear bands. Following this, it was observed in the 1910's that plastic deformation took place on rational crystallographic planes. The modern era of research on plastic deformation began in 1934, with the introduction of the dislocation to account for the low observed levels of stress to achieve permanent deformation. This important development was followed by proposals of mechanisms of dislocation motion and models by which dislocations interact with obstacles. Thereby evolved the relationship between theories of dislocation dynamics and the form of the observed stress–strain curve for a large variety of crystalline materials.

The chapters in this volume review the great profusion of ideas on plastic deformation which have been put forward over the years. The overall goal here has been to analyze the important developments and to place them in a modern perspective. Plastic deformation of metals, crystalline non-metals and polymers are covered in the seven chapters of this volume. Attesting to the continuous activity in this area of research are the more than 1000 references contained herein, which must in truth be considered to be a modest number.

We have attempted to present a global view of plastic deformation in solids, and in such a venture certain areas and references will be unavoidably left out. For this the Editor apologizes.

It is hoped that this volume will contribute to the understanding of this very active and diverse area of research and will enable the reader to more readily focus on the pressing current and future problems in plastic deformation—both in the research laboratory and in industrial practice.

R. J. ARSENAULT

Preface

Materials limitations are often the major deterrents to the achievement of new technological advances. In modern engineering systems, materials scientists and engineers must continually strive to develop materials which can withstand extreme conditions of environment and maintain their required properties. In the last decade we have seen the emergence of new types of materials, literally designed and processed with a specific use in mind. Many of these materials and the advanced techniques which were developed to produce them, came directly or indirectly from basic scientific research.

Clearly, the relationship between utility and fundamental materials science no longer needs justification. This is exemplified in such areas as composite materials, high-strength alloys, electronic materials, and advanced fabricating and processing techniques. It is this association between the science and technology of materials on which we intend to focus in this treatise.

The topics to be covered in *Treatise on Materials Science and Technology* will include the fundamental properties and characterization of materials, ranging from simple solids to complex heterophase systems. The *Treatise* is aimed at the professional scientist and engineer, as well as at graduate students in materials science and associated fields.

The Editor would like to express his sincere appreciation to the members of the Editorial Advisory Board who have given so generously of their time and advice.

H. HERMAN

Contents of Previous Volumes

Treatise on Materials Science and Technology

VOLUME 6

Plastic Deformation of Materials

Low Temperature of Deformation of bcc Metals and Their Solid-Solution Alloys†

R. J. ARSENAULT

Engineering Materials Group, and
Department of Chemical Engineering
University of Maryland
College Park, Maryland

I. Introduction

During the past decade the number of investigations undertaken for the purpose of understanding the nature of plastic deformation in bcc metals and their alloys exceeds the total number of such investigations initiated prior to 1960. Most of the earlier investigators were more interested in studying fcc and hcp metals and their alloys primarily because they were a lot easier to prepare and to test.

† This work was supported by the United States Atomic Energy Commission under contract No. AT(40–1)–3612.

1

One of the several reasons why there has been an increased interest in bcc metals and their alloys is that the apparatus necessary for the production of high purity single crystals has been perfected. Also, deformation characteristics of bcc metals are in many instances quite different from those of fcc and hcp metals and their alloys. For example: (1) the resolved shear stress for yielding exhibits asymmetry, i.e. Schmid's law is not obeyed; (2) the presence of a small amount of interstitial impurities may or may not affect the temperature dependence of the yield stress; (3) only at very low interstitial concentrations are three stages of work hardening observed; (4) reduction (versus an increase, as generally observed in fcc metals and alloys) of the yield stress can occur as a result of a substitutional or interstitial alloy addition, and also as a result of neutron and electron irradiations.

Christain (1970) has prepared an excellent review of the low deformation characteristics of single crystal bcc metals and their alloys; therefore only the most recent work will be covered on the topics reviewed by him. The main additional areas to be covered are superconductivity and the effect of irradiation damage. Conrad et al. (1961; Conrad, 1963) have also written extensive reviews of properties of polycrystalline bcc metals. Some of the areas which will not be covered are twinning, fracture, and grain size effects.

II. Asymmetry of Slip and Yielding

A unique slip direction, but no unique slip plane, has been observed in bcc and their alloys. Transmission electron microscopy (TEM) investigations of channel planes in neutron-irradiated molybdenum and vanadium have indicated that the dislocations move on the {110} and {211} planes (Huang and Arsenault, 1973b, 1972) as shown in Fig. 1. The observed microscopic slip plane, however, is in general an irrational plane depending on: crystal orientation, mode of testing, temperature, and impurity content. The orientation dependence can be conveniently expressed in terms of ψ–χ curves, where χ is the angle between the $(\bar{1}01)$ plane and the maximum resolved shear stress plane (MRSS), and ψ is the angle between the (101) plane and the observed slip plane (Fig. 2). Rogausch and Mordike (1970) have investigated tantalum and tantalum alloys. In their investigations they avoided sample orientations near the edges of the standard stereographic triangle. Therefore, they did compression tests for $\psi < -8°$ which were inverted at $\psi = \chi = 0°$. This is justified if normal stresses do not have to be considered, because slip in compression for $\chi > 0°$ is crystallographically equivalent to slip in tension for $\chi < 0°$. The measured ψ–χ curves are discussed as tensile ψ–χ curves; but

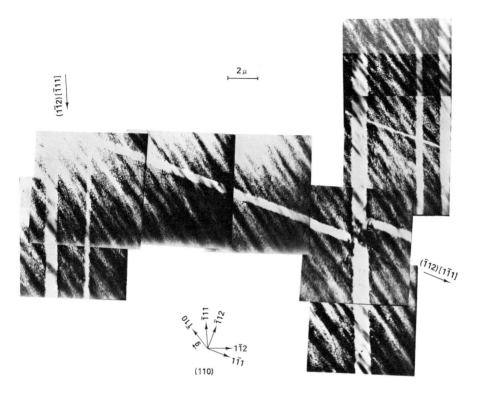

Fig. 1. Transmission electron micrograph of neutron-irradiated and deformed single crystals of molybdenum.

it should be kept in mind that the left branch, i.e. for $\chi < -8°$, is derived from compression tests.

The results obtained by Rogausch and Mordike (1970) are plotted in Fig.3, and can be summarized by the following generalizations: (1) The assumption regarding inverting compression and tensile results is largely confirmed by the smooth continuation of both branches in the overlap region, $-5° < \chi < 5°$. This is even true for pure tantalum in compression at 77° K and $\chi = 5°$, the only case in which the $[\bar{1}11]$ slip direction was observed. The only obvious failure is observed in pure tantalum at 295° K. Whenever there is a tendency to deviate, it is such that the compression branch tries to avoid the MRSS plane; whereas the tension branch favors it. This might indicate that normal stresses have an influence on the asymmetry of slip. (2) A definite influence of temperature on the ψ–χ curve is observed only for pure tantalum. The crystallographic $(\bar{1}01)$ slip, observed in tantalum

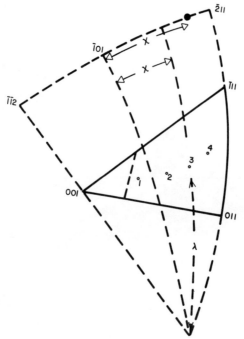

Fig. 2. Definition of ψ and χ.

at 295° K and $0° < \chi < 27°$, changes to noncrystallographic slip at lower temperatures. For $\chi < 0°$ the effect of temperature is reversed, i.e. $(\overline{1}\overline{1}2)$ slip at 77° K deviates towards MRSS slip at higher temperatures. (3) Alloying tends to promote slip on the MRSS plane, but to a much lesser extent for $\chi > 0°$ than for $\chi < 0°$. No crystallographic slip is observed. Rhenium has a slightly stronger influence on the ψ–χ curve than molybdenum. If there was no asymmetry of the slip plane, ψ would be zero.

Slip lines in neutron-irradiated molybdenum were investigated by Hasson *et al.* (1973), and slip lines in neutron-irradiated vanadium was investigated by Huang and Arsenault (1973b). The reason for investigating neutron-irradiated molybdenum and vanadium was based on the assumption that the slip lines would be straight, course, and well defined due to the process of channeling (Arsenault, 1971b). Figure 4 shows scanning electron micrographs of a neutron-irradiated single crystal of vanadium deformed at 300° K in compression. The micrographs are from the central portions of the samples. For the vanadium sample, the micrograph is taken from the side where there would be zero slip lines if only dislocations of the primary Burgers moved, i.e. the side parallel to the primary Burgers vector. As is evident in Fig. 4 the slip lines are not straight lines and the slip lines within

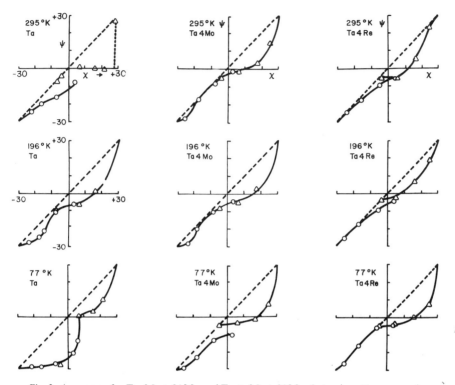

Fig. 3. ψ–χ curves for Ta–3.8 at. % Mo and Ta + 3.8 at. % Mo. $\triangle$, tension; $\bigcirc$, compression. Rogausch and Mordike (1970).

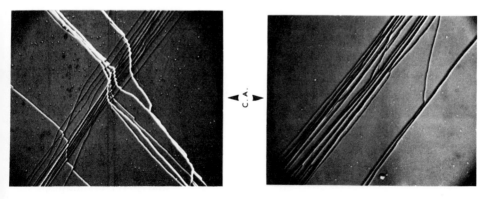

Fig. 4. Scanning electron micrograph of slip lines on a neutron-irradiated and deformed vanadium single crystal. × 375.

the slip bands are not straight. There is a significant amount of cross-slip within a slip band, and also cross-slip from one slip band to another (Fig. 4). The slip bands are even less well defined in the case of molybdenum. A two-surface slip line analysis was performed on molybdenum compression samples of the [$\bar{4}$91], [011], and [001] orientations. If the most predominant slip bands were used, and the slip within the slip bands was ignored, then in all cases the slip plane was a ($\bar{1}$01) type and the slip direction was [$\bar{1}$11]. This agrees with the results of Guiu (1969) for molybdenum.

The reason for the above boundary conditions is that even at small strains there are extraneous slip lines. It is assumed that other investigators have used a similar boundary condition.

Anomalous (0$\bar{1}$1) slip has been observed in very pure niobium (Foxall *et al.*, 1967). The reason for defining it as anomalous slip is that there are six other slip systems which have a larger resolved stress acting on them. Bolton and Taylor (1972) have continued an investigation of this anomalous slip and have shown that this system contributes a significant portion of the strain. Reed and Arsenault (1973) have also observed anomalous slip in high purity niobium from 45° K to 77° K. However, prestraining appears to eliminate anomalous slip. G. Taylor (private communication, 1973) has also observed this anomalous slip in high purity vanadium. As stated by Bolton and Taylor (1972), the origin of anomalous slip is a mystery. In concluding a discussion of slip geometry a quote by Rogausch and Mordike (1970) might be appropriate. "In conclusion it can be said that the geometry of slip is closely related to the mechanism of slip in bcc metals. On the other hand, the morphology of slip is the result of the dynamics of yielding and can give only limited information about the basic deformation mechanism."

A mechanism for the orientation dependence of the yield or flow stress is not well defined. The temperature variation of the effective stress and the proportional limit for the [001], [$\bar{4}$91], and [011] unirradiated molybdenum samples tested in compression are shown in Fig. 5 (Hasson *et al.*, 1973). The proportional limit curves show very little orientation dependence in contrast to the results of Stein (1967, 1968) and Sherwood *et al.* (1967). They found from their proportional limit results that the stress for the [011] orientation is a factor of 3 higher than for the [001] orientation. The temperature variation of the effective yield stress for the present study, however, does show an orientation dependence for temperatures from 77 to 400° K. The present results show the [001] and [$\bar{4}$91] to be harder than the [011] orientation. The differences in the effective stress for the [001] and [011] orientations, however, are not as great as the other investigators (Stein, 1967; Sherwood *et al.*, 1967) have reported. Also, the previous investigators found the [011] orientation to be stronger than the [001], whereas the present results indicate the opposite. The rapid decrease in the effective stress in the temperature range from 300 to 340° K shows an almost monotonic decrease and a shift of

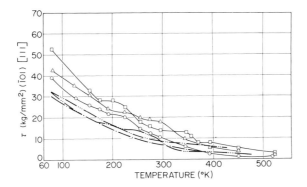

Fig. 5. Effective stress (open symbols) and proportional limit (dotted lines) versus temperature for unirradiated molybdenum single crystals of [100], ○, — · —, [491], □, — ·· —, and [110], △, — ··· —, orientations.

the inflection point to lower temperatures as the orientation changes from [001] to [$\overline{4}$91] and then to [011]. The results in Fig. 5 suggest that the proportional limit should *not* be used to determine the orientation dependence of the yield stress.

Lachenmann and Schultz (1972) have conducted an interesting investigation of the effect of interstitial nitrogen on the asymmetry of the yield stress of tantalum. Figure 6 is a plot of tensile yield stress versus ψ for the high purity samples (total interstitials < 4 at. ppm) in the region of temperature dependent yield stress. A pronounced asymmetry exists, characterized by three inequalities: (1) $\sigma(\chi)$ tension $\neq \sigma(\chi)$ compression, (2) $\sigma(\chi) \neq \sigma(-\chi)$, and (3) $\sigma(\chi)$ tension $\neq \sigma(-\chi)$ compression.

The inequality (3) appears less pronounced than (1) and (2). The yield stress is highest in tension for $\chi = +30°$, corresponding to ($\overline{2}$11) as the MRSS plane, and in compression for $\chi = -30°$, i.e. when ($\overline{1}$12) is the MRSS plane. This result is in accordance with observations on other bcc metals which indicate that for a glide on [112] there exists a "hard" and "soft" direction depending on whether the stress is applied to the twinning direction or not.

The asymmetry of the yield stress disappears at $T = 354°$ K. Above this temperature, the yield stress is *independent* of temperature.

An increase in the interstitial concentration (250 wt ppm) considerably reduces the asymmetry. The slip plane can be approximated as the MRSS. However, these samples do not follow Schmid's law. If the Schmid yield stress law were obeyed, then the yield stress would be that shown as the dashed line with open circles in Fig. 6.

The main points of Lachenmann and Schultz's investigation can be summarized as follows:

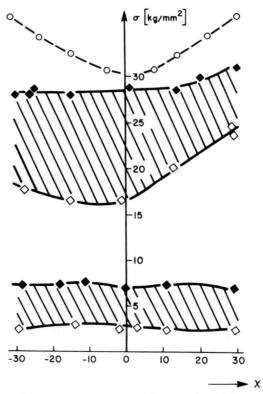

Fig. 6. Orientation dependence of interstitial solution hardening for deformation in tension (Lachenmann and Schultz, 1972). $\bigcirc$, Schmid's Law prediction; $\blacklozenge$, 250 at. ppm N; $\diamondsuit$, <4 at. ppm C + N + O; $\varepsilon = 1 \times 10^{-4}$, tension. Upper shaded area: $T = 197°$ K; lower shaded area: $T = 354°$ K.

1. High purity tantalum crystals show an asymmetry of yield stress and slip geometry which disappears at the same temperature as the thermal component of the yield stress.

2. The asymmetry of the yield stress is continuously *reduced* by adding impurity interstitials, and slip planes tend toward the MRSS planes.

3. The hardening due to interstitial impurities in the low temperature region is orientation dependent. The increase of the thermal component of the yield stress is most pronounced for orientations near $\chi = 0$. For samples with orientations near $\chi = +30$, the additional hardening is nearly athermal.

These results strongly indicate that the asymmetry of the yield stress is *caused by the properties of the pure tantalum lattice.* They also support the idea that the same intrinsic dislocation property is responsible for the temperature dependent yield stress and for the asymmetry effects.

III. Dislocation Configurations

The dislocation configuration most generally observed in bcc metals deformed at large effective stresses is straight screw dislocation (Keh, 1965; Foxal *et al.*, 1967; Arsenault and Lawley, 1967; Huang and Arsenault, 1973a). This observation, i.e. straight screw dislocation, has been used as evidence that the double-kink mechanism is the rate-controlling mechanism of dislocation motion at low temperatures in bcc metals (Arsenault, 1966a; Conrad, 1963). If interstitial atoms were the srb, then edge dislocations would be present (Fleischer, 1964; Gibbs, 1969; Ravi and Gibala, 1970b). Figure 7 is representative TEM of screw dislocation configuration.

Huang and Arsenault (1973a) have conducted an investigation of the effect of oxygen on the dislocation configuration in vanadium. An advantage

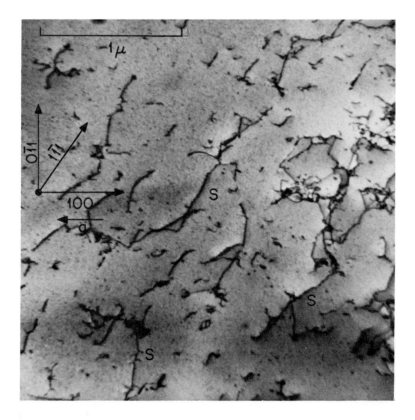

Fig. 7. Electron transmission micrograph form a tantalum single crystal deformed in tension $\gamma = 0.15\%$ strain at 273° K. Plane of foil slip plane defined as (011).

of this system is that the solubility limit of oxygen in vanadium is high, i.e. ≈ 2 wt %. In this investigation the level of strain was small, i.e. 2–4% shear strain. The reason for this small amount of strain was due to the fact that at higher oxygen concentrations, i.e. greater than 150 wt ppm, the density of dislocations became very large for larger strains, and it became difficult to resolve and analyze individual dislocations. The lower test temperature was chosen in order to avoid twinning which occurred at temperatures below 173° K in the high oxygen content samples. In general, the dislocation density was greater in samples tested in compression than in samples tested in tension, independent of the oxygen concentration. The probable reason for this is that the stress on the specimen is not as uniaxial as in the case of the tension test.

The dipoles and dipole clusters observed in samples deformed in both compression and tension at 300° K are very similar to those observed in other bcc metals (Fig. 8). The individual dipoles vary in length and width,

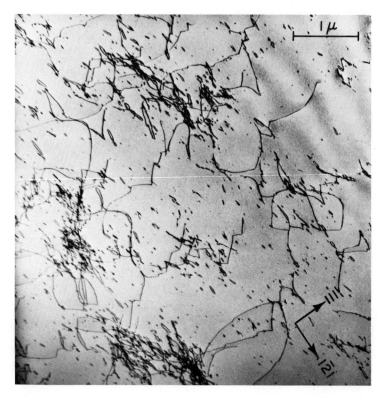

Fig. 8. Electron transmission micrograph from a low oxygen vanadium single crystal deformed at 300° K.

but are very comparable (approximately $1/2\ \mu$ in length to 400 Å in width) to those observed in tantalum. There are some secondary dislocations. The dislocation structure resulting from low temperature deformation is very similar to that observed in other bcc metals deformed at low temperatures, i.e. a predominance of straight screw dislocations. However, if the deformation was conducted in the compression mode at low temperatures, then there was evidence of secondary slip and tangling. In Figs. 9a and b, it is evident

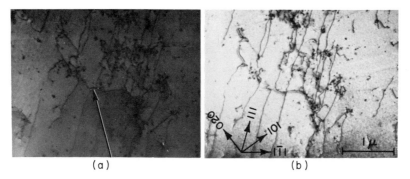

(a)   (b)

Fig. 9. Dislocation structure in high purity vanadium which contains less than 40 ppm O_2 and deformed 4% in compression at 123° K. Orientation is [$\bar{1}$01]. (a) **g** = (020); (b) **g** = (101).

that there is a reaction between the secondary and primary dislocations as follows:

$$a/2[111] - a/2[\bar{1}11] = a[010]$$

The arrow in Fig. 9 is pointed at the [010] dislocation. The length of the dislocation product, i.e. the length of the [010] type dislocation, was, in general, very short. The segment in Fig. 9 is by far one of the largest ones. The length of the [$\bar{0}$10] type dislocation is shorter than that observed in tantalum (Arsenault and Lawley, 1967) and in iron (Ohr and Beshers, 1963).

The extinction condition of $g \cdot b = 0$ is only valid for an isotopic elastic material, but vanadium is slightly anisotropic. Computer calculations by J. Hern and A. Head (private communication, 1972) indicate that for the conditions employed to determine the [010] Burgers vector, the intensity for $g \cdot b = 0$ should give a residual contrast of $\sim 15\%$ over background. A careful examination of Fig. 9b indicates that there is slight residual contrast. The computer calculations for $g \cdot b = 2$ gives a single dark band, not a double line, which is again consistent with the image in Fig. 9.

An increase in the oxygen content results in an increase in the degree of tangling and the density of dipoles in samples deformed both in tension and compression at 300° K. The dislocation structure observed at 4% strain was

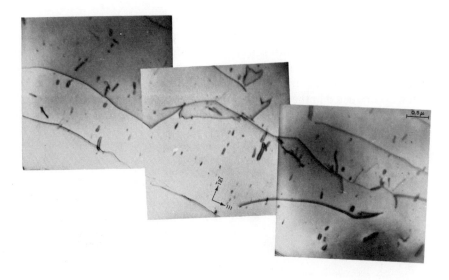

Fig. 10. Dislocation structure in vanadium which contains 300 ppm O_2 and deformed 4% in tension at 123° K. Orientation is [$\bar{1}01$].

comparable to the dislocation structure observed at 10% strain for the lower oxygen content samples. However, a difference in the dislocation structures was seen between samples tested at the lower temperatures. Dipole loops were evident, as shown in Fig. 10. The density of secondary dislocations greatly increased due to the increase in oxygen concentration, and dislocations representing almost every slip system are present. In Fig. 11 there are dislocations at approximately right angles to each other. It is possible to mistakenly identify one set as edge dislocations and others as screw dislocations. However, from a $g \cdot b$ analysis it was shown that one set of dislocations consisted of primary screws and the other of secondary screws.

In a few micrographs, parts or segments of a dislocation line could be identified as edge. However, in no micrographs could an entire length of dislocation line be identified as an edge dislocation. Therefore, the ratio of edge to screw dislocations is almost zero, since there are only a few segments of dislocation line which are edge in character.

An increase in the oxygen content has a significant effect on dislocation structure in samples deformed at 300° K in both tension and compression. The dislocation structures, when observed in a foil parallel to the primary slip plane ($\bar{1}01$), are a tangled array of dislocations; and it is almost impossible to observe individual dislocations, as shown in Fig. 12. Well-defined bands (slip bands) are observed in foils cut perpendicular to the primary slip

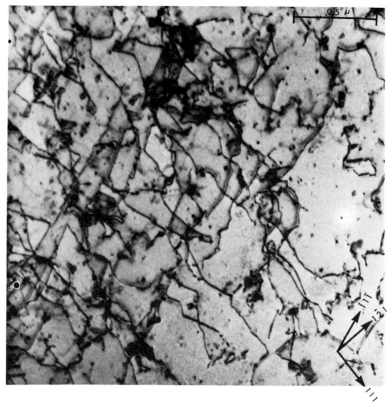

Fig. 11. Dislocation structure in vanadium which contains 300 ppm O_2 and deformed 10% in compression at 123° K. Orientation is [$\bar{1}01$].

plane (Fig. 13). In the case of low-temperature deformation, the dislocation density increased in comparison with the intermediate and low oxygen content samples deformed at an equivalent strain. Again, there was a predominance of screw dislocations with a few edge dislocation segments. Also, the density of dipole loops had increased.

The most significant observation is the predominance of screw dislocations in samples deformed at low temperatures independent of oxygen content. If the oxygen atoms *were* the srb to dislocation motion, then there should be a higher density of edge dislocations then screw dislocations; but this is not the case. There may be screw dislocations due to the presence of super logs.

There are differences in the dislocation structure due to a change in the oxygen concentration. An increase in the oxygen concentration results in a

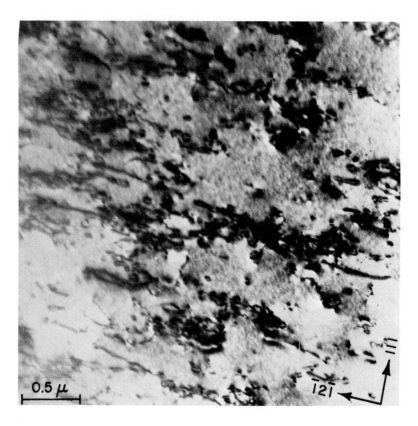

Fig. 12. Dislocation structure in vanadium which contains 1000 ppm O_2 and deformed 4% in tension at 300° K. Orientation is [$\bar{1}$01], i.e. the slip plane.

marked increase in the density of dipole loops; and there is some evidence of segments of edge dislocations in samples deformed at low temperatures. The increase in dipole density, with an increase in oxygen concentration, can be explained in terms of a cluster or a short-range order (SRO) arrangement of interstitial atoms. A cluster has to be considered because a single oxygen atom will not produce a torque on a screw dislocation (Arsenault and de Wit, 1974). If a cluster forms, then there is a dilatation strain associated with that cluster. A torque can then develop on the screw dislocation causing it to cross-slip (Weertman and Weertman, 1964). This torque is due to the following shear stress on the cross:

$$\tau = -6\mu\varepsilon xz r_0^3/\gamma^5 \tag{1}$$

where μ is the shear modulus, r_0 the radius of the defect, $r^2 = x^2 + y^2 + z^2$,

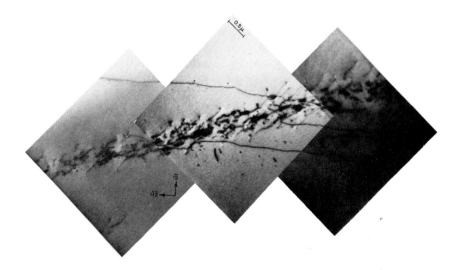

Fig. 13. Dislocation structure in vanadium which contains 1000 ppm O_2 and deformed 4% in compression at 300° K. Orientation is [011], i.e. perpendicular to the slip plane.

and ε is the dilatational strain of the cluster. This is an additional stress on the cross-slip plane aiding in the cross-slip of the screw dislocation; and as a result of the double cross-slip, we end up with a super jog which then leads to a dipole. In order to calculate the additional stress due to the cluster it is necessary to estimate ε. If it is assumed that there are at least three oxygen atoms in a cluster, then the following energy condition must be satisfied for the cluster to exist:

$$E_{3O_2} < \sum_1^j E_{1O_2} \qquad (2)$$

where E_{3O_2} is the strain energy of the cluster and E_{1O_2} is the strain energy due to a single oxygen atom. The strain energy of a single oxygen atom is related to the tetragonal strain ($\Delta\varepsilon$) which is 0.4 (Hasson and Arsenault, 1971) for oxygen in vanadium. A conservative value for ε for the cluster would be 0.7. The value of stress on the cross-slip plane is a maximum near the cluster, but at $20b$ ($\sim$ length of a double kink, Arsenault, 1967a), the stress is 11.2 kg/mm^2 for $\varepsilon = 0.7$. The density of the cluster and the number of oxygen atoms in the cluster increases with oxygen concentration. Therefore, there would be an increase in the dipole density as the oxygen concentration increases.

Increasing the oxygen content resulted in an increase in the density of primary and secondary dislocations. In the low oxygen content material deformed at low temperatures, secondary dislocations are almost nonexistent. This increase in the presence of secondary dislocations is probably due to two causes: (1) the grown-in dislocations become severely locked as the oxygen concentration increases. These grown-in dislocations then act as stress concentrations so that other slip systems can operate; they are also the nuclei of tangles (Arsenault and Lawley, 1967). (2) The interstitial clusters can also produce results similar to those obtained from the locked grown-in dislocations.

There is strong evidence that the interaction between screw dislocations results in [010] type dislocations.

The structure at 300° K is also dependent upon oxygen concentration. The structure in the low oxygen concentration samples is similar to that observed in stage I of other bcc metals deformed at small effective stresses. An increase in the oxygen concentration results in an increase in tangling which is similar to that observed in other bcc alloys. However, for the higher oxygen concentration, extreme tangling takes place and the deformation is confined to well-defined slip bands. The reason for tangling could be due to the clusters, but the reason for the well-defined slip bands is not obvious.

Arsenault and Lawley (1967) have determined that the changes in dislocation configurations are a function of work-hardening stages and substitutional alloy additions. In order to better understand this function, it is necessary to discuss work hardening in bcc metals, secondary slip, and the "overshot" phenomenon. It is also necessary to convert the load time curves into shear stress–shear strain curves.

In converting the load–time curves into shear stress–shear strain curves, the assumption was made that $[\bar{1}11]$ was the operative slip direction. This assumption was justified by observations of the rotations of the tensile axis during deformation, and by the nature of the optically observed slip traces. The slip plane was chosen to give the highest Schmid factor. The operative primary slip plane was approximately (001), as determined from slip traces and from the dislocation structure in foils cut perpendicular to $[1\bar{1}1]$. The curves of samples of Ta and Ta–Nb with tensile axis orientations inside the stereographic triangle can be divided into various states, as shown in Fig. 14. The designation of the stages is that proposed by Mitchell *et al.* (1963). Separation of the stress–strain curves into specific stages is only possible at temperatures in the range 296–501° K. As the test temperature is decreased, the extent of stage II decreases; whereas, stage 0 increases, reaches a maximum at approximately 15% shear strain at 243° K, and then decreases. By further reducing the test temperature ($\lesssim 77°$ K), the stress–strain curves exhibit a negative slope of stress against strain. Crystals of Ta and Ta–9.2 wt % Nb with a tensile axis near the (001)–(101) symmetry line have a small stage I, and as a result the stress–strain curves are nearly parabolic.

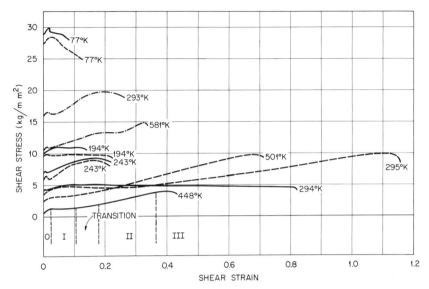

Fig. 14. Representative shear stress–shear strain curves for Ta, ——, Ta–9 at. % Nb, - - -, and Ta–9 at. % W, — · — · —, crystals as a function of temperature.

This general description of the stress–strain curves does not apply to the Ta–9.1 wt % W crystals. In this case, the stress–strain curves are parabolic in shape at high temperatures ($\gtrsim 293°$ K), but exhibit a negative slope of stress against strain at low temperatures ($\lesssim 194°$ K). There is an anomaly in that at high temperatures there is what appears to be a small stage II. Surprisingly, this increase in the rate of work hardening occurs after the sample has begun to neck down.

The separation of the stress–strain curves of tantalum into various stages was found to be a sensitive function of the interstitial concentration. Increasing the total interstitial concentration by a small amount (from 30 to 50 wt ppm) resulted in a parabolic stress–strain curve. Since a parabolic curve was obtained for the Ta–W alloy, an alloy containing a total interstitial concentration of about half this amount was prepared, and again, the stress–strain curve had a parabolic shape. Lachenmann and Schultz (1973) have shown very convincingly that interstitials affect θ_{II}, as is shown in Fig. 15. The correlation between secondary slip and stage II can be studied in another manner. In investigating the overshoot phenomenon in bcc metals, it becomes immediately obvious that the degree of overshoot is much greater in bcc metals than in fcc metals or alloys. Figure 16 shows the tensile axis rotation of a niobium single crystal (Arsenault, 1966b). Secondary slip lines are visible and the transition region begins at the strain designated as point A in Fig. 16. The large degree of overshoot observed in bcc metals had led to

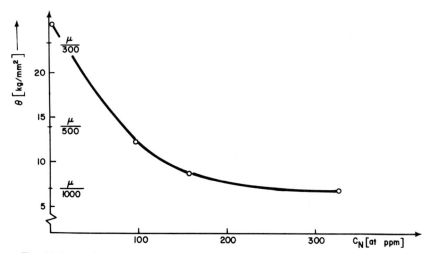

Fig. 15. Rate of work hardening in stage II as a function of nitrogen concentration. $T = 473°$ K; $\dot{\varepsilon} = 5.5 \times 10^{-5}$ sec^{-1}; $\mu = 6979$ kg/mm^2.

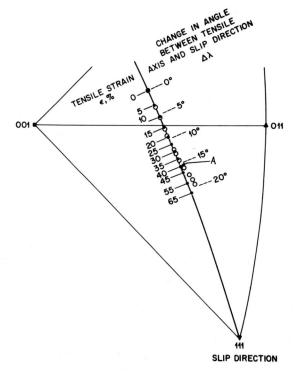

Fig. 16. Tensile axis rotation of a Nb single crystal deformed at 300° K. ○, Measured direction of tensile axis, ●, calculated direction of tensile axis for single slip on ($\bar{1}$01)[111].

the application of dislocation dynamics as an explanation of the overshoot phenomenon (Arsenault, 1966b). It can be justifiably assumed that the mobile dislocation density on the primary slip plane exceeds that on the secondary slip plane. For the strain rate to be the same in the secondary system and the primary system, the stress on the secondary slip plane must be greater than that on the primary slip plane. This difference in stress is directly proportional to the degree of overshoot and can be expressed as follows:

$$\tau_s^* - \tau_p^* = \Delta\tau^* = [kT/v^*(\tau^*)][\ln (N_p/N_s)] \tag{3}$$

where N_p and N_s are the mobile dislocation densities on the primary and secondary slip planes, respectively; τ_s^* and τ_p^* are the shear stresses on the secondary and primary slip planes, respectively; and $v^*(\tau^*)1^*$ is the activation volume. The temperature dependence of the degree of overshoot is related to parameter $T/v (\tau^*)$ which goes to zero as τ^* goes to zero. Thus the data (Fig. 16) indicates that there is only a limited amount of secondary slip.

The separation of the stress–strain curve into various stages is strongly dependent on the test temperature. Of primary interest is the temperature dependence of the work hardening rate θ_{II} in stage II. It is well known that in fcc metals and solid solutions the rate of work hardening in stage II is only slightly temperature dependent. Figure 17 shows the temperature dependence of θ_{II} for Ta and Ta–Nb. Below a given temperature (approximately 300° K for Ta, approximately 275° K for Ta–9 at. % Nb), stage II does not exist; as the temperature increases, θ_{II} increases and saturates at higher temperatures. The temperature at which the rate of work hardening θ_{II} reaches a maximum is lower for Ta–Nb than for pure Ta. The maximum value of θ_{II} is less in Ta–Nb than Ta.

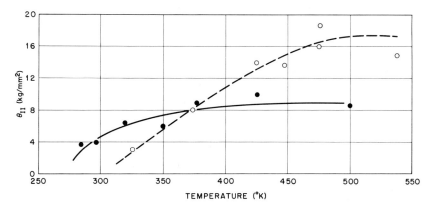

Fig. 17. Rate of work hardening in stage II versus temperature; the samples were tested in tension. ○, Ta; ●, Ta–9 at. % Nb.

In order to determine the stability of the obstacle or barrier present in stage II, a series of differential (stress) temperature tests was conducted. A sample was first deformed into stage II, unloaded except for a small aligning load, then reloaded at a lower temperature. The results of these tests are plotted in Fig. 18. The curve designated 541° K defines the initial test temperature. The extent of the solid line shows the amount to which the sample was initially strained: the continued dashed line is the curve that would have been produced if the sample had been strained to fracture at

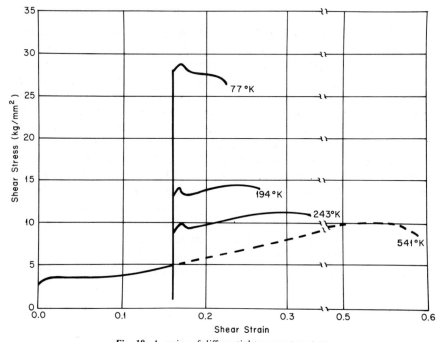

Fig. 18. A series of differential temperature tests.

541° K. The curve designated 243° K is that produced by first prestraining at 541° K and then reloading at 243° K. If a nonprestrained sample were deformed at 243° K, the slope of the stress–strain curve would be approximately zero. As a result of prestraining, the slope after the yield drop is similar to that which resulted when the prestrain test was stopped. The extent of this work-hardening region at 243° K is about 10% strain; if the reloading temperature is 194° K, the extent of the work-hardening region is only about 5% strain. Again, if a nonprestrained sample were deformed at 194° K, the slope of the stress–strain curve would be approximately zero. If the retesting temperature is decreased to 77° K, there is no work-hardening region similar to a nonprestrained sample. However, the strain to fracture of a prestrained sample is greater than that of a nonprestrained sample.

It is apparent from these results that the barriers introduced in stage II are still present at macroscopic yielding in the low-temperature reloaded samples; the yield stress has increased by an amount equal to the increase in stress due to work hardening during the prestraining. These barriers continue to increase in effectiveness when the reloading temperature is 195° K and 243° K, but the total extent of the work-hardening region (i.e. stage II) is less than that which would have been obtained if the sample had been strained to fracture at 541° K. At 77° K there is no work-hardening region. The question arises as to why the barriers do not continue to increase in effectiveness once they have formed. This question is related to why stage II ends and stage III begins. Explanations based on cross-slip are unlikely since this occurs even at low strains in bcc metals.

A possible reason why the barriers stop increasing in effectiveness could be that the secondary slip sources are stopped as a result of latent hardening from the primary slip plane.

Now, it is time to consider the dislocation configurations. At the beginning of stage I, patches of dipolex ("multipole" clusters) develop; all the dipoles are found to have the $[1\bar{1}1]$ primary Burgers vector (Fig. 19). In

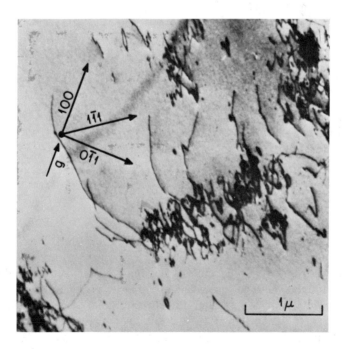

Fig. 19. Structure of deformed Ta–Nb. $\gamma = 0.09$ at 373° K, dark field with $g = 200$. Electron beam along [011].

general, these dipole clusters are separated by distances $< 10 \mu$ on the primary slip plane. Diffraction contrast experiments using the [211], [200], [011], and [211] reflections respectively show that secondary dislocations [11$\bar{1}$], [111], [$\bar{1}$11] Burgers vectors comprise less than 5% of the total dislocation density. More of the [111] and [$\bar{1}$11] vectors are present than [11$\bar{1}$]; the former correspond to the Burgers vector of the conjugate and critical slip systems, respectively. Individual dipoles within the clusters are elongated along [21$\bar{1}$] with loop lengths $\leq 1/2 \mu$, and loop widths up to 400 Å. Overall dimensions of the dipole patches vary; typically, patches extend 2μ in the [1$\bar{1}$1] direction, but have a variable overall length in the [21$\bar{1}$] direction. Dislocation densities within the dipole clusters are 2.8×10^9 cm cm^{-3}.

The onset of stage II is characterized by secondary slip: this slip results in the development of dislocation networks in the regions *between* the dipole clusters, Figs. 20a and b. Diffraction contrast experiments reveal extensive

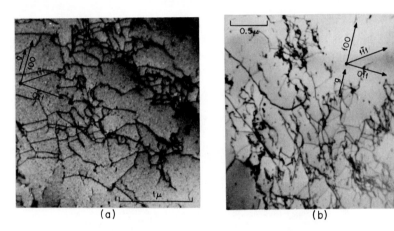

(a)   (b)

Fig. 20. (a) Structure of deformed tantalum on the (011) plane. $\gamma = 0.15$ at 455° K, bright field. Operative reflection g is 200; (b) structure of deformed Ta–Nb. $\gamma = 0.28$ at 373° K, bright field with $g = 200$. Electron beam along [011].

interactions between primary and secondary dislocations. Forest dislocations having a Burgers vector of [111] or [$\bar{1}$11] now constitute $> 10\%$ of the total density: numerous dislocations having the [11$\bar{1}$] vector in the (011) plane are present. A large percentage of the dislocations in the networks lie close to the [100] direction. Reactions between primary and secondary dislocations can give rise to the $\langle 100 \rangle$ dislocations lying in the $\langle 100 \rangle$ direction. In this study, the Burgers vector of the dislocations lying along [100] was not unambiguously determined. At the end of stage I, the dislocation density is fairly uniform and is in the range $(2.2-2.8) \times 10^9$ cm cm^{-3}.

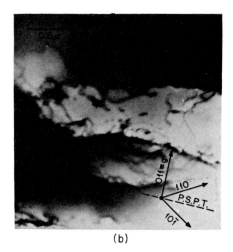

(a)

(b)

Fig. 21. Structure of deformed tantalum. $\gamma = 0.44$ at 463° K. (a) Electron beam along [011] with $g = 200$ in dark field. (b) Electron beam along [111] with $g = 011$ in dark field.

At the end of stage II, extensive dislocation activity had taken place over all of the primary slip plane, Fig. 21a. Networks (N) and intense dislocation tangles (T) are visible. The alignment of the tangles is much more irregular, and the overall substructure approaches a cellular structure. Dislocation densities are 9.3×10^9 cm cm^{-3}.

In stage II, dislocation walls are observed in foils cut perpendicular to $[1\bar{1}1]$ (Fig. 21b); the walls run along the direction of the primary slip plane trace. The dark–light contrast across the walls shows that lattice misorientation has occurred. In this orientation, the electron beam is parallel to the direction of the primary slip vector, i.e. $[1\bar{1}1]$. As a consequence, for all the possible reflecting planes, the plane normal g (in reciprocal space) is perpendicular to b (primary), and $g \cdot b = 0$. Under these conditions, the primary dislocations will be extinct except for an anomalous contrast effect at the top and bottom foil surfaces. Thus, the dislocations visible in Fig. 21b have vectors other than $\pm a/2$ $[1\bar{1}1]$. Optical examination of the crystal surface reveals secondary slip (on the $(0\bar{1}1)$ [111] system) at this level of strain. Slip line observations (optical) at various levels of strain in stage I and stage II establish that secondary slip is a necessary condition for the existence of stage II hardening.

These substructures in stage I and stage II are very similar to those reported by Hirsch (1963) for fcc copper. The exact mechanism of dipole formation in the tantalum and tantalum base alloys is not known. The observations on foils cut perpendicular to the primary vector $[1\bar{1}1]$ (stage I and the beginning of stage II) show that the dipole clusters form in bands along the projection of the line of intersection of the primary slip plane with the conjugate or critical secondary planes. [In this orientation, the direction of the projected line of intersection in ($1\bar{1}1$) is the same for the conjugate and critical planes.] This suggests that nucleation of the dipoles occurs at forest intersections.

If the testing temperature is reduced or an alloy addition is made so that stage II hardening does not exist, the dislocation configuration is generally a relatively uniform distribution of screw dislocations (Fig. 7) or a combination, i.e. the screws and tangles. The dislocation configuration in Fig. 7 is in marked contrast to Fig. 20: both are of samples deformed the same amount.

In addition to the primary screw dislocations (Fig. 22), operation of the secondary slip vectors $\pm a/2$ [$11\bar{1}$], $\pm a/2$ [$\bar{1}11$] has been established: secondary slip constitutes 30% of the total dislocation density (2.4×10^9 cm cm^{-3}) at this level of strain ($\delta = 0.15$). However, optical examination of the

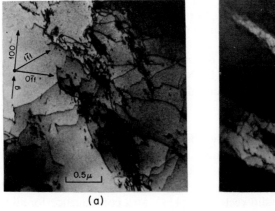
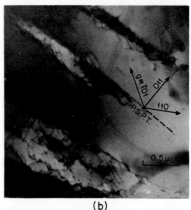

(a)   (b)

Fig. 22. (a) Structure of deformed Ta–Nb. $\gamma = 0.66$ at 273° K. Electron beam along [011] with $g = 200$ in bright field. (b) Structure of deformed Ta–Nb. $\gamma = 0.66$ at 273° K. Electron beam along [111] with $g = \bar{1}01$ in dark field.

slip traces, and a knowledge of the direction of rotation of the tensile axis, indicated that gross deformation on the secondary system ($0\bar{1}1$) [111] had not taken place. In some cases the amount of overshoot may be as much as 10° (Arsenault, 1966b) before there is any indication of secondary slip. It must then be concluded that either the dislocations produced on the secondary system move only short distances, or that the positive and negative dislocation strains from secondary slip are canceled.

High levels of strain at temperatures which do not lead to stage II hardening produce a structure which is a combination of that observed at the lower temperatures (relatively straight primary screw dislocations) and that produced in stage I at the higher temperatures (dipole clusters), Fig. 22a. Dislocation densities are now 5.2×10^9 cm cm^{-3} with the density of non-primary dislocations approaching that of the primary dislocations. Again, optical examination of the slip traces indicated that there was *no* extensive secondary slip. The dislocation structure observed in foils prepared perpendicular to [1$\bar{1}$1] again shows the subboundaries (Fig. 22b).

The dislocation structure in Ta–9 at. % W crystals, deformed ($\delta = 0.14$) at 581° K, is illustrated in Figs. 23a and b. Consistent with a parabolic form

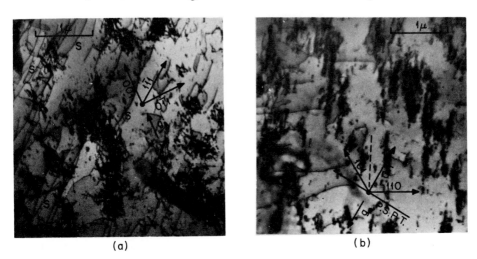

(a)   (b)

Fig. 23. (a) Structure of deformed Ta–W. $\gamma = 0.14$ at 581° K. Electron beam along [011] with $g = 200$ in bright field; (b) Structure of deformed Ta–W. $\gamma = 0.14$ at 581° K. Electron beam along [111] with $g = [0\bar{1}\bar{1}]$ in bright field.

of stress–strain curve, the structure on the primary slip plane (Fig. 23a) is characterized by long screw dislocations (S) and regions of intense dislocation tangling (T). The latter are associated with grown-in immobile dislocations. In contrast to Ta and Ta–Nb, subboundaries are not observed along the direction of the primary slip plane trace (Fig. 23b).

Foxall and Statham (1970) investigated the dislocation configurations in Nb–Mo and Nb–Re alloys. The observations were quite similar to those made by Arsenault and Lawley (1967) on Ta–W, and by Huang and Arsenault (1973a) on vanadium–oxygen alloys.

The most consistent configurations, as stated before, are screw dislocations in samples deformed at large effective stresses. Interstitial and substitutional additions do not change the fact that there is still a predominance of screw dislocations in samples deformed at large effective stresses. This can mean one of two things: (1) The motion of screw dislocations is the controlling dislocation species in both the pure bcc metal and the alloys, or (2) the predominance of screw dislocations does not have anything to do with the rate-controlling mechanism of low-temperature deformation of bcc metals and their alloys.

IV. Core Structure of a Screw Dislocation

Hirsch (1968) suggested that the core of a screw dislocation in a bcc metal could dissociate with a threefold symmetry on three intersecting (112) planes. If a reasonable stacking fault energy is assumed for a given bcc metal, the width of the stacking fault is very small, $1b$ or possibly $2b$. Since then a large number of investigators (see Christain, 1970; Basinski et al., 1971) have employed various interatomic force potentials and have shown that there is a threefold symmetry about a screw dislocation corresponding to three (112) planes. Basinski, Duesbery, and Taylor have also determined the effect of an applied stress on the core structure and have shown that it changes slightly. They also determine a Peierls stress; however, the value is approximately an order of magnitude too large. They also concluded that a normal stress component to the slip plane would reduce the Peierls stress, and since most experiments are done in unioxidal tension the observed extrapolated yield stress at $0°$ K would be lower than the calculated value.

Recently Vitek and Christain (1973) have considered the effect of a shear stress and the initial core width on subsequent displacements about a screw dislocation in a bcc lattice. They have arrived at the following conclusions: (1) Atomistic calculations show that screw dislocations in bcc metals have much higher Peierls stresses than dislocations with some edge component. The relative immobility of screws is probably mainly responsible for the low-temperature strength. (2) The core structure of the screw dislocation and its motion under stress show an asymmetry associated with the sense of the twinning–antitwinning shear across {112} planes. This is consistent with the asymmetry effects observed in $\psi-\chi$ curves. (3) The deformation geometries found in various bcc materials may result from differences in the strength of binding. A general explanation of the temperature dependences of $\psi-\chi$ curves can be given for higher temperatures. The extreme asymmetry which is sometimes observed at low temperatures is not fully explained. (4) A satisfactory theoretical description of anomalous slip and of twin nucleation is not yet available.

V. Thermal Activation Analysis of Dislocation Dynamics

Plastic deformation is the motion of dislocations; and it is a generally accepted fact that thermally-activated motion controls plastic deformation over a large segment of the stress–temperature region. However, there has been and probably still are objections to the application of equilibrium thermodynamics and absolute reaction rate theory to plastic deformation, which is an irreversible process. Probably the most concise explanation of the reaction rate theory is given by Darken and Gurry (1953). Only the main points will be covered here. For any conceivable way that the reaction might occur [i.e. the dislocations jumping the short-range barrier (srb)], it is possible to imagine a plot of energy versus a distance coordinate: the path taken is any conceivable path. Of all the possible paths, there will be one path for which the maximum energy is lower than for all other paths: this path is known, in general, as the reaction coordinate. For dislocation motion, this reaction coordinate is generally in units of Burgers vectors. A plot of energy against the reaction coordinate is shown schematically in Fig. 24. If a dislocation is free to move on a path, which has a srb whose height (energy) is

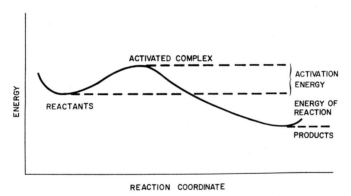

Fig. 24. Schematic of an energy curve for a reaction. (Darken and Gurry).

given by the curve, it is obvious that the dislocation will rest in the position defined as "Reactants," for this is a position of minimum energy (even though it may be a metastable position). The dislocation, of course, would never of its own accord move to the minimum designated as "Products," for it must first acquire the activation energy necessary to surmount the peak designated as "Activated Complex." At absolute zero temperature there is no energy available, so the dislocation stays at the "Reactants."

At any finite temperature, however, there may be a few dislocations of the 10^6–10^8 dislocations/cm^2 which may have a finite probability of acquiring the necessary activation energy from the energy of thermal agitation of the entire system; it is obvious that the higher the activation energy, the lower the probability.

There are two main points of the reaction rate theory that are concerned with this activated complex. They are as follows:

1. The activated complex, though exceeding short life, may be treated as having a definite set of thermodynamic functions and *is in equilibrium with the "Reactants."* This means that the dislocations at the "Reactants" are in equilibrium with the dislocations at the "Activated Complex." The thermodynamic equilibrium constant K may be written. The free energy of formation of the activated complex in the standard state is written as ΔG (τ^*, T structure). The two are related in the usual way:

$$\Delta G = -kT \ln K \qquad (4)$$

where k is Boltzman's constant.

2. The specific rate of decomposition of the activated complex into products is a universal rate independent of the nature of the *particular reaction* or particular *activated complex*. This means that the motion of the dislocation from the activated complex state to the products state can be treated as any other thermally-activated process. The rate is kT/h, where h is Planck's constant. The average stay of the dislocation at the activated complex state is very short, $\sim 10^{-13}$ sec.

The dislocations in the "Reactants" position or state can be designated as R^1, and the "Activated Complex" as A; then the reaction for the function of activated complex state becomes simple.

$$R^1 = A \qquad (5)$$

and the equilibrium constant is

$$K = C_{A_1}/CR \qquad (6)$$

There is no need to account for the activity. The reaction rate, or the number of dislocations which move from the activated complex state to the products state is:

$$\text{Reaction rate} = (kT/h)C_A \qquad (7)$$

which can be written as

$$\text{Reaction rate} = (kT/h)KC_R \qquad (8)$$

and the expression relating ΔG and K can be rewritten in the form:

$$K = e^{-\Delta G/kT} \qquad (9)$$

Now inserting Eq. (9) into Eq. (8), the reaction rate can be written as

$$\text{Reaction rate} = (kT/h)C_R e^{-\Delta G/kT} \qquad (10)$$

It is now possible to more specifically define C_R as the total number of dislocations per unit area or volume which have the possibility of moving (Pm). Also it is possible to approximate $kT/h \cong \nu_D$ where ν_D is the Debye frequency; and further, if it is assumed that distance between the "Reactants" and "Products" state is one Burgers vector, it is possible to define a specific velocity.

$$v = \nu_D h^1 e^{-\Delta G/kT} \qquad (11)$$

where h^1 is the distance covered per thermally activated event, and recalling that

$$\dot{\gamma} = \rho_m b v \qquad (12)$$

inserting Eq. (11) into Eq. (12) gives

$$\dot{\gamma} = \rho_m b h^1 \nu_D e^{-\Delta G/kT} \qquad (13)$$

These equations (which are starting equations for the consideration of thermally-activated plastic deformation) were obtained by using classical equilibrium thermodynamics and absolute reaction rate theory.

The main concern of most investigators who have studied thermally-activated deformation has been to determine the obstacle to dislocation motion and the mechanism by which the dislocation overcomes this obstacle. The method usually employed to determine the obstacle and the rate-controlling mechanism (RCM) is to experimentally determine the activation parameters; i.e. activation energy (ΔH), free energy (ΔG), and activation areas (i.e. activation volume v^*) with calculated parameters of specific mechanisms.

Before beginning a discussion of the activation parameters, it is necessary to consider several other factors associated with both the derivation and experimental determination of these parameters. First, it is assumed that the flow stress can be divided into two components: an athermal component (τ_μ) and a thermally activated component (τ^*). Figure 25 is a schematic showing this division. The dislocation barrier giving rise to the athermal component is a long-range barrier 10 Å, i.e. it affects the dislocation over this range. When these barriers are present, thermal fluctuations cannot assist the external stress in helping the dislocation overcome this barrier. If the long-range barriers exist at a high temperature, they should also exist at lower temperatures. These particular assumptions have been discussed in great detail elsewhere (Arsenault, 1967c). This dislocation barrier giving rise to the thermal component of the flow or yield stress is a short-range barrier, and thermal

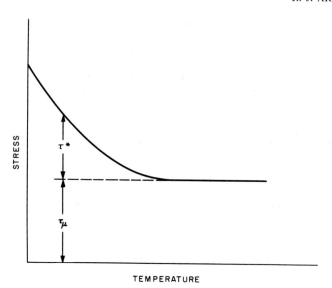

Fig. 25. Schematic yield or flow stress versus temperature curve illustrating the division of stress.

fluctuations can assist dislocations in overcoming these barriers. Another factor that needs to be considered is the energy obtained from a calculation of defect–dislocation interaction. This energy has been labeled the change in potential energy, the appropriate Gibbs free energy, and implicitly the activation enthalpy. However, upon examination, the energy calculated is really a work term. There is another work term which results from the work done during the thermally activated event by the effective stress. Generally there is an implicit assumption made that the dislocation is in thermodynamic equilibrium at the ground and activated state. This leads to the classical definition of the Gibbs free energy, since

$$(\Delta G)_{P,\,T} = DW_{\text{net, rev}} \tag{14}$$

where P is the hydrostatic pressure. This description is misleading since the pressure applied is not hydrostatic but a shear stress. A more accurate description would be that this energy results from reversible work which is free energy (Zenner, 1942). However, using Schoeck's (1965) terminology this quantity will be referred to as the Gibbs shear free energy or Gibbs free energy.

Two thermodynamic systems have been developed to theoretically determine the activation parameters (Conrad and Wiedersich, 1960; Schoeck, 1965). These thermodynamic systems are shown schematically in Fig. 26.

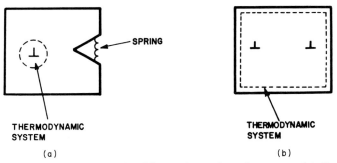

Fig. 26. Schematic illustrating the different thermodynamic systems. (a) Conrad and Wiedersich 1960 system. (b) Schoeck 1965 system.

The spring in Conrad–Widersich's model is outside the thermodynamic system, but it is of the same material as that inside the thermodynamic system. This spring gives rise to the internal stress. The thermodynamic system is a small volume around the dislocation. Schoeck's thermodynamic system contains at least one stationary dislocation which gives rise to the internal stress. Hirth and Nix (1969) have shown that self-consistent results can be obtained from both thermodynamics systems, and that one is more useful than the other. Surek *et al.* (1973) have shown that for a rigorous treatment, the temperature and stress dependence of A^* (activation area) should be considered. They have also shown that compatibility equations exist between the two thermodynamic systems, but the fact that two systems are used should be kept in mind.

There is still the problem of comparing the experimentally determined activation parameters with those predicted by theory. The following is a general equation for the Gibbs free energy for a general srb:

$$\Delta G = \int_{y_1(T,\,\tau*)}^{y_2(T,\,\tau*)} F(y)\,dy - bA^*(\tau_a - \tau_\mu) \tag{15}$$

where $F(y)$ is the force–distance relationship of the particular srb under consideration. Two representative force–distance diagrams are schematically shown in Fig. 27. Most investigators in determining ΔG_{Th} (i.e. the theoretical Gibbs free energy) make the following assumptions:

$$d\mu/dT = 0 \quad\text{and}\quad \tau_\mu = 0 \tag{16}$$

where μ is the shear modulus. Therefore $\tau_a = \tau^*$. Now Eq. (15) is rewritten as follows:

$$\Delta G_{Th} = \Delta G_0 - bA^*\tau^* \tag{17}$$

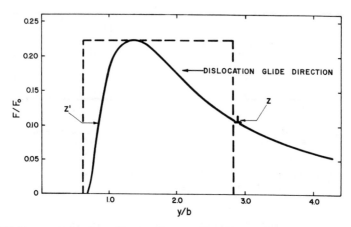

Fig. 27. Representative force–distance diagrams. The solid curve is that predicted by Gibbs for the tetragonal distortion defect.

where ΔG_0 is the total area under the force–distance diagram. Also

$$\Delta S_{Th} = -\partial \Delta G_{Th}/\partial T = 0 \tag{18}$$

then

$$\Delta G_{Th} = \Delta H_{\tau*, Th} \tag{19}$$

As shown by numerous investigations and by Schoeck (1965) the activation energy Q is given by the following equation:

$$Q_{\tau*}^* = -k[\partial \ln (\dot{\gamma}/\dot{\gamma}_0)/\partial(1/T)]_{\tau*} = \Delta H_{\tau*, exp} \tag{20}$$

It is *very* tempting to use the following equation:

$$\Delta H_{\tau*, Th} = \Delta H_{\tau*, exp} \tag{21}$$

However, the above equation is *wrong*, due to the fact that the $\Delta H_{\tau*, exp}$ is, in general, not obtained under the conditions of $\partial \mu/\partial T = 0$ and $\tau \mu = 0$. The satisfactory way of making the comparison is to determine ΔG_{exp} and compare it with ΔG_{Th}. This can be done by experimentally determining $T \Delta S$ for

$$T\Delta S_{exp} = \Delta H_{\tau* exp} - \Delta G_{exp} \tag{22}$$

Recently there has been considerable discussion of how to determine $(\Delta S)_{\tau* exp}$ or $(\Delta S)_{\tau_a exp}$ by Arsenault (1971a, 1972a) and Jones *et al.* (1971, 1972). There are at least two methods of experimentally determining $(\Delta S)_{\tau* exp}$ or $(\Delta S)_{\tau_a exp}$. In the first method to be considered for experimentally determining S there is one critical factor which has to be considered, and that is the temperature range in which $(\partial \Delta S/\partial T)_{\tau *}$ is a maximum. ΔS can be expressed as follows:

$$\Delta S = \Delta S_0 + \int_0^T (1/T)(\partial \Delta H/\partial T)_{\tau * \text{ or } \tau_a} dT + \int_0^\tau (\partial A^* b/\partial T)_{\tau * \text{ or } \tau_a} dT \quad (23)$$

Li (1965) has argued that ΔS_0 which is the change in entropy at 0° K is zero (the third law of thermodynamics). Also, as a consequence of accepting the third law of thermodynamics, $\lim (\partial \Delta S/\partial T) = 0$ as $T \to 0$. Because of experimental limitations it is not possible to determine $(\partial \Delta H/\partial T)_{\tau *}$ over the entire temperature range; however, it is possible to determine it over a narrow temperature range ($100-200^\circ$ K). The same applies for $(\partial Ab^*/\partial T)$. As pointed out by Arsenault (1972a), if $(\partial \Delta H/\partial T)_{\tau * \text{ or } \tau_a}$ is determined in the temperature range where $(\partial \Delta S/\partial T)$ is the largest, then it is possible to define a maximum upper limit of ΔS at any given temperature. Unfortunately, in order to do this, it is necessary to make a few assumptions. If the assumption is made that a major contribution to a possible ΔS arises from the temperature dependence of the shear modulus (this is an assumption made by most investigators), then the activation entropy can be expressed as

$$\Delta S_{\tau_a} = -(1/\mu)(d\mu/dT)(\Delta G + bA^*\tau_a) \quad (24)$$

where

$$\Delta G = [\Delta H_{\tau_a} + bA^*\tau_a(T/\mu)(d\mu/dT)]/[1 - (T/\mu)(d\mu/dT)] \quad (25)$$

and $(\Delta S)_{\tau_a}$ is the entropy of activation when considering the thermodynamic system as proposed by Schoeck (1965). It is also possible to define $(\Delta S)_{\tau *}$ (Surek et al., 1973). The next result of both solutions for $(\Delta S)_{\tau * \text{ or } \tau_a}$ is that $(\partial \Delta S/\partial T)_{\tau * \text{ or } \tau_a}$ is largest in the temperature range of $77-300^\circ$ K. This temperature range obviously depends upon the metal under consideration, e.g. for iron the range is $200-300^\circ$ K; for magnesium it is a little lower. Therefore the temperature range over which measurements of $(\partial \Delta H/\partial T)_{\tau * \text{ or } \tau_a}$ should be made is from 100 to 300° K. The net result of all of these measurements (Arsenault, 1964, 1966b; Conrad et al., 1961; Koppenaal and Arsenault, 1965) is that $(\partial \Delta H/\partial T)_{\tau * \text{ or } \tau_a}$ is zero, or has a small positive or negative value. Jonas et al. (1972) have argued that if the value of $(\partial \Delta H/\partial T)_{\tau * \text{ or } \tau_a}$ was experimentally measurable then ΔS would be very large. Therefore, they state that determining $(\partial \Delta H/\partial T)_{\tau * \text{ or } \tau_a}$ is not a satisfactory method for determining ΔS_{exp}.

There is a second experimental method for determining ΔS_{exp}. The basis of this analysis is a consideration of the following equation, which is just a different version of Eq. (13):

$$\Delta G = kT \ln (\dot{\gamma}_0/\dot{\gamma}) \quad (26)$$

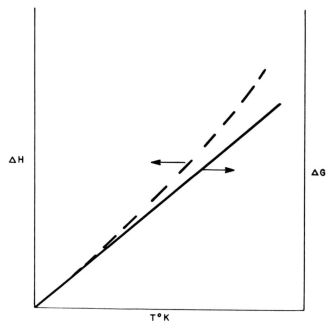

Fig. 28. Schematic representation of ΔG vs T and ΔH vs T for a positive value of ΔS.

where $\dot{\gamma}_0 = \rho_m bh^1 v_D$. Now if we assume $\dot{\gamma}_0$ is a constant, i.e. independent of stress or temperature over the entire range of consideration, then at constant $\dot{\gamma}$ a linear relationship should exist between ΔG and T as shown schematically in Fig. 28. For the previous method of determining $(\Delta S)_{exp}$ it is only necessary to assume that $\dot{\gamma}_0$ is independent of stress and temperature over a limited range of stress and temperature. If there is a positive value of ΔS associated with the dislocation motion, then $(\Delta H)_{\tau^* exp}$ would have a positive curvature as shown by the dashed line in Fig. 28. If experimentally a nonlinear $(\Delta H)_{\tau^*}$ vs T relationship is observed, then it is possible to determine the $(\Delta G)_{exp}$ curve by making the assumption that $T\Delta S$ is small at the lowest temperature at which $(\Delta H)_{\tau^* exp}$ is measured. The difference between the $(\Delta H)_{\tau^* exp}$ and $(\Delta G)_{exp}$ can be logically assumed to be $T(\Delta S)_{exp}$, as most recently reported by Cagnon (Arsenault, 1973). However, before making this simple subtraction it is necessary to consider the fact that a positive curvature of ΔH vs τ^* can arise due to the presence of τ_μ, as shown clearly by Arsenault and Li (1967). After making this correction the difference can be $T\Delta S_{exp}$. An analysis of a large amount of the experimental data for bcc metals and their alloys (Conrad, 1963) reveals that there is a linear relationship between $\Delta H_{\tau^* exp}$ and T which would indicate once more that $T\Delta S_{exp}$ is zero, or has small positive or negative value. If $T\Delta S$ is believed to be large, then $\dot{\gamma}_0$ must be temperature dependent in an alternate way, i.e. compensate for the $T\Delta S$ term, which seems highly unlikely.

If the $T\Delta S$ is calculated by the methods of Schoeck (1965) and Surek et al. (1973), the values obtained from bcc metals are small. For example, in the case of iron determining $\Delta S_{\tau*}$ at $\tau* = 0$, the value of $T\Delta S = 0.01$ eV using the $\Delta H_{\tau*}$ obtained by Arsenault (1964), and the elastic constants of Lord and Beshers (1965).

In the above analysis of the activation parameters it was assumed that the preexponential $(\dot{\gamma}_0)$ is independent of temperature of stress and temperature. If it is now assumed that $\dot{\gamma}_0 = \dot{\gamma}_0(T\tau)$, then the expressions for activation enthalpy and volume change.

$$\Delta H = -k(\partial \ln \dot{\gamma}/1/T)_{\tau} + k(\partial \ln \dot{\gamma}_0/1/T)_{\tau} \tag{27}$$

$$v(\tau) = kT(\partial \ln \dot{\gamma}/\partial \tau)_T - kT(\partial \ln \dot{\gamma}_0/\partial \tau)_T \tag{28}$$

where the first term on the right-hand side of the above equations is the experimentally determined ΔH_{exp} and $v(\tau)_{exp}$. Now in order to obtain the correction term it is necessary to go through an iteration process as devised by Jonas et al. (1973). Briefly it is necessary to assume some relationship for ΔS, where the force–distance diagram is a function of μ, which is temperature dependent. This is the same assumption used to obtain Eqs. (24) and (25).

Then ΔG is determined, and then a corrected ΔH is determined, and the processes repeated until an asymmetric value of ΔH is obtained. Since the magnitude of ΔS determines the magnitude of the stress dependence of $\dot{\gamma}_0$, and since ΔS is small as determined by Eq. (25), then the stress dependence of $\dot{\gamma}_0$ must be small.

However, there is an additional factor which has to be considered and that is related to density of mobile dislocations. The following is used to define the applied yield or flow stress:

$$\tau_a = \tau_\mu + \tau^* \tag{29}$$

where

$$\tau_\mu = \tau_0(\mu) + \alpha\mu b\sqrt{\rho_m}$$

$$\tau^* = \tau_0^*(\dot{\gamma}, T, \rho_m)$$

and ρ_m is the mobile dislocation density. Now if ρ_m increases τ_μ increases and τ^* decreases; therefore, there is a given ρ_m to obtain a minimum τ_a for a constant ε. The minimization of Eq. (29) leads to a nearly linear relationship between ρ_m and T, i.e. ρ_m increases with increasing temperature. This means that τ_μ increases with increasing temperature. The magnitude depends upon the parameters chosen, but in general the change in τ_μ due to a change in ρ_m

is much larger than the change in τ_μ due to a change in modulus with temperature. Hence Eqs. (24) and (25) should be modified to take into account the change in ΔG due to a change in ρ_m with temperature.

The experimental methods for determining the activation parameters have been reviewed by Arsenault (1967c). The relationships between the activation parameters and τ^* and T have been comprehensively discussed in past reviews (Christain, 1970; Conrad, 1963). However, the one relationship which is distinctly different from fcc metals is that the activation volume is in general independent of strain for bcc metals and their alloys. This is a strong indication that the density of obstacles does not increase with strain. At present three models are being considered to account for the magnitude and temperature dependency of τ^*. These are: the double-kink model, the sessile-glissile transition model, and the tetragonal distortion model.

The double-kink models are based on the assumption that the intrinsic lattice, i.e. the Peierls stress, is the srb to dislocation motion. The model of the movement (Seeger, 1956; Weertman, 1956) of a dislocation segment over the Peierls potential energy barrier (double-kink formation) has been used to explain several low-temperature deformation phenomena. Seeger (1956) made the first detailed calculations of the energy of formation of a double kink; these calculations were applied to the problem of explaining the low-temperature internal friction peak (the Bordoni peak) in fcc metals. The double-kink model was proposed by Conrad et al. (1961; Conrad, 1963) as an explanation of the large temperature and strain rate dependence of bcc metals deformed at low temperatures. This model can be used to explain the solid-solution effects in bcc alloys and other materials in which the formation of the double kink is the rate-controlling mechanism.

The experimentally determined activation parameters (activation energy and activation volume) can only be compared with those predicted from calculations if the calculated parameters are determined as a function of stress. In his theory, Seeger calculated the energy to form an isolated single kink at zero applied stress. Seeger suggested that the attractive force between the single kinks is proportional to ln d where d is their separation. This force later was found by Seeger and Schiller (1962) to vary as $1/d^2$. There is a critical separation of kinks at a given externally applied stress. If the separation is greater than the critical value, the kinks will continue to move apart. At smaller spacings they will move together and collapse. The configuration that was assumed to result in the maximum energy consisted of two half-kinks at the critical separation connected by a dislocation segment. The position of this segment was such that its Peierls' energy was a maximum (that is, it was sitting at the top of the Peierls' barrier). Therefore, the (activation) energy of formation of a double kink consists of the energy of a single kink and the energy to move a segment of dislocation corresponding to separation of the kinks over the Peierls' potential energy barrier, less the work done by the applied stress in forming the double kink. Arsenault (1964) has discussed in detail the attempts to use the models of Seeger (1956) and Seeger and Schiller (1962) to explain the data obtained from the low-temperature deformation of bcc metals. It was shown that these models lead to an activation volume that is stress independent, in contradiction to experimentally determined activation volumes (Arsenault, 1964).

Seeger's theory may have failed because it was developed under the assumption that the applied stress is equal to zero. Dorn and Rajnak (1964) and Celli *et al.* (1963) have calculated the energy of formation of a double kink. Some of their results will be mentioned later. However, these investigators did not consider the interaction energy between kinks and calculated the formation energy only for infinite dislocation lengths. In this review an attempt will be made to demonstrate a theory which can be used to calculate the energy of formation of a double kink as a function of applied stress for various dislocation lengths, and take into consideration the interaction energies associated with the kinks. The consequence of more than one double kink forming, i.e. the movement of the dislocation center over a second potential energy barrier or more, will also be considered.

The Peierls' potential energy barrier (Peierls' stress) and the activation parameters can be affected by solute additions. Of interest is the fact that these effects on the Peierls' stress due to solute atoms can result in solid-solution weakening, whereas all previous theories of solid-solution effects predict strengthening due to solute additions.

The energy of formation of a double kink will be considered when the dislocation is initially parallel to the potential energy barrier, i.e. the Peierls' stress hills. It can be argued that not all the dislocations lie parallel to the Peierls' stress hills in as-grown or plastically deformed crystals. The dislocations that lie at an angle to the Peierls' stress hills contain geometrical kinks. These kinks can move at very small stresses, and their motion is considered to be not thermally activated (Brailsford, 1961, 1962). Upon application of a stress, all dislocation segments will move into positions that lie parallel to the potential energy barriers. Their further motion will require the formation of double kinks. A double-kink configuration is shown schematically in Fig.29. In calculating the formation energy of a double kink it was assumed

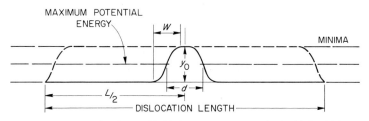

Fig. 29. Schematic of a double kink at zero stress. *W* is the kink width.

that (1) the dislocation can be treated as a line and (2) the continuum elasticity theory can be used. The solution of the following partial differential equation will give the displacement of the dislocation line as a function of the distance χ measured along its length:

$$E(y)\, d^2y/dx^2 = dE(y)/dy - b_\tau \tag{30}$$

This equation is obtained by setting the force on each element of a dislocation equal to zero. It is the same as the one proposed by Seeger (1956) and by Read (1953). In Eq. (30), $E(y)$ is the line energy of the dislocation. This energy is a periodic function of y with a period equal to a lattice constant a. The term τ^* is the external stress and b is the Burgers vector of the dislocation.

The line energy $E(y)$ can be represented by a Fourier series (Seeger, 1956):

$$E(y) = E_0 - (\tau_p^0 ba/2\pi)\cos(2\pi y/a) + \cdots \tag{31}$$

Here $E_0 = \mu b^2/n$ where n varies from 1 to 5 depending upon the theory (Cottrell, 1953; Friedel, 1964) used to determine the line energy of the dislocation, μ is the shear modulus, and τ_p^0 is the Peierls' stress at absolute zero. Since E_0 is much larger than $\tau_p^0 ab/2$, $E(y)$ can be replaced by E_0 in the left side of Eq. (1). Equations (1) and (2) thus reduce to

$$E_0\, d^2y/dx^2 = \tau_p^0 b \sin(2\pi y/a) - \tau b \tag{32}$$

The solution of Eq. (32) was obtained with the following set of boundary conditions:

$$y = 0 \quad \text{at} \quad x = 0 \tag{a}$$

$$y = y_0 \quad \text{at} \quad x = L/2 \tag{b}$$

$$dy/dx = 0 \quad \text{at} \quad x = L/2 \tag{c}$$

The values of y_0 at $L/2$ are shown in Fig. 29 and symmetry is assumed with respect to the y axis.

The following expression for distance x as a function of y is obtained from Eq. (32):

$$x(y) = (A/2)^{1/2} \int_0^y (K - \cos y - cy)^{-1/2}\, dy \tag{33}$$

where $A = E_0 a/2\pi\tau_p^0 b^3$, $c = \tau/\tau_p^0$, and K is an integration constant.

The equation for the energy of the configuration defined by Eq. (33) is again similar to that proposed by Seeger (1956).

$$E = 2\int_0^{L/2} \{\tfrac{1}{2}E_0(dy/dx)^2 + (\tau_p^0 ab/2\pi)[1 - \cos(2\pi y/a)] - \tau by\}\, dx \tag{34}$$

which can be written in the form

$$E(y_0, K) = B(A/2)^{1/2} \int_0^{y_0} [2(K - \cos y - cy)^{1/2}$$

$$+ (1 - K)(K - \cos y - cy)^{-1/2}] \, dy \qquad (35)$$

where $B = \tau_p^0 ab^2$ since

$$x(y_0, K) = (A/2)^{1/2} \int_0^{y_0} (K - \cos y - cy)^{-1/2} \, dy = L/2 \qquad (36)$$

Equation (35) can be simplified to

$$E(y_0, K) = B(A/2)^{1/2} \left[\int_0^{y_0} 2(K - \cos y - cy)^{1/2} \, dy + (1 - K)L/2 \right] (37)$$

Several implicit assumptions are involved in the above solution, i.e. $d\mu/dT = 0$ and $\tau_\mu = 0$. Equation (37) can be evaluated numerically for the boundary condition given by Eq. (36) by two different methods. One method is to fix $L/2$ and compute the value of K from Eq. (36) for each value of y_0 required to satisfy the boundary conditions (a) and (b). The potential energy of the configurations is next computed with these values of K and y_0. Results of such calculations for various values of external stress are shown in Fig.30. At a given stress, the potential energy first decreases to minimum energy and then begins to increase for larger values of y_0, i.e. further displacements of the center of the double kink, until a maximum energy is attained. On further increasing y_0 the energy decreases. The dislocation configurations that occur at the minimum and maximum energies will be called the maximum and minimum configurations. For increases in stress the distance between the maximum and minimum energy position decreases and at $\tau = \tau_p^0$ there is no maximum or minimum energy, only an inflection in the curve.

The second method involves the use of the boundary conditions (a) and (c), and only the maximum and minimum potential energies are calculated as a function of stress for dislocation lengths from $30b$ to $2000b$. The difference in energy between the maxima and minima can be defined as the Gibbs free energy ΔG. The difference in ΔG was found to be independent of dislocation length for lengths greater than $40b$ for iron. The dependence of ΔG on the dislocation length at small dislocation lengths is due to end effects imposed by the boundary conditions.

For a dislocation of infinite length, and with $\Delta G = E_{max} - E_{min}$ where

$$E_{max} = B(A/2)^{1/2} \left[\int_0^{y_{max}} (2K - \cos y - cy)^{1/2} \, dy + (1 - K)L/2 \right] (38)$$

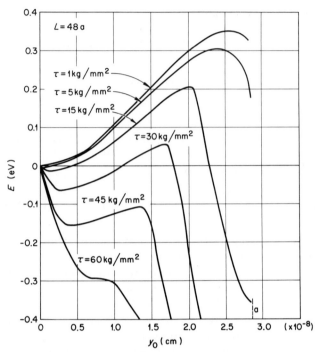

Fig. 30. Energy of the dislocation configuration as a function of the displacement of the center of the double kink. $\mu = 7.73 \times 10^3$ kg/mm^2 and $\tau_p^0 = 60$ kg/mm^2.

and

$$E_{\min} = B(A/2)^{1/2}\left[\int_0^{y_{\min}} 2(K - \cos y - cy)^{1/2}\, dy + (1 - K)L/2\right] \quad (39)$$

we have

$$\Delta G = B(2A)^{1/2} \int_{y_{\min}}^{y_{\max}} (K - \cos y - cy)^{1/2}\, dy \quad (40)$$

The activation energy is actually independent of the dislocation length and is only a function of stress. The activation energies calculated by the second method are in agreement with those obtained by the first method.

The calculations mentioned above were made for the formation of only one double kink on a dislocation line. The successive maximum and minimum potential energies for the movement of a dislocation over a second Peierls' stress hill, i.e. the formation of a second double kink in the center of the dislocation line of finite length, can be calculated using Eq. (35). The results from a typical calculation are shown in Fig. 31. After the center of the dislocation line has passed over a given number (determined by the stress

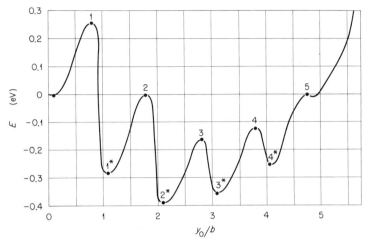

Fig. 31. Energy of the dislocation configuration of the formation of more than one kink, for a stress of 10 kg/mm² or $C = 1/6$ and $L = 70b$.

and dislocation length) of potential energy barriers, the energy increases monotonically. An example of this can be seen when the dislocation has passed over five potential energy barriers as shown in Fig. 31.

The dislocation can reach an equilibrium position, such as position 2* in Fig. 31, after several double kinks have formed. An increase in the stress results in an increase in the number of potential energy barriers which are overcome before the energy begins to increase monotonically. This situation is shown in Fig. 32. The number of potential energy barriers that the dislocation must overcome before the energy begins to increase monotonically is a function of both stress and dislocation length. This dependence is shown in Fig.33. The difference in energy between the equilibrium minimum energy and minimum energy on each side is small for long dislocation lengths as compared to shorter lengths. This effect is seen by comparing Figs. 31 and 32. Ono and Sommer (1970) have recently considered the effect of finite line lengths. These finite dislocation line lengths are due to dispersed interstitial atoms and have shown effectively that an increase in τ_μ occurs. Physically this is the result of the fact that the dislocation line after forming a double kink at small τ^* may have a higher energy than prior to the formation as shown in Fig. 30, for a $\tau^* = 1$ kg/mm² and a line length of 48a. The dislocation line has a larger energy after the double kink has formed than before. Therefore the equilibrium position or configuration is the straight dislocation, i.e. it does not move. For the parameters used to determine the curves in Fig. 30, a stress between 5 and 15 kg/mm² is required for dislocation motion. This is, therefore, an increase in τ_μ of 5 to 15 kg/mm².

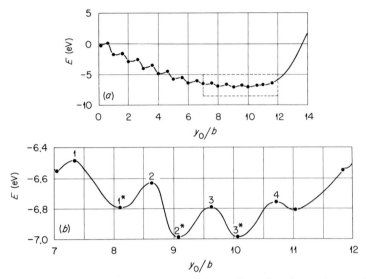

Fig. 32. Energy of the dislocation configurations for the formation of more than one double kink, for a stress of 30 kg/mm² or $C = 1/2$ and $L = 70b$. Part (b) is an expanded view of the section enclosed by the dashed lines in part (a).

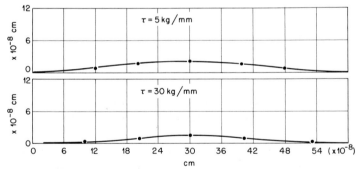

Fig. 33. The configuration of the dislocation at the maximum and minimum energies for $C = 1/12$ and $C = 1/2$.

The activation volume is related to the work done by the external stress when a dislocation overcomes the barrier. This work (energy) is the product of three length terms and the stress. The three length terms are called the activation volume which can be calculated using two approaches. The first involves calculating the difference in area of maximum and minimum configurations and multiplying it by the Burgers vector.

$$v^* = b \left[\int_0^L y_{\max}(x)\, dx - \int_0^L y_{\min}(x)\, dx \right], \qquad (41)$$

which can be transformed into

$$v^* = (ab^2/2\pi)(2A)^{1/2}\left[\int_0^{y0\;max} y(K - \cos y - cy)^{-1/2}\,dy\right.$$

$$\left. - \int_0^{y0\;min} y(K - \cos y - cy)^{-1/2}\,dy\right], \tag{42}$$

where again v depends only on the stress and is independent of the dislocation length.

The second approach to calculating the activation volume is based on the following expression:

$$v^* = -\partial\Delta G/\partial\tau = -\partial(E_{max} - E_{min})/\partial\tau \tag{43}$$

The negative of the differential of $(E_{max} - E_{min})$ with respect to stress results in the same expression as Eq. (43), as would be expected if all the transforms are correct.

In the above calculation of the energy of formation of a double kink the interactions between the kinks and between the kinks and the remaining portions of the dislocation line were not taken into consideration. To determine whether these interaction energies are important, the interaction energy was calculated for a double-kink configuration corresponding to the maximum potential energy. The approach used was similar to that of Seeger and Schiller (1962): the mutual interaction energies of a segmented dislocation containing the double kink and a corresponding segmented straight dislocation were calculated. The difference in energies between the two dislocation configurations then corresponds to the interaction energy of the double kink. The formation of Lothe (1962) and Jøssang et al. (1965) was employed rather than Kroner's (1958). The total energy of a segmented dislocation made of c_1 segments with a Burgers vector b can be expressed as

$$W = \sum_i W_{si} + \sum_{i<j} W_{ij}, \tag{44}$$

where W_{si} is the self-energy of the segment c_i. Lothe (1962) used Blin's (1955) formula for the interaction energy

$$W_{ij} = -(2\mu/4\pi)\int_{c_i}\int_{c_j}(b_i \times b_j)(dl_j \times dl_i)/R$$

$$+ (\mu/4\pi)\int_{c_i}\int_{c_j}(b_i \times dl_i)(b_j \times dl_j)/R$$

$$+ \frac{\mu}{4(1-v)}\int_c\int_{c_i}(b_i \times dl_i)\tilde{T}(b_j \times dl_j) \tag{45}$$

where v is Poisson's ratio, dl_i is a dislocation segment, and $\tilde{T}$ is a tensor with components

$$\tilde{T}_{ij} = \partial^2 R/\partial x_i\ \partial x_j \tag{46}$$

where $R^2 = (x_i - x_j)^2 + (y_i - y_j)^2 + (z_i - z_j)^2$. The above equation can be simplified if it is assumed that the dislocation is either edge or screw. The equation for the interaction energy can be written for a screw dislocation as follows:

$$W_{ij} = z \int_{x_i} \int_{x_j} \frac{\left[1 + \dfrac{1}{1-v}\left(\dfrac{dy_i}{dx_i}\right)(dy_i/dx_j)\right]\ dx_j\ dx_i}{[(x_i - x_j)^2 + (y_i - y_j)^2]^{1/2}} \tag{47}$$

where $z = \mu b^2/4$. The maximum configuration for $\tau = 1/2\tau_p^0$ was used, and the interaction between segments was computed numerically using a double precision technique. It turned out that the sums of the interaction energies of the two configurations were large numbers and that their differences were small (0.03 eV); thus the interaction energy of the double kink cannot be stated with any accuracy. However, the results do indicate that the interaction energy contribution can be disregarded when considering the formation energy of a double kink since the interaction energy is 0.03 eV, whereas the formation energy is 0.55 eV.

Most of the contribution to the difference in energy results from the interactions of segments which are defined as the widths of kinks (as shown in Fig. 29). Therefore several simplifications can be made, and the following equation for the differences in energy can be formulated:

$$\Delta W_{ij} \approx z[R_k\ \ln R_k - R_s\ \ln R_s) - (R_k - R_s)] \tag{48}$$

where R_k is the product of the length of the kink segment and distance between the other kink segment with which it is interacting, and R_s is the corresponding straight segment. According to Eq. (48) the interaction energy decreases as the separation increases. As a result there is an attractive force between the kinks.

The magnitude of the energy of formation of the double kink is strongly dependent on the Peierls' stress, as is evident from Eq. (34). Unfortunately, the theoretical calculations of the Peierls' stress can be described only as rather poor estimates of this quantity; but all of these estimates are related to the same parameters. Although an exact calculation cannot be made of the Peierls' stress, relative changes in the Peierls' stress can be predicted from changes in parameters such as lattice spacing. The following is a typical expression for the Peierls' stress of a screw dislocation found in the literature (Weertman and Weertman, 1964):

$$\tau_{ps}^0 \approx \alpha(\mu b/2c) \exp{(-2\pi y/c)} \tag{49}$$

where α is a constant related to the periodic force law used (it can vary from 1/2 to 1); δ, the "width" of the dislocation, is equal to $a/2$; a is the lattice constant; c is approximately equal to b for bcc metals. These parameters are related to the positions and interactions of the lattice atoms surrounding the dislocation; however, these parameters can be modified by replacing some of the lattice atoms with foreign (solute) atoms.

There are some specific ways in which a solute can affect these parameters that in turn change the Peierls' stress: (1) change in modulus, (2) local distortion in the lattice near the core of the dislocation and/or change in dislocation width, and (3) change in the density of occupied "d" states.

Arsenault has discussed these effects in detail elsewhere, and the general conclusion is that any disturbance to the lattice results in a reduction in the Peierls' force. Arsenault et al. (1972) have considered the effect of varying the line vector of a dislocation from edge to screw and have shown a change in the τ_p^0.

Recently Chou and Sha (1972) have considered the effect of anisotropic elastic constants on the Peierls' stress, and have again shown that in general $\tau_{pe}^0 < \tau_{ps}^0$.

In order to compare the calculated values of activation energies and activation volumes with those obtained experimentally, it is necessary to first consider several additional factors. In calculations of the formation energy of a double kink, the potential energy of the dislocation configuration was determined, and the activation energy was actually the difference between a minimum and maximum potential energy. Another factor to be considered is the separation (lateral motion) of the kinks once the double kink has formed. There have been several calculations of the velocity and stress necessary for the lateral motion of kinks. Only for an abrupt kink (Cottrell and Bilby, 1951) is there a possibility that the motion may be a thermally-activated process. As a result lateral motion does occur under small stresses and is an athermal process. Hirth (1970) has again considered lateral kink motion and, contrary to his conclusion, the lateral motion of kinks does not affect the overall dislocation motion. The mean lifetime of a kink is $\sim 10^{-7}$ sec, whereas the time required to nucleate a kink is $\sim 10^{-2}$ sec.

In the above calculations it was assumed that if other double kinks are forming in the same dislocation line, there will be no interaction effects. This assumption can be justified by the rapid decrease in the interaction force with distance between kinks. Only for short separations is there an effect upon the activation energy. In Fig. 34 the calculated Gibbs force energy and experimental activation enthalpy energy for iron is shown. In calculating the activation energies the Peierls' stress at $0°$ K, $\tau_p^0 = 60$ kg/mm^2, was obtained by extrapolating the experimentally determined flow or yield stress curves to absolute zero.

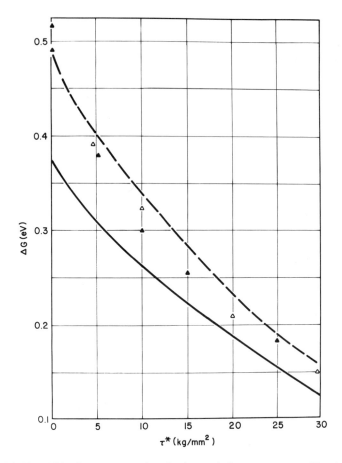

Fig. 34. The Gibbs free energy and activation enthalpy versus stress. The dashed line represents a line energy of $\frac{1}{2}\,\mu b^2$, $\tau_p^0 = 60$ kg/mm^2, $L = 148b$, and the solid line represents a line energy of $\frac{1}{5}\,\mu b^2$, $\tau_p^0 = 60$ kg/mm^2, $L = \infty$. ▲, Arsenault; △, Conrad and Fredrick.

The two sets of calculated values shown in Fig. 34 are for dislocation line energies of $\frac{1}{2}\,\mu b^2$ and of $\frac{1}{5}\,\mu b^2$. The larger line energy results in better agreement with the experimental data. The same procedure was also used to calculate the activation energies for tantalum (using a line energy of $\frac{1}{2}\,\mu b^2$ and a τ_p^0 of 60 kg/mm^2) (Arsenault, 1967a). The agreement was not as good; the experimental value of ΔH_0 at $\tau^* = 0$ is 0.8 eV. Dorn and Rajnak (1964) also obtained the best agreement for iron in their comparison of theory and experiment of data for some six bcc metals.

The calculated activation volumes also can be compared with experimental data. Figure 35 shows the variation of activation volume with respect to

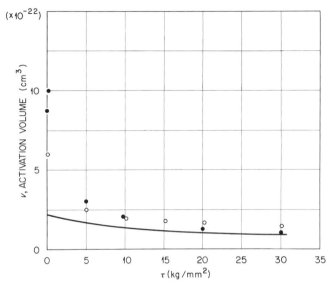

Fig. 35. Activation volume versus stress. The solid line represents the activation volume calculated for a line energy of $\frac{1}{2}\,\mu b^2$. ●, Arsenault; ○, Conrad and Fredrick; ——, calculated.

stress. The calculated values are less than those obtained experimentally (Arsenault, 1964; Conrad and Fredrick, 1962). The disagreement increases at the smaller stresses. Tantalum exhibits similar behavior.

The discrepancy between the experimental and the calculated v^* determined at small τ^* can be accounted for in terms of the periodic internal stress (Arsenault and Li, 1967). As a result of a periodic τ_μ, the v^* goes to infinity as $\tau^* \to 0$.

The double-kink mechanism described above cannot account for the asymmetry of the yield stress. However, Dorn and Mukherjee (1969) have proposed a modification of the Peierls' potential and as a result have predicted some asymmetry of the yield stress.

The theories based on the assumption that the mobility of the screw dislocations is controlled by the thermally-activated sessile–glissile transformation can account for the asymmetry of the yield stress (Vitek, 1966; Vitek and Kroupa, 1966; Kroupa and Vitek, 1967; Duesbery and Hirsch, 1968; Escaig, 1970; Duesbery, 1969), and all of those models are based on dissociated core models (Christain, 1970). The bases or mechanisms of these theories are very similar to the double-kink mechanism; however, there is one additional step, i.e. the constriction of the screw dislocation.

The theory outline by Duesbery (1969) will be considered here. Figure 36 shows the successive stages of formation of the double kink. At very low stresses, the saddle point is determined by the elastic interaction between the

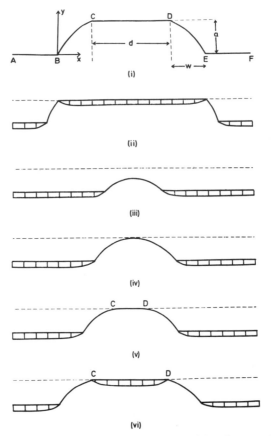

Fig. 36. Successive stages in motion of dissociated screw dislocation according to Duesbery (1969).

two kinks as shown in Fig. 36 (ii): this is also true of the Peierls' double-kink model, which was treated by Seeger (1956), but at most stresses the kink–kink interaction can be neglected. For intermediate stresses, the dislocation loop bows out until it reaches the next position at which it can dissociate [Figs. 36 (iii)–(v)]. It can reextend in sessile form [Fig. 36 (vi)] when CD exceeds a critical length. This is the saddle point configuration since the kinks then move apart with continuous decrease of energy. However, at high stresses the bowed-out loop becomes unstable before the stage of Fig. 36 (iv) is reached.

This model predicts three saddle point configurations, i.e. three ΔG depending upon the stress level. In order to consider ΔG, it is necessary to consider the configuration of dislocation. As shown in Fig. 37, the dislocation can exist in six distinct orientations in the lattice. Each configuration

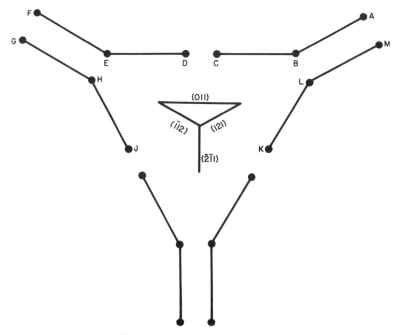

Fig. 37. Possible dislocation configuration (Duesbery, 1969).

shown can slip on either of its two composition planes, but in general with different activation energies. Since there are so many possibilities, it is necessary to consider the plastic strain rate in order to arrive at the activation parameters.

The plastic strain rate ($\dot\gamma$) resulting from the movement of a mobile screw dislocation density ρ_s with velocity v_s and a mobile edge dislocation ρ_E with velocity v_E is given by

$$\dot\gamma = \rho_s b \,|v_s| \,+\, \rho E^b \,|v_E| \tag{50}$$

Two extreme interpretations of this equation can be made.

(a) For a single dislocation loop with screw and edge components moving with different velocities it is clear that the screw dislocation density, for example, must be dependent on the edge dislocation velocity. If the velocities are uniform, then while the loop is freely expanding

$$\rho_{s(L)}/\rho_{E(L)} = |v_E| / |v_s| \tag{51}$$

where the subscript (L) indicates that the subscripted quantity refers to a single loop. The contribution of this loop to the total plastic strain rate may consequently be written:

$$\dot\gamma_{(L)} = 2\rho_{s(L)} b \,|v_s| \tag{52}$$

(b) In a real crystal, in addition to the term due to expanding loops, there will be a contribution from loops whose edge or screw parts have been blocked. In this case the screw dislocation density in Eq. (47) will depend on the mean free path in the crystal of edge dislocations, Λ_E, and the edge dislocation density on the screw mean free path, Λ_s. Any relation between ρ_s and ρ_E would be determined by the interaction, if any, between Λ_s and Λ_E. An estimate of this interaction would require consideration of the work-hardening of the system, and is beyond the scope of the present treatment.

In the steady state it may be expected that the relative contributions of (a) and (b) to the total strain rate will remain the same, so that the total strain rate will be of the form:

$$\dot{\gamma} = 2\left[\sum_L \rho_{s(L)} + \rho_s(\Lambda_E)\right]b\,|v_s| + \rho_E(\Lambda_s)b\,|v_E| \tag{53}$$

In the treatment below it will be assumed that Eq. (53) can be written approximately as

$$\dot{\gamma} \simeq \rho_s b\,|v_s| \tag{54}$$

where ρ_s is to be interpreted as an "effective" screw dislocation density.

Christain (1970) has arrived at the same conclusion, i.e. it is only necessary to define the screw velocity.

Each of the six configurations of the dissociation can move in four different ways. The resultant velocity of each configuration will be given in magnitude and direction by an expression of the form:

$$v = \sum_j (v_D b/l_j)\cdot(L/l_j)\cdot a_j \exp{(-\Delta G_j/kT)}, \tag{55}$$

where v_D is the Debye frequency, $v_D b/l_j$ the vibration frequency of a dislocation segment of length l_j, the activation length, L the free length of a screw dislocation, L/l_j the number of sites per dislocation available for activation, and a_j is the slip distance for a single activation; entropy is neglected and ΔG_j is the activation energy given by the appropriate mechanism. The summation is taken over the four possible slip modes. The free length of screw dislocation, L, which is the length of dislocation swept out by the kinks during a single activation, can depend on several factors, as shown by Duesbery and Hirsch (1967); we consider here only the case in which L is effectively constant, as for example when limited by defect interactions.

The resultant velocity of a screw dislocation is given by the summation of Eq. (55) over the six possible configurations. If ρ_i, v_i are the density and velocity, respectively, of the ith configuration, and ρ_s is the total screw dislocation density, then the total resultant screw velocity is

$$v_s = \sum_i \rho_i v_i/\rho_s \tag{56}$$

A basic difficulty in comparing Eqs. (54) and (55) with the experimental data is the lack of information on the variation of mobile dislocation density with strain. It will be assumed in the present treatment that the screw density ρ_s remains constant, and that the configuration densities ρ_i of ρ_s are equal.

The great majority of the available experimental data on the deformation of bcc materials relates to tests performed at constant plastic strain rate. Combining Eqs. (54) and (55) with this condition gives

$$6\dot{\varepsilon}/\rho_s v_D b^2 L = \text{const} = \left| \sum_j (a_j/l_j^2) \exp(-\Delta G_i/kT) \right| \tag{57}$$

with the summation taken over all possible configurations and slip modes. It must be stressed that while the experimental results may accurately reflect the condition of constant strain rate, the theoretical solutions of Eq. (57) represent only the condition of constant screw dislocation velocity.

Equation (20) can be rewritten as follows:

$$\Delta H_{\tau* \exp} = -kT^2 (\partial \ln \dot{\varepsilon}/\partial \tau^*)_T (\partial \tau^*/\partial T) \tag{58}$$

Implicit in the derivation of each ΔG_i is the following: $d\mu/dT = 0$ and $\tau_\mu = 0$; therefore $\Delta H_{\tau* \exp} = \Delta G_i$. Duesbery numerically differentiated Eq. (58) and obtained a $\Delta H(\tau^*)$ vs τ^* relationship. Recalling that the τ^* can be obtained as follows:

$$v^* = kT[\partial(\ln \dot{\gamma}/\dot{\gamma}_0)/\partial \tau^*]_T \tag{59}$$

he again performed a numerical differentiation of Eq. (57) and the results are shown in Fig. 38.

The intriguing thing about the curve in Fig. 38 is the local maximum at A. This local maximum has been observed in high purity niobium and molybdenum (Reed and Arsenault, 1975; Hasson et al., 1973).

If this analysis is carried to the point of considering the τ^* vs T relationship at constant strain rate, then the curve should not be a smooth function as Escaig (1972) has also shown. However, the data he uses to correlate with the theoretical prediction do not show a pronounced change in slope. A pronounced change in slope was indeed observed by Hasson et al. (1973) (Fig. 5).

The third model is based on the fact that interstitial solutes are effective hardeners because they produce tetragonal distortions, and that they can be the srb to dislocation motion. Fleischer (1962a,b) was the first to develop an analytical mechanism based on the fact that the interstitials can be the short-range barriers. This model is discussed in detail by Koppenaal and Arsenault (1971), and Gibbs (1969) has proposed a correction to Fleischer's original analysis which results in a reduction in τ_0^*, the effective stress at $0°$ K. Contrary to statements by Fleischer (1967), the magnitude of τ^*

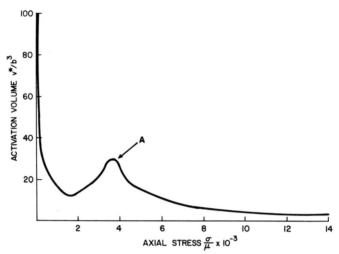

Fig. 38. The stress dependence of the activation volume for slip in pure iron. The curve represents the theoretical activation volume for stacking fault energies. $\Gamma_{110} = 242$ ergs/cm^2; $\Gamma_{112} = 473$ ergs/cm^2; $\ln (6\dot{\gamma}/\rho_s \nu_D b^2 L) = -28$. (Duesbery, 1969).

predicted for a given concentration of interstitials is much less than that observed. Even though it may be argued that the formulation of interaction energy between the solute and dislocation predicts too small of a force, a larger empirical value can be chosen. However, as Christain (1968) has pointed out, there is a limit to the magnitude of this force based on the simple fact that if the force is increased above a given amount depending upon the concentration, then the yield becomes independent of temperature, i.e. the Orowan mechanism becomes operative.

The analyses by Fleischer and Gibbs are analytical solutions in which several simplifying assumptions have to be made, one of which is that a regular array distribution is assumed for the interstitials, i.e. srb. Arsenault and Cadman (1975) have considered the problem of a random array of srb. This solution will be discussed here.

A random distribution of srb was placed on a plane, i.e. the slip plane; and the internal stress was assumed to be zero. The minimum separation between srb was $1b$. Two different force distance profiles (Fig. 27) were used. One was that obtained by Gibbs (1969) as follows:

$$F(y) = F_0[(y/b)^2 + (1/\sqrt{2})(y/b) - 1]/[(y/b)^2 + 1]^2 \qquad (60)$$

where $F_0 = \mu \Delta \varepsilon b^2/3.86$, $\Delta \varepsilon$ is the tetragonal strain, μ the shear modulus, and the value of ν for vanadium, 4.67×10^{11} dyn/cm^2, was used, b is the Burgers vector, and the value of b for vanadium, 2.63×10^{-8} cm, was used. This force–distance relationship is zero at $y = \infty$. However, thermal fluctuations cannot occur over distances greater than $10b$, so a cutoff of $10b$ was assumed.

A recent investigation by Barnett and Nix (1973) has shown that the above equation is wrong. The constant $1/\sqrt{2}$ should be replaced by $2\sqrt{2}$. The second force–distance diagram used was a delta function where the integrated area under the delta function was equal to the integrated area given by Eq. (60), and the maximum force of the delta function was again equal to the maximum force defined by Eq. (60).

Initially, a dislocation enters the slip plane and moves at a drag velocity defined by the following equation (Frost and Ashby, 1971):

$$v_D = B\tau \tag{61}$$

where B is a drag coefficient and τ_a is the applied stress. Since the long-range internal stress is zero, $\tau_a = \tau^*$ (the effective stress). When the dislocation makes contact with the srb, a force balance is determined by the following equation (Foreman and Makin, 1966):

$$\mu b^2 \cos \phi/2 = F(y) \tag{62}$$

where ϕ is the included angle about the srb; and it is assumed that the line tension is $\mu b^2/2$, and $F(y)$ is defined by Eq. (60) or by the delta function. The bowout between the initial points of contact occurs as a circular arc and is defined as follows: $R = \mu b/2\tau$. When the dislocation is bowing out, a search is carried out to determine if contact is made with other srb. If contact is made, then a new force balance has to be obtained at each point of contact with a srb. Then, again, the dislocation bows out between each srb where contact is made; and once more a search is carried out. This process is continued until an equilibrium force balance is obtained. The position of equilibrium is defined as Z (Fig. 27), and the other position of force equilibrium is defined as Z_1. The positions of force equilibrium are defined directly by the delta function. The main reason for using a delta function force–distance diagram was to reduce the computer time involved in making the calculations. The Gibbs force–distance profile was initially used for some of the investigations concerning the configuration of the dislocation as it moved across the slip plane. However, an immense amount of computer time was required to obtain a force balance due to the fact that the force on a particular srb was dependent upon the position of the dislocation on the neighboring srb.

If the applied force (F_A) is greater than the strength of the barrier (F_{max}), the dislocation is allowed to move over the barrier and the whole process is repeated. When a force balance is obtained and there are no points of contact where F_A is F_{max}, then the time required to overcome each srb where contact is made is determined by the following equation:

$$t_i = t_0 \exp \left[\Delta G_i(\tau)/kT\right] \tag{63}$$

where $t_0 = L/bv_0$, L is the average spacing of the srb, v_D the frequency, and

$\Delta G_i(\tau^*)$ is the activation energy of the ith barrier. The activation energy is determined by the following equation:

$$\Delta G_i = \int_z^{z'} F(y)\, dy - F(y)\bigg|_z \Delta y \qquad (64)$$

where the limits of integration are obtained from the force balance and $F(y)$ is again defined by Eq. (60) or the delta function.

The dislocation is then allowed to move over the srb which has the smallest jump time. By allowing the dislocation to jump only the srb with the smallest jump time, it is implicitly assumed that the probability of the dislocation jumping any other srb is very small. The smallest time is then stored after the jump has been made. This time is also accumulated at other points of contact and a linear relationship is assumed between the probability of overcoming a srb and time. The time accumulated by a dislocation at srb reduces the time necessary to overcome one of the srb. For example, if the time of one srb (No. 1) is 1 sec, and another srb (No. 2) is 10 sec, then the dislocation has increased its probability of overcoming the srb No. 2 by 10% when it has jumped srb No. 1. Now if, as a result of the first jump and a new force balance, the jump time at srb No. 2 is decreased 2 sec (and this is the minimum jump time), then only 1.8 sec of time is stored and not 2 sec. This procedure is continued until the dislocation first emerges from the array. What is neglected is the possibility that the dislocation might jump the barrier during the time it is bowing out to the static equilibrium position (Frost and Ashby, 1971). However, the time involved in the bowout for the small stress case is $\sim 10^{-8}$ sec, whereas the controlling jump time may be 10^3 sec. For larger stress, the bowout times are shorter, i.e. 10^{-11} sec, while the controlling jump times are still $\sim 10^3$ sec. It was therefore assumed that the dislocation would not jump the srb during the time required for the bowout to the static equilibrium position. The average velocity is determined as follows:

$$\bar{v} = y_L / \sum t_i \qquad (65)$$

where y_L is the length of the slip plane in a direction of motion of the dislocation and t_i is the time stored for each jump.

Determination of the boundary conditions and parameters which result in a steady state condition is *the critical factor* in determining the τ^* vs T relationship, and the activation parameters vs τ^* and T relationships.

The simplest manner by which the steady state condition can be defined is as follows: the steady state condition exists for a given temperature, stress, and concentration of srb if the average $\bar{v}$ does not change with an increase in the width or length of the slip plane. Figure 39 is an indication of a *non-steady* state condition even though the dislocation has overcome 21,000 srb: Fig.40 again indicates that a steady state condition does not exist. The

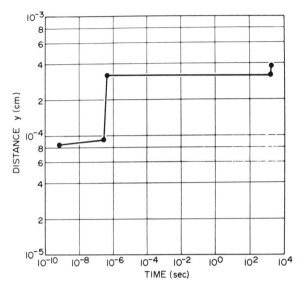

Fig. 39. Distance traveled by a dislocation as a function of time. $\Delta\varepsilon = 3$; $N_x = 66$; conc. = 143 ppm; $\tau = 3.45$ kg/mm^2; $T = 40°$ K; srb $\approx 21,000$.

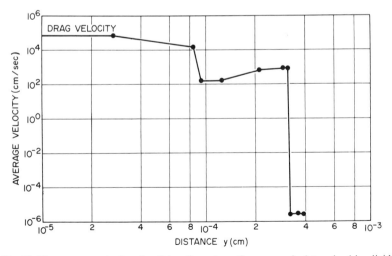

Fig. 40. The average velocity of a dislocation where the average is determined by dividing the distance traversed by total time to cover that distance as a function of distance. $\Delta\varepsilon = 3$; $N_x = 66$; conc. = 143 ppm; $\tau = 3.45$ kg/mm^2; $T = 40°$ K; srb = 21,000.

motion of the dislocation across the slip is controlled effectively by the time required to overcome one hard spot. Figures 39 and 40 may be extreme ones, i.e. for all practical purposes the total time required for the dislocation to travel across the slip is controlled by *one* jump. However, for numerous other conditions the time required for traversing the slip plane was controlled by a few jumps, 4 or 5 "hard spots." The nature of these hard spots can be defined in the following manner: 5 srb are in a rectangular area of $100b$ in the x direction and $20b$ in the y direction. (These dimensions are for a concentration of 143 ppm.) After defining the hard spot in this manner a search could be made for the hard spots of the slip plane. After locating the position of the hard spots, the dislocation was placed at a variable distance behind a hard spot and then allowed to move. The dislocation was removed from the slip plane when it had jumped a hard spot, and then placed behind the next hard spot. A correlation of the total times needed to traverse the slip plane by letting the dislocation travel the length of slip plane completely or picking it up and moving it from one hard spot to the next did *not* exist.

From these negative results it appears that there is a difference between a "dynamic hard spot" and a "static hard spot."

Therefore, the procedure chosen was to let the dislocation traverse the entire slip plane. Figure 41 is an example of a case where the dislocation motion has reached a steady state condition; changing the y dimension from 1.5×10^{-5} to 1.1×10^{-4} cm does result in a change in the average velocity $\bar{v}$, thereby satisfying one boundary condition. The second boundary condition which has to be satisfied is that the $\bar{v}$ does not change as a result of

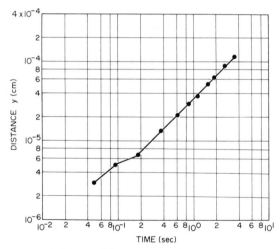

Fig. 41. Distance traveled by a dislocation as a function of time. $\Delta\varepsilon = 1.5$; $N_x = 66$; conc. = 2283 ppm; $T = 650°$ K; $\tau = 0.25$ kg/mm^2.

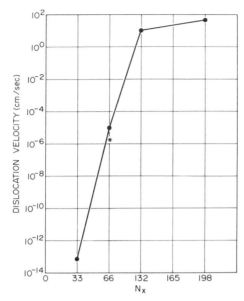

Fig. 42. Average dislocation velocity as a function of the width of the slip plane. N_x represents the average number of srb in the x direction, i.e. the width of the slip plane. $\Delta\varepsilon = 3$; $\tau = 3.35 \text{ kg/mm}^2$; $T = 40°$ K; conc. $= 143$ ppm; $*$, $\tau = 3.45 \text{ kg/mm}^2$.

changing the y dimensions, i.e. width of the slip plane. Figure 42 is a plot of v vs N_x which is the average number of srb in the x direction. As N_x is increased from 33 to 198 the $\bar{v}$ increased by $\sim 10^{16}$. For N_x's greater than 132 the $\bar{v}$ has reached a steady state condition. A very convincing point which indicates that a steady state condition does not exist is the fact that for $N_x = 66$ an increase in τ^* resulted in a decrease in $\bar{v}$. A simple explanation as to why the width of slip plane has a great effect on the velocity is based on two points: mirror boundary conditions were used, and the probability of having a large spacing between points of contact increases as the width of the slip plane increases. There is still a third boundary condition to be satisfied before a steady state can be determined and that is that ΔH does not change with width, as shown in Fig. 43.

The conditions, i.e. the width and length of the slip plane, required at a given temperature and stress do not necessarily hold for another temperature and stress condition. In general, as the stress is increased for a given concentration, the width and length have to be increased to obtain steady state conditions.

After defining the barrier it is necessary to define the mechanism by which the dislocation overcomes this barrier. Then it is possible to obtain expressions for the activation parameters as a function of stress, concentration of

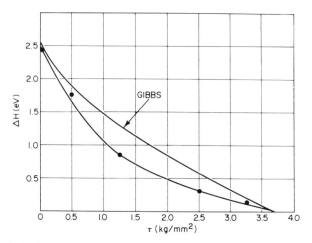

Fig. 43. Activation energy as a function of stress. ●, $\Delta\varepsilon = 3$; srb conc. = 143 ppm.

srb, temperature, etc. Finally, it is possible to obtain an expression for τ^* vs T. Therefore, activation energy will be considered first. The activation energy (ΔH) was obtained by determining the change in $\bar{v}$ due to a change in temperature at a constant τ^*, according to the following formula:

$$\Delta H_{\tau*} = k \ln (\bar{v}_1/\bar{v}_2)/(1/T_1 - 1/T_2) \tag{66}$$

Figure 44 is a plot of ΔH vs τ^* for a srb concentration of 143 ppm, and even though a square force–distance diagram was employed, ΔH is *not* a linear function of τ^*.

The value of ΔH_0 obtained by extrapolating ΔH to zero τ^* is in good agreement with the calculated value of the area under the force–distance diagram. For example, at a concentration of 143 ppm and $\Delta\varepsilon = 3$, the calculated value of ΔG_0 is 2.42 eV and the experimental ΔH_0 value is 2.45 eV. The temperature dependence of ΔH was determined by obtaining ΔH at various temperatures at a constant τ. Within the accuracy of the calculations, ΔH is *not* a function of temperature; i.e. ΔH is independent of temperature.

The Fleischer type analysis of ΔG for a *square* force distance diagram is quite simple:

$$\Delta G = \Delta G_0 - \tau^* b l \tag{67}$$

where ΔG_0 is the area under the force–distance diagram, d the width of the force–distance diagram, l the mean spacing between srb, and $l = C^{-1/2}$, where C is the concentration of srb. This analysis obviously results in a

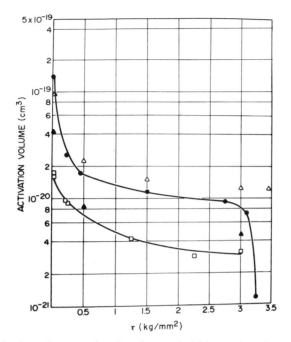

Fig. 44. Activation volume as a function of stress. △, Gibbs; ●, $\Delta\varepsilon = 3$, conc. = 143 ppm; ▲, Gibbs; □, $\Delta\varepsilon = 1.5$, conc. = 2283 ppm.

linear relationship between ΔG and τ. It is assumed that $\Delta G = \Delta H$, i.e. zero entropy.

The Gibbs modification of the Fleischer analysis, which employs the Freidel correction for the dislocation length between srb, is as follows:

$$\Delta G = \Delta G_0 - \tau^* b \, d\lambda \qquad (68)$$

where ΔG_0 and d are the same as in the original Fleischer analysis; but λ is defined as follows:

$$\lambda = K(\mu b l^2/\tau)^{1/3} \qquad (69)$$

where K is a constant equal to 1, and $l = C^{-1/2}$. The Argon (1972) analysis results in a value of K between 1 and 2. Combining Eqs. (68) and (69) results in the following:

$$\Delta G = \Delta G_0 - K^1 \tau^{*2/3} \qquad (70)$$

where $K^1 = Kb^{4/3} \, d(\mu l^2)^{1/3}$, or, for the purposes of comparing the ΔG vs τ^* relationship, K^1 is a normalizing constant. In Fig. 43 the Gibbs relationship

between ΔG and τ^* is shown, and there is fair agreement with the numerically determined results.

The activation volume (activation area times the Burgers vector) was
obtained by determining the change in $\bar{v}$ due to a change in τ^* at constant T,
as follows:

$$v^* = kT \ln (\bar{v}_1/\bar{v}_2)/(\tau_1 - \tau_2) \tag{71}$$

Figure 44 illustrates the large effect τ^* has on v^* for both concentrations of
srb.

The Fleischer type analysis predicts that v^* is independent of stress, i.e.
$v^* = b \, dC^{-1/2}$.

The Gibbs formulation of v^* is as follows:

$$v^* = b \, d\lambda \tag{72}$$

where λ is given by Eq. (69) and $d \cong 3b$. For a srb concentration of 143 ppm
and 2283 ppm, the values of l are 2.5×10^{-6} cm and 6.25×10^{-7} cm, respectively. The v^* vs τ^* relationships obtained using Eq. (72) are shown in
Fig. 44 and, surprisingly, there is remarkable agreement. There are a few
problems with this formulation such as the fact that v^* does not go to zero at
τ_0^*, which occurs in the numerical solution.

If the formulation of Fleischer is used for v^*, then l can be obtained from
v^* (v^* at zero τ^*). If the extrapolated values of v_0^* of 1.5×10^{-19} (v^* at zero
τ^*) are used, then the value of $l = 7 \times 10^{-5}$ cm. If $l \approx C^{-1/2}$ then the value
of $l \approx 2.5 \times 10^{-6}$ cm, i.e. almost two values of magnitude smaller than the
numerically calculated l. A similar analysis for a larger concentration case,
i.e. 2283 ppm, again produces similar results. The value of l obtained numerically is approximately two orders of magnitude larger than that obtained
from a $l \approx C^{-1/2}$ relationship.

The stress versus temperature relationship was obtained at approximately
constant $\bar{v}$ conditions. Figure 45 is a plot of the stress required to obtain an
approximately constant $\bar{v}$, as a function of temperature for two values of τ_0^*
and three different concentrations. The stress at $0°$ K (τ_0^*) varies as $C^{1/2}$, and
in all cases the values of τ_0 obtained by extrapolating to $0°$ K are in good
agreement with those predicted by Foreman and Makin (1966).

It is also possible to compare the temperature dependence of τ^* vs T as
predicted by Fleischer and Gibbs with numerical results by the use of the
following equation:

$$\bar{v} = v_0 \exp \left[- \Delta G(\tau^*)/kT \right] \tag{73}$$

where $\Delta G(\tau^*)$ is defined by Fleischer and Gibbs and v_0 is related to the
attempt frequency and the average distance traveled by the dislocation per

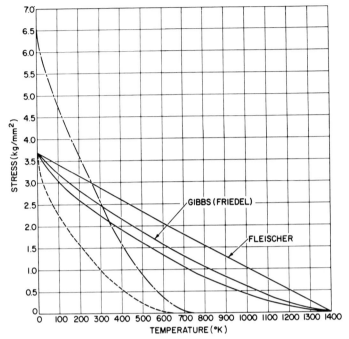

Fig. 45. Stress as a function of temperature at approximately average dislocation velocity, $\bar{v} \cong 10^{-5}$ cm/sec. ——, $\Delta\varepsilon = 3$, srb conc. = 143 ppm; – – – –, $\Delta\varepsilon = 1.5$, srb conc. = 571 ppm; — – —, $\Delta\varepsilon = 1.5$, srb conc. = 2283 ppm.

thermally-activated event (*h*). If the Fleischer and Gibbs expressions for ΔG are substituted into Eq. (73), then it can be rewritten as follows:

$$\tau_{\text{Fleischer}} = [\Delta G_0 - kT \ln (v_0/\bar{v})]/b \, dl \qquad (74)$$

or

$$\tau_{\text{Fleischer}} = B - CT$$

$$\tau_{\text{Gibbs}} = [\Delta G_0 - kT \ln (v_0/\bar{v})]^{3/2}/K' \qquad (75)$$

or

$$\tau_{\text{Gibbs}} = [D - ET]^{3/2}$$

The above analysis only assumes that the force–distance relationship is positive, i.e. opposing the motion of the dislocation. Arsenault and de Wit (1974) have shown if a carbon in an iron lattice is considered, then the force–distance relationship can be either positive or negative, i.e. the carbon atom can aid the motion of a dislocation. Arsenault and Cadman (1973c) have developed a computer simulation model for dislocation motion on a slip plane which has defects having both positive and negative portions of the force–distance relationship. If there is Peierls stress and the dislocation motion is controlled by double-kink nucleation, then the presence of low carbon, i.e. < 100 at. ppm has very little effect. Increasing the carbon concentration can result in a significant *reduction* in the yield stress.

VI. Interstitial Alloys

It has been argued by some workers (Ravi and Gibala, 1969, 1970b) that there have been no investigations where the true characteristics of the bcc metals have been observed, i.e. in all bcc metals, even the highest purity bcc

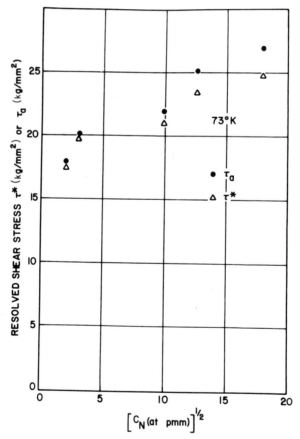

Fig. 46. Stress versus concentration to the 1/2 power (Lachenmann and Schültz, 1970).

metals produced to date, the dislocation kinetics are still controlled by impurity interstitials. The earliest strong evidence that interstitials were the rate-controlling srb in iron–carbon alloys was in the work of Stein and Low (1966; Stein et al., 1963). However, later work by several investigators has almost completely invalidated Stein and Low's data (Christain, 1970).

There are many published reports and reviews (Conrad, 1963) of the variation of the yield stress with interstitial concentration, C, for alloys with relatively high interstitial content. The indication is that the strengthening produced is approximately athermal. In addition, there are cases where interstitials, i.e. oxygen in niobium (Ravi and Gibala, 1969, 1970a,b), have reduced the temperature dependence of the flow or yield stress. This observation certainly contradicts the simple theories of interstitial hardening.

However, there are some recent results on interstitial doped niobium and tantalum which appear to indicate that some of the temperature dependence of the flow stress is due to very small amounts of interstitials. Lachenmann and Schultz (1970) have investigated nitrogen doped tantalum single crystals. If the data obtained at 73° K are replotted against $C^{1/2}$ then a reasonable linear relationship is obtained for single crystals with $\chi = 0$, as shown in Fig.46. This is rather strong evidence that interstitials are affecting the temperature dependence of the yield stress in some manner. However, the data in Fig. 10 for a [110] orientation indicate that nitrogen only affects the athermal component of the yield stress.

Smialek and Mitchell (1970) have also investigated nitrogen doped single crystals of tantalum. Figure 47 is a plot of τ_a versus nitrogen concentration for two different temperatures. For nitrogen concentrations greater than 200

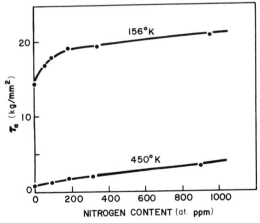

Fig. 47. Yield stress versus concentration of nitrogen for tantalum–nitrogen alloys (Smialek and Mitchell, 1970).

at. ppm, the data indicate that the interstitial nitrogen only affects τ_μ. It should be pointed out that the solubility of nitrogen in tantalum at $300°K$ is 5–10 at. %. Therefore, the linear region cannot be explained in terms of a precipitation phenomenon. For nitrogen concentration < 300 at. ppm, the data indicate that nitrogen atoms are affecting τ^*. However, if the nitrogen atoms are the srb's to dislocation motion, then τ^* should be related to $C^{-1/2}$. The data of Smialek and Mitchell, as shown in Fig. 47, indicate that τ^* is *not* a $C^{1/2}$ function of nitrogen concentration.

Taylor and Bowen (1970) investigated nitrogen doped niobium single crystals of two different orientations. They also observed that for the [491] orientation, nitrogen had a larger effect on the $(\partial\tau^*/\partial T)$ then for an orientation near [111]. However, they also observed zero effect of nitrogen on $(\partial\tau^*/\partial T)$ for differential temperature tests, with τ^* for high purity increasing from $\simeq 12$ kg/mm² to 30 kg/mm² due to the differential temperature testing at 77° K. Reed and Arsenault (1975) have also determined the effect of differential temperature testing on τ^*. Figure 48 is a plot of τ vs γ at various temperatures for the purpose of showing how a plot of τ^* vs T is obtained by the differential temperature technique. There is no straining during the time required to make the temperature change. Now the data points are obtained by summing the τ^* to the particular temperature.

Reed and Arsenault (1975) have observed differences between prestrained or incremental tests and nonprestrained samples from 4.2 to 77° K in the tensile yield stress of high purity Nb, i.e. 60 vs 48 kg/mm² incremental and nonprestrained, respectively. If the angle of the slip plane is taken into account, the difference in resolved shear stress is large at 4.2° K, i.e. 30 vs 13 kg/mm² incremental and nonprestrained samples, respectively.

At 77° K there is no difference in the tensile yield stress of incremental and nonprestrained tested samples. However, taking into account the difference in slip plane, then nonprestrain samples have a yield stress of 6.5 kg/mm² and the incremental tested samples have a yield stress of 12–15 kg/mm².

Another factor which has been observed in high purity Nb and Mo is a localized maximum, in plots of v^* vs τ^*, as shown in Fig. 49. The height of the localized maximum, and even a discontinuity of the v^* vs τ^*, has been shown to be dependent on the prior history of the sample (Fig. 50). Prestraining reduces this localized maximum. Also associated with this difference in the v^*–τ^* relationship at $\simeq 12$ kg/mm² is that prestraining also affects the nature of the stress–strain curve (Fig. 51). Therefore, there appears to be a strong correlation between the slip and the v^*–τ^* relationship. A possible cause of the discontinuous slip in the nonprestrained samples is adiabatic heating. However, the analysis presented by Kubin and Jouffery (1971) does not appear to fit the present results.

The localized maximum in the v^*–τ^* relationship has been predicted by

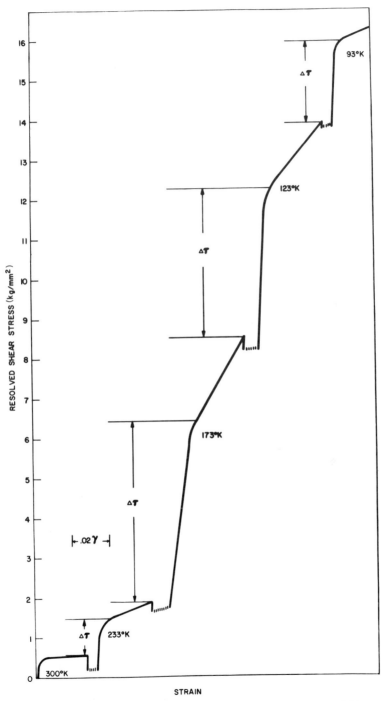

Fig. 48. Stress versus temperature for a differential temperature test on niobium single crystals.

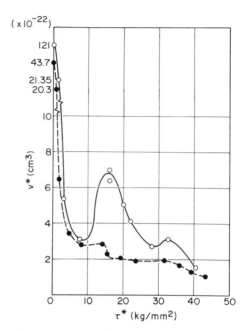

Fig. 49. Activation volume versus stress for molybdenum single crystals. ○, nonirradiated Mo; ●, irradiated Mo (fluence = 1.6×10^{19} n/cm^2).

Duesbery (1969) for the dissociated screw mechanism. By a very simple argument it can be shown that the localized maximum cannot be due to a change in mechanism, for example from the tetragonal distortion to the double-kink mechanism. Consider the schematic diagrams of Fig. 52 and the fact that $v^* \propto \partial\tau/\partial T$.

The vast majority of the results indicate that impurity interstitials only affect the athermal component of yield stress. However, there are some results on high purity niobium and tantalum which indicate that impurity interstitials may affect the effective stress (τ^*); but only in the high purity samples is the asymmetry very pronounced and, more importantly, a localized maximum of the v^* vs τ^* is observed. This maximum can be readily explained in terms of the intrinsic lattice as the srb to dislocation motion. A theoretical problem which has to be solved is how or why a small amount of impurity interstitial appears to affect τ^* in niobium and tantalum. The simple analytical theories and the computer simulation models predict that these small interstitial concentrations should have about zero effect. Also interstitial impurities can lead to a reduction in τ^* if it assumed that the lattice, i.e. Peierls stress, is the srb to dislocation motion.

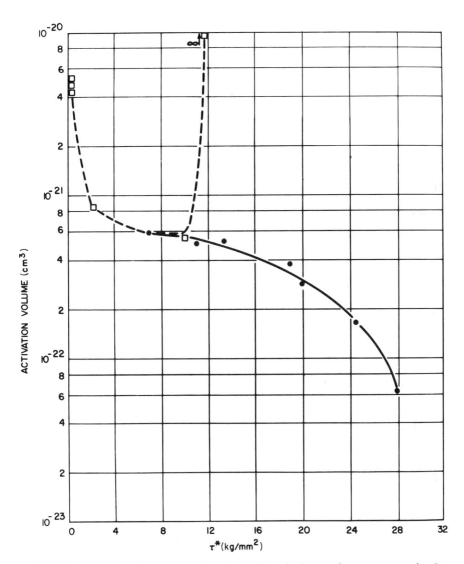

Fig. 50. Activation volume versus stress for niobium single crystals. □, nonprestrain; ●, prestrain.

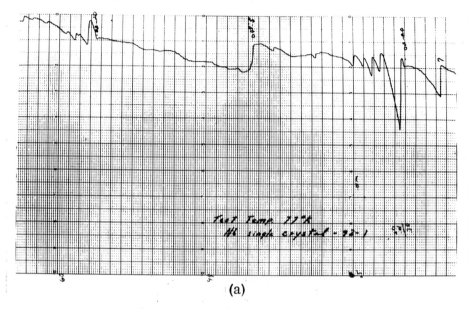

(a)

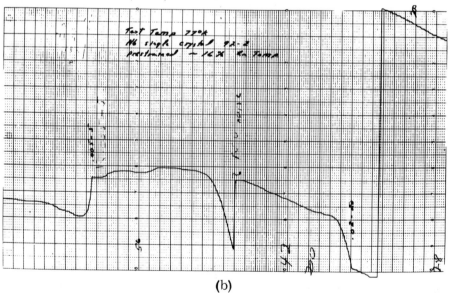

(b)

Fig. 51. Instron charts of load versus time for the deformation of niobium. (a) nonprestrain; (b) prestrain.

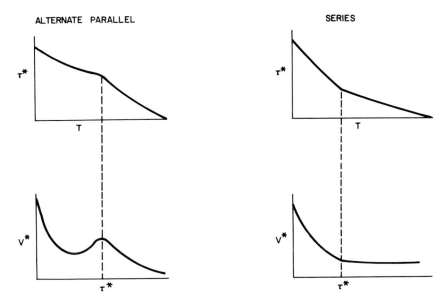

ALTERNATE PARALLEL **SERIES**

Fig. 52. Schematic diagrams showing the interdependence of the activation volume–effective stress and effective stress–temperature relationship.

VII. Hydrogen Alloys

The effect of hydrogen on the dislocation kinetics in bcc metals is treated separately for two reasons. The solubility of hydrogen in several bcc metals (i.e. vanadium, niobium, tantalum) decreases to a very small amount in the temperature range of 100–150° K (Westlake, 1967, 1969a,b,c; Owen and Scott, 1970, 1972). The mobility of hydrogen at these temperatures is sufficient so that precipitation can occur.

There have been numerous studies of the effect of small amounts of hydrogen (i.e. less than 2 wt ppm) on the ductility of iron base alloys. Also Scott and co-workers and Westlake have shown that small amounts of hydrogen result in a loss of ductility of vanadium, niobium, and tantalum.

There have been a few investigations of the effect of hydrogen on the low-temperature yield and flow stress of bcc metals. Eustice and Carlson (1961) observed a "bump" in the plot of yield stress versus temperature of a V–H alloy.

An investigation was undertaken by Chen and Arsenault (1975) to determine the influence of hydrogen on the low-temperature deformation characteristics of high purity vanadium single crystals in the temperatures ranging from 77 to 473° K. Vanadium containing hydrogen exhibits a local discontinuity in the temperature dependence of yield stress (Fig. 53). The local discontinuity can not be explained in terms of normal or stress-induced precipitation hardening due to the fact that the pronounced increase in yield stress disappears at lower temperatures. However, it must be related to the precipitates because the local maximum occurs below the solvus temperature. On the basis of an analysis of the activation parameters, a mechanism which could explain the local maxima is the stress-induced reorientation of the hydride platelets due to the stress fields of the moving dislocations.

In order to characterize more generally the hydrogen effect on the low-temperature deformation behavior of Group Va metals, an investigation has been undertaken by Chen and Arsenault (1975) on high purity niobium and tantalum single crystals in the temperature range of 20 to 300° K. Similar to that observed in V–H, the addition of hydrogen causes a local discontinuity in the temperature dependence of yield stress. The local maximum in yield stress occurs at a temperature below the solvus and the effect disappears with a further decrease in temperature. The measured strain rate sensitivity and activation parameters are in good agreement with those obtained from the yield stress versus temperature results. The strengthening phenomena

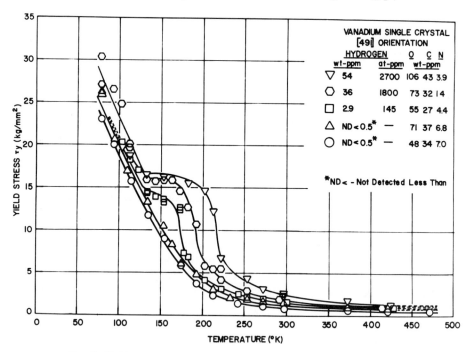

Fig. 53. Yield stress versus temperature for various vanadium–hydrogen alloys.

occurring in Nb–H and Ta–H alloys at low temperatures can be attributed to the stress-induced reorientation of hydrides due to the stress field of the moving dislocations. Additional observations are made on the effect of isothermal aging and oxygen content. It is demonstrated that neither precipitation nor trapping phenomena can account for the local discontinuity in the temperature dependence of yield stress.

Ravi and Gibala (1971) did not observe a "bump" in their investigation of Nb–H alloys most likely because they did not perform tests at sufficiently low temperatures. Ravi and Gibala (1971) analyzed the increase in τ^* in terms of a dislocation interstitial interaction mechanism.

VIII. Substitutional Alloys

In a discussion of the effect of alloy additions on dislocation dynamics of bcc solid solutions it is important to reconsider the fact that the yield stress or the flow stress consists of two components, i.e. τ^* and τ_μ. A general statement which is made in the introduction of most review papers on the

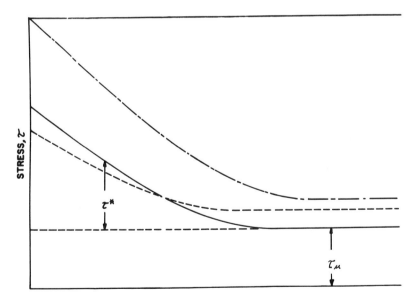

Fig. 54. Schematic diagram illustrating the various possible changes in the yield stress due to an alloy addition. ——, pure bcc; – – – –, solid solution No. 1; — – —, solid solution No. 2. $\tau = \tau^* + \tau_\mu$.

effect of alloy additions is as follows: When element A is added to element B the resulting solid solution is stronger than either A or B. This statement is not necessarily correct for bcc solutions, i.e. the solid solution may be weaker than either A or B. However, in general the solute addition results in an increase in τ_μ, but the solute addition may result in a decrease or an increase in τ^*. Figure 54 is a schematic diagram illustrating the possible ways in which the solute addition can affect the yield stress. Since solute additions affect the yield stress in several different ways, the discussion of alloy effects will be divided into two major categories: strengthening and weakening. The category of strengthening is further subdivided into an increase in τ_μ and an increase in τ^*.

A. Solid-Solution Hardening

1. SOLID-SOLUTION STRENGTHENING OF THE ATHERMAL COMPONENT

Bcc solid solutions in general exhibit a linear relationship between strengthening and solute concentration when strengthening occurs (Fig. 55). The strengthening considered here is the increase in τ_μ due to the solute addition.

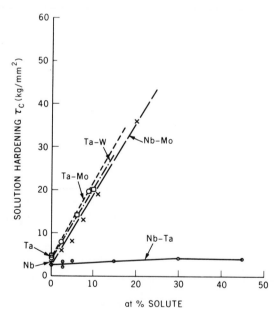

Fig. 55. BCC substitutional solid-solution hardening of Ta and Nb at 273° K. Results for Ta–W and Ta–Mo from Mitchell and Raffo and for Nb–Mo and Nb–Ta from Peters and Hendrickson.

Labusch (1972) has given an excellent review of most of the existing solid-solution strengthening theories and has shown that they all predict a $C^{1/2}$ or $C^{2/3}$ dependency. This is not surprising since these theories were developed to explain the solid-solution strengthening observed in fcc solid solutions. In general, these theories predict the strengthening at $0°$ K, and it is implicitly assumed that the strengthening is independent of temperature. However, even in fcc metals the experimental results indicate that the strengthening is temperature dependent.

There are only two theories which have been proposed to account for the linear hardening observed in bcc metals, those of Suzuki (1971) and Arsenault (1973).

Suzuki's theory employs a statistical determination of "hard spots," but Suzuki states that the entire strengthening is due to an increase in the strength and density of srb. This means that alloy addition results in only an increase in τ^*; there is no change in τ_μ. In the high temperature, i.e. low τ^* range, the rate-controlling step is related to the lateral motion of a single kink. More specifically, a statistical determination is made of the distribution and the strength of the "hard spots" along the dislocation line. The rate-controlling step is the jump of the kink over the hardest of the "hard spots" in a given length of dislocation line. In the low-temperature range, the rate-controlling step is the nucleation of the double kink. There are a few difficulties when applying this theory. Since the strengthening is due to a change in the thermally activated motion of the dislocation, it is possible to define the strain rate by Eq. (13). Taking typical values of ΔG_0 (the activation energy at $\tau^* = 0$) of 1 eV, and $v_0 = 10^8$ cm/sec and $v = 10^{-4}$ cm/sec, the value of T_c, the temperature at which τ^* is zero, is approximately $400°$ K. This means that above $400°$ K there would be zero strengthening due to the alloy addition. The experimental results of Rudolph and Mordike (1967) conclusively demonstrate that the strengthening continues to a temperature of $1400°$ K.

Suzuki (1971) has a rather complete survey of references which demonstrate the linear relationship between hardening and C.

As a first approximation, Predmore (1975) assumed a random solution for the purposes of accounting for the linear hardening. This model will be outlined below.

A random solid solution containing a range of solute contents was constructed as the model to simplify the calculation of the hardening effect. The random solid solution models were divided into small groups of solute atoms containing from one to ten solute atoms per group. These groups correspond to locally ordered regions in the alloys. The distribution of groups containing various solute contents were determined by using a random number table. Three random arrays of 2500 atoms and three arrays of 2304 atoms were selected using the random number tables and containing 10%, 20%, and 30% solute atoms, respectively. The arrays containing 2500 atoms were divided into 250 groups of 1 by 10 atoms in a row. The number of groups of each solute concentration were counted for each array. The total number of groups containing 2 or more solute atoms through 6 or more solute atoms is shown in Fig. 56. These groups are termed "hard spots." The arrays containing 2304 atoms were divided into 256 groups of 3 × 3 atoms in a plane. The total number of groups of each composition was counted.

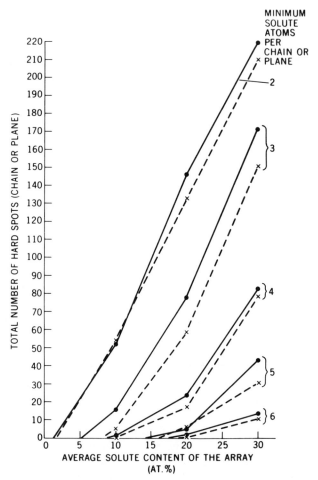

Fig. 56. Total number of hard spots per random array when each hard spot contains from 2 or more to 6 or more solute atoms per 10 atom chain or 9 atom plane above nominal array composition. ●, 2500 atom array, 10 atoms per chain; ×, 2304 atom array, 9 atoms per plane.

The total number of groups containing 2 or more solute atoms through 6 or more solute atoms was plotted for arrays containing 10%, 20%, and 30% solute atoms in Fig. 56. The number of groups containing 9 atoms per plane is smaller than the number containing 10 atoms per chain because they correspond to a lower composition for an equal number of solute atoms in the group. The total force exerted on the dislocation by the obstacle was calculated using Eq. (62). The dislocation is assumed to break away from the obstacle or hard spot when the angle between the arms of the dislocation reaches the value ϕ.

The interpretation of the short-range order and the atomic displacement parameters shows that the strain around a solute Ta atom in Mo has a strain of 5% of the Ta atomic diameter and 10% for Mo dissolved in Ta. The total force exerted on an edge dislocation by the solute atom strain field can be calculated from Weertman and Weertman (1964). Equating the obstacle–dislocation interaction force [(Eq. (62)] to the interaction force of an edge dislocation and the solute atoms strain field leads to

$$\cos \phi/2 = \varepsilon b^2 (xy)/R^4 \tag{76}$$

where ε is the strain, x and y are the distance coordinates of the solute atom, and R is the distance between the solute atom and the dislocation. The strength of the obstacle (per solute atom) was calculated using $gy = x = a$, $R = 1.4a$, $\varepsilon = 0.05$ for Ta solute atoms and $\varepsilon = 0.10$ for Mo solute atoms. This obstacle strength, corresponding to one solute atom, was multiplied by the number of solute atoms in each group in order to determine the strength of the hard spots.

The increase in strength due to solution hardening was calculated from the relation for weak obstacles given as (Foreman and Makin, 1966):

$$\tau = (\mu b/L)(\cos \phi/2)^{3/2} \tag{77}$$

where L is the distance between hard spots located on a square grid and $(\cos \phi/2)^{3/2}$ characterizes the strength of the hard spots. The shear modulus was taken as a linear relation between 13.4×10^{11} dyn/cm^2 for Mo and 5.17×10^{11} dyn/cm^2 for Ta (Armstrong and Mordike, 1970). The value for $(\cos \phi/2)^{3/2}$ was taken to be the weighted average of the number of hard spots of each composition and the obstacle strengths corresponding to the solute concentration in the hard spot. The solution hardening was calculated for each array using groups containing 3 or more solute atoms per group as hard spots for the 10% alloy, 4 or more solute atoms per group as hard spots for the 20% alloy, and 5 or more solute atoms per group as hard spots for the 30% alloy, and then plotted in Fig. 57. Similar solution hardening results were obtained by choosing hard spots as groups with 22 or more solute atoms for the 10% alloy, with 3 or more solute atoms for the 20% alloy, and

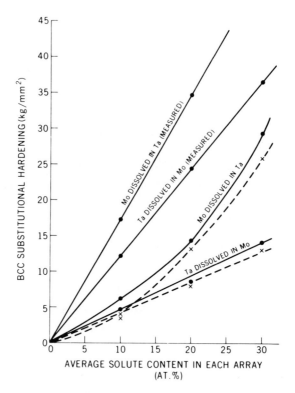

Fig. 57. Measured solution hardening and calculated substitutional solution hardening obtained from an edge dislocation interaction with the hard spot point strain. Hard spots have 2 or more solute atoms above nominal array composition. ●, 2500 atom model; ×, 2304 atom model.

with 4 or more solute atoms for 30% alloy. The solution hardening is similar for both the 2500 atom arrays, which shows the statistics are consistent.

The solution hardening calculation predicts a nearly linear solution hardening rate, an absence of solution hardening with an absence of size effect, and that the Ta rich alloys will have greater solution hardening than the Mo rich alloys. The hard spots are large enough in size to adequately prevent thermal motion of the dislocation. The hard spots form long-range barriers to dislocation motion and induce plateau substitutional hardening. These substitutional hardening calculations based on a random distribution of atoms explain about half of the observed hardening.

Theories reviewed by Labusch (1972), and Suzuki's (1971) and Arsenault's (1973) theories are based on a random solid solution. However, Predmore (1975) has shown that where there is an atomic size difference and substantial hardening there is also short-range ordering. Figure 58 is an X-ray intensity map of the $h_1 h_2 0$ plane in reciprocal space from a Mo–91%

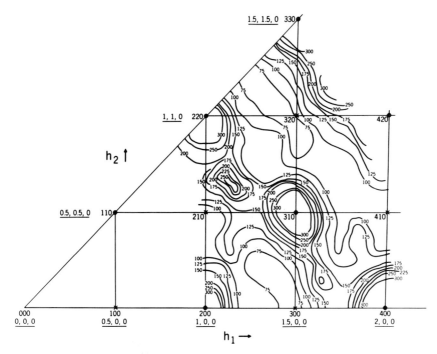

Fig. 58. Diffracted X-ray intensity distribution in the $h_1 h_2 0$ plane of reciprocal space for Mo at ambient temperature counts per 30 sec.

Ta alloy. The local intensity maximum near 210 and 300 represents short-range order. Unfortunately the existing theories of short-range order strengthening do not predict a linear relationship between hardening and composition (Fischer, 1954). Simulating a large array of atoms to agree with the measured SRO parameters, such as Gehlen and Cohen (1965) have done, may increase the separation distance and strength of the hard spots due to higher solute concentration, and in turn increase the calculated solution hardening.

Rudman (1965) attributed the large increase in solution hardening of concentrated bcc solutions to the SRO mechanism of Fischer (1954). The moving dislocation increases the energy of the solution as it passes through by deforming it to a random solution. The mechanism requires a large force to pass the dislocation through the ordered lattice, thus producing a large increase in solution hardening. This hardening mechanism which disorders the lattice could account for the additional solution hardening observed in the Ta–Mo alloys.

Mordike and Rogansch (1970) over the past few years have conducted extensive investigations of the strengthening of tantalum base alloys. They have shown that very good agreement can be obtained with Labusch's (1972) theory by choosing the appropriate misfit parameters.

1. The size misfit parameter

$$\varepsilon_b = (1/a)(da/dc)$$

2. The modulus misfit parameter

$$\varepsilon_\mu = (d\mu/\mu \ dc)[1 - (1/a\mu) | d\mu/dc |]^{-1}$$

The size misfit is measured by the change in lattice parameter, a, due to alloying. The definition of ε_μ is not unique, as the value depends upon the combination of elastic constants used. The modulus interaction arises due to the difference in the energy of the dislocation as a result of a change in modulus. The strength of the interaction is thus determined by the modulus which describes the stress field of the dislocation. The modulus for a screw dislocation in bcc metals is K_{33}, as defined by Head (1964) as $K_{33} = C_{44}$. The following equation was derived by Labusch (1972):

$$\Delta\tau_0 = Z\mu\varepsilon^{4/3}C^{2/3} \tag{78}$$

where $\Delta\tau_0$ is the difference in yield stress between the alloy and the base metal, $Z = 2 \times 10^{-3}$, and $\varepsilon = \varepsilon_\mu + \alpha\varepsilon_b$. The constant α is an adjustable parameter within limits. For a screw dislocation, it is less than 4. Mordike et $al.$ used a value of $\alpha = 2$. The $\alpha\varepsilon_b$ was obtained in the following manner. Recall from the simple isotropic elasticity theory that there should be zero total force interaction between a spherical dilatational field and a screw dislocation. However, Chou (1965) in a very interesting paper has shown that by considering the anisotropic elastic constants, a screw dislocation in a bcc lattice has a dilatational field; and for tantalum its strength is only a factor of 10 less than that of an edge dislocation; whereas in the case of iron the difference is even less—a factor of 6. If it is assumed that $\alpha = 30$ for an edge dislocation, then a reasonable value of α for a screw would be less than 3 for tantalum.

Figure 59 is a plot showing the correlation between $\Delta\tau$ and the misfit parameter. All of the alloy data points are in good agreement, and the proportionality constant, $Z = 0.7 \times 10^{-3}$, is in reasonable agreement with the predicted value. Mordike and co-workers further concluded that since the line passes through the origin, the interaction between the screw dislocation and a solute atom is purely elastic, i.e. there are no valance (no electronic) effects.

If the alloy data used by Mordike and co-workers are analyzed in terms of an edge dislocation–misfit interaction then there is no agreement. A second factor which should be considered is the force on a screw dislocation which arises from the torque term (Urakami and Fine, 1970). Arsenault and Kuo (1973) have made some preliminary calculations and have shown that the

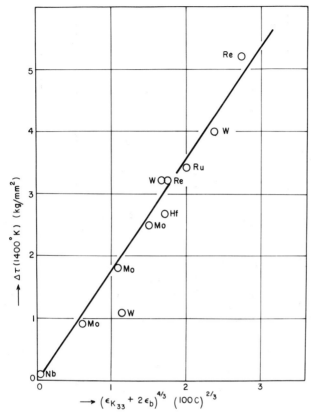

Fig. 59. Correlation between $\Delta\tau_0$ and the combined misfit parameter (Mordike and Rogausch, 1970).

force extended per unit of dislocation due to solute atoms is approximately the same, regardless of whether it is screw or edge dislocation, as is evident in Fig. 60.

Christian and co-workers (Statham and Christain, 1970; Statham et al., 1972) have conducted extensive investigations of niobium–molybdenum and niobium–rhenium alloys. They did not perform tests at a sufficiently high temperature to reach the athermal region, but concluded that at the highest temperatures studied there was a large thermal component. However, at their highest test temperature, 373° K, they observed a nearly linear behavior between strengthening and molybdenum concentration up to molybdenum concentrations of 50% Mo; whereas at lower test temperatures, there is approximately a $C^{1/2}$ relationship between strengthening and molybdenum. For a limited molybdenum concentration (i.e. 0–8.5% Mo) they obtained a linear dependence between strengthening and alloy composition.

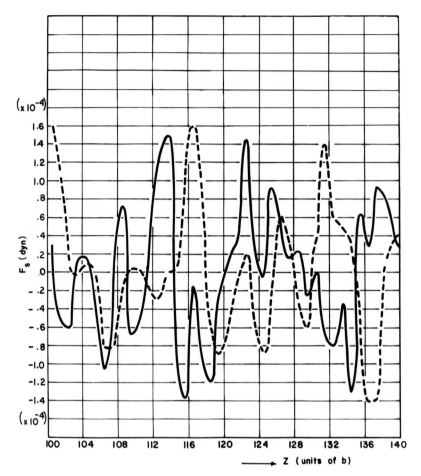

Fig. 60. The force on a screw and edge dislocation due to a ditational strain center; isotropic elastic constants were employed. 30% solute concentration; ——, screw; - - -, edge; $\varepsilon = 0.0556$; $\mu = 10^{12}$ dyn/cm^2; $b = 3 \times 10^{-8}$ cm; active volume $15b \times 15b \times 14b$.

Christain and co-workers concluded that the actual source of the strong solution hardening is uncertain. They further concluded that the theories of Fleischer (1962), Friedel (1959), and Suzuki (1969) cannot be applicable since they are based on the interaction of an individual solute atom and a dislocation.

In a recent paper by Slining and Koss (1973) they observed a linear relationship between the strengthening of τ_μ with alloy additions.

Mitchell and Raffo (1967) have investigated tantalum base alloys, i.e. Ta–Re and Ta–W. Again at their highest test temperature, 623° K, they obtained a linear relationship between strengthening and alloy composition.

They analyzed their data in terms of the Fleischer (1962) mechanism which is only applicable for dilute solutions, i.e. in the ppm range. They obtained surprisingly good agreement between theory and experimental results, even when considering the fact that the Fleischer theory predicts a $C^{1/2}$ dependence between strengthening and alloy composition. Also they explain all of strengthening as a consequence of a size mismatch.

There are some exceptions to the linear concentration dependence, one of which is found in the data obtained by Pink and Arsenault (1972) on a vanadium–titanium alloy, where they obtained a $C^{1/2}$ relationship.

2. Solid-Solution Strengthening of the Thermal Component

The fact that the yield should be divided into two components has not been realized by many investigators of bcc solid solutions, and as mentioned before most theories only predict the increase in strength at $0°$ K.

The concentration dependence of the strengthening for the thermal component is $C^{1/2}$ (Pink and Arsenault, 1972; Mitchell and Raffo, 1967; Statham *et al.*, 1972).

The changes in the activation parameters due to alloy additions have been measured by Arsenault (1969), Pink and Arsenault (1972), and Statham *et al.* (1972). A general result is that the alloy addition produces an increase in ΔH_0 and also increases the activation volume. Statham *et al.* (1972) also observed discontinuities in the v* vs τ* relationship and stated that this was rather strong evidence that two parallel mechanisms of slip were operating. However, it is not clear whether the solute modifies the existing slip mechanisms or whether one mechanism which is controlled by the solutes operates in the low τ* range and the intrinsic bcc mechanism operates at the higher τ*. The data obtained by Pink and Arsenault from V–Ti alloys indicate that the solute atoms result in a slip mechanism at low τ*, and then there is a reversion to the intrinsic mechanism at large τ*. However, what is not clearly understood is the mechanism of solute strength of τ*. Pink and Arsenault have put forward the concept that within the array there can be hard spots (i.e. similar to Predmore and Arsenault), but these obstacles are thermally activable and their strength does not change with composition.

B. *Solid-Solution Weakening*

Solid-solution weakening refers to a reduction of τ* due to a solute (substitutional or interstitial) addition. The solute addition may also produce a reduction in τ_a. Arsenault (1966a) was the first to consider solid-solution weakening in detail. This came about from the difference in theories proposed to account for τ* in the unalloyed bcc metals. One theory is based on the scavenging mechanism which assumes that the impurity interstitials are the srb to dislocation motion. Hahn *et al.* (1963) were the first to expand this idea. The substitutional solute addition reacts with the impurity interstitial concentration. Recently Ravi and Gibala (1970a,b) have restated this theory and have applied it to interstitial–interstitial reactions, but, as Christain (1970) has stated, this does not appear to be a likely theory. Hasson and Arsenault (1972) have reviewed the evidence of substitutional–interstitial interactions, but the results are not conclusive.

The second theory assumed that the solute addition affects the intrinsic lattice or aids in the formation of a double kink. In a previous section a discussion was presented on how the solute addition can affect the Peierls' stress.

Recently Sato and Meshii (1973) have used a concept originally proposed by Urakami and Fine (1970). They demonstrated that if a solute atom has a *dilatational* strain field associated with it, a torque is developed on a screw dislocation and this torque would aid in the formation of a double kink. Sato and Meshii applied this concept to oxygen in niobium. First of all, oxygen has a tetragonal strain field, where $\Delta\varepsilon = 0.26$, and to assign a dilatational strain to this would be difficult; but by applying a simple elasticity theory where the dilatational strain is

$$V = \tfrac{1}{3}(\varepsilon_1 + \varepsilon_2 + \varepsilon_3)$$

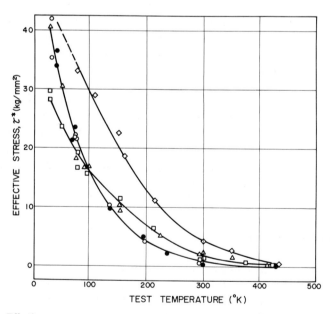

Fig. 61. Effective stress versus temperature for vanadium–titanium alloys. ●, unalloyed V; ○, 0.75 wt % Ti; △, 1.6 wt % Ti; □, 4.0 wt % Ti; ◇, 20.0 wt % Ti; $\dot\varepsilon = 8.3 \times 10^{-5}$/sec.

for the case of carbon in iron, $V = 0.10$ and for oxygen in niobium $V = 0.08$, not 0.35 as assumed by Sato and Meshii. The work of Arsenault and de Wit (1974) and Arsenault and Cadman (1973b) has shown that if tetragonal defects are present then solution weakening depends on the concentration of defects in a manner completely different from that predicted by Sato and Meshii.

The concept proposed by Urakami and Fine (1970) can be used to explain the solid-solution softening observed by Pink and Arsenault (1972) in low titanium content vanadium–titanium alloy, as shown in Fig. 61.

As in the case of unalloyed bcc metals the phenomenon of solid-solution weakening can be interpreted by either of two mechanisms, but the majority of the data supports the concept that solute additions affect the intrinsic barrier or aid in the formation of a double kink.

IX. Radiation Effects

The number of fundamental investigations of the effect of irradiation produced damage on the dislocation dynamics of bcc metals is limited in comparison to fcc metals. A possible deterrent, in addition to those mentioned previously, is the fact that the early investigations of the effect of neutron damage revealed that there was no change in τ^* due to neutron irradiation (Smidt, 1965; Arsenault, 1967b; McRickard, 1968). Arsenault (1967b) put forward an explanation as to why the neutron damage does not result in an increase in τ^*. However, later data by other investigators (Soo, 1969; Diehl et al., 1968; Kitujima, 1968; Ohr et al., 1970) indicated that impurity interstitials had a very large effect on how the neutron-produced damage affected τ^*.

Arsenault and Pink (1971) conducted a thorough investigation of the effect of impurity interstitials on the change in τ^* due to neutron irradiation of vanadium. However, before considering the effect of neutron damage on τ^* it is necessary to consider a few other factors first. Arsenault and Pink (1971) observed that for interstitial concentrations above $\simeq 200$ wt ppm oxygen, there was a change in the RCM of deformation. In addition, Hasson and Arsenault (1974) have shown that the impurity interstitial concentration has a large effect on the τ_μ, as is shown in Fig. 62.

It is also necessary to consider the ability of the irradiation damage (vacancies, interstitials, divacancies, diinterstitials, and clusters) to act as trapping sites or sinks for impurity interstitials. Conversely, the impurity interstitials can also act as trapping sites for the neutron-produced defects

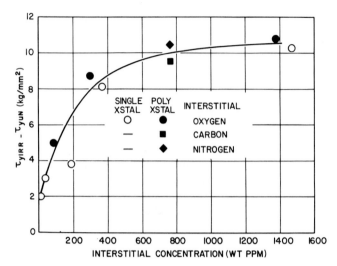

Fig. 62. The change in the athermal component of yield versus oxygen concentration, due to neutron irradiation.

(Evans, 1970). Internal friction results, obtained from vanadium which contained 40 wt ppm oxygen, indicate that a dose of 1.6×10^{19} neutrons/cm^2 (without a post-irradiation anneal) results in at least a 40% reduction in the random interstitial concentration (Hasson, 1973).

In a discussion of the effect of neutron damage on the dislocation motion, the results should be considered in terms of the amount of interstitial impurities in the samples. At the higher level of oxygen concentration (1800 wt ppm), the irradiation damage can be completely saturated with oxygen without appreciably decreasing the random oxygen concentration. The rate-controlling mechanism is still due to the oxygen atoms, and τ^* does not change since the amount of oxygen in random solid solution remains approximately the same. The irradiation-produced defects, which have been saturated with oxygen, result in long-range barriers to dislocation motion. Therefore, the irradiation only causes an increase in τ_μ.

For intermediate oxygen content (i.e. 100–300 wt ppm) the neutron-produced defects are large enough sinks so that the random oxygen concentration is appreciably reduced. Therefore, it is possible that neutron irradiation can cause a decrease in τ^* (Fig. 63). This reduction can be deduced from the following simple equation (Koppenaal and Arsenault, 1971):

$$\tau^* = \{\tau_0^* - \tau_0^{*1/2}[(kT/F_0 b) \ln (A/\dot{\gamma}_0)]^{1/2}\}^2 \tag{79}$$

where $\dot{\gamma}$ is the strain rate and $\dot{\gamma}_0$ is the product of the attempt frequency, the

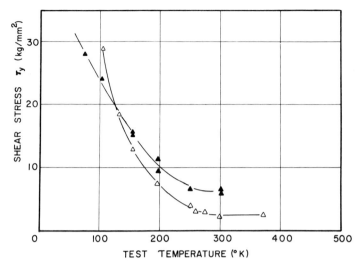

Fig. 63. Yield stress versus temperature of intermediate vanadium oxygen alloys both irradiated and nonirradiated. Vanadium single crystal, intermediate oxygen 200 wt ppm; △, annealed; ▲, irradiated 1.6 × 10^19 neutrons/cm².

number of dislocations involved in the thermally activated event, and the strain produced per event; F_0 is defined by the following equation:

$$F_0 = \mu \Delta \varepsilon b^2 / 3.86 \tag{80}$$

and

$$\tau_0^* = F_0 / Lb \tag{81}$$

where $L = b/(4C)^{1/2}$ and C is the oxygen concentration.

Therefore, if there is a reduction in the oxygen concentration there will be a reduction in τ^*. As suggested by the fact that ΔH_0 is still about 1 eV, a complete removal of the random oxygen atoms does not occur. This suggestion is also supported by the internal friction results. Therefore, oxygen interstitials are still the srb's to dislocation motion in the irradiated intermediate oxygen samples.

In the case of low oxygen concentration material, the rate-controlling mechanism of slip is probably the double-kink mechanism. The neutron-irradiation-produced defects in this material are not saturated with oxygen, and as a result the neutron-produced defects act as srb's to dislocation motion. Before analyzing the activation parameters to determine the nature of the neutron-produced defects, it is necessary to consider whether there is a single obstacle size or strength. In the case of neutron-irradiation strengthening of fcc metals, it was proposed (Koppenaal and Arsenault, 1965) that there is a spectrum of obstacle sizes, and thus a spectrum of activation energies (Koppenaal and Arsenault, 1965; Arsenault, 1966a). It was demonstrated that if the following equation

$$\Delta H_0 = \Delta H(\tau^*) + \int_0^{\tau^*} v(\tau^*)\, d\tau^* \tag{82}$$

is considered, and if it is demonstrated that ΔH_0 is independent of τ^*, then there is strong evidence for a single rate-controlling mechanism. From a determination of ΔH_0 vs τ^* for the irradiated low oxygen content vanadium, it was found that H_0 was not a function of τ^*, within the range of scatter as observed in the case of neutron-irradiated Cu (Koppenaal and Arsenault, 1965). Therefore, it may be concluded that there is a single rate-controlling mechanism for the neutron-irradiated low oxygen content vanadium.

The spacing of the neutron-produced defects, which are the srb's, can be calculated from the following equation:

$$v^* = (-\partial \Delta G/\partial \tau^*)_T = (F_0 b/\tau_0^*)[(\tau_0^*/\tau^*)^{1/2} - 1] \tag{83}$$

where F_0 and τ_0^* are defined in Eqs. (80) and (81); and with $F_0 b/\tau_0^* = b^2 L$, where L is a defect spacing. If v^* is equal to 10^{-21} cm^3 at $\tau^* = 1$ kg/mm^2, then L is $\simeq 10b$.

Now, recalling the results of the computer simulation discussed in a previous section, the spacing between defects may be a factor of 4 larger.

A consideration of ΔH makes it possible to determine both the strain caused by the irradiation-produced defect and its approximate size. The following equation can be used to make these approximations (Gibbs, 1969);

$$\Delta G = F_0 b[1 - (\tau^*/\tau_0^*)^{1/2}] \tag{84}$$

This equation was derived for a defect of a single atomic volume. If it is assumed that the irradiation-produced defect, which is causing a change in the rate-controlling mechanism, is of one atomic size, then the associated tetragonal strain can be calculated. The calculated strain is small: $\Delta \varepsilon = 0.48$. There is an alternative manner in which $\Delta \varepsilon$ of the irradiation-produced defect can be obtained since there is a linear relationship between $\Delta \varepsilon$ and ΔH. In the case where the oxygen atom is the srb to dislocation motion, the value of H_0 is 1.0 eV and $\Delta \varepsilon = 0.53$. In the neutron-irradiated case, $H_0 = 0.65$ eV; therefore, $\Delta \varepsilon = 0.34$ for the neutron-produced defect. If the value of $\Delta \varepsilon = 0.39$ (Hasson and Arsenault, 1971) for oxygen is used, then $\Delta \varepsilon = 0.25$ for the neutron-produced defect. According to the fair agreement between $\Delta \varepsilon$ obtained by both methods, it would appear that the defect is between 1 and $2b$ in size, e.g. a divacancy. The defect cannot be a collapsed vacancy loop because in such a case $\Delta \varepsilon = 1$ (Fleischer, 1962). Isolated single vacancies have been observed in bcc tungsten which neutron-irradiated at approximately 70° C (Attardo and Galligan, 1966).

The above explanation assumes an elastic interaction between the neutron-produced defect and the dislocation. However, a modulus interaction energy between a vacancy and an edge dislocation could be 0.5 to 0.9 eV for a vacancy in vanadium.

The fact that the defect remains at ambient temperatures further suggests that the defect is small, probably a single or divacancy, for its migration energy is higher than the migration energy of a diinterstitial (Johnson, 1964, 1966). Another point which must be considered is the relationship between the irradiation-produced defects and impurity interstitials such as the manner in which the impurity interstitials eliminate, cancel, or prevent the formation of the irradiation-produced srb's. One possible explanation is that a single vacancy traps an oxygen interstitial resulting in a large energy of motion of this complex (Evans, 1970). Therefore, these complexes do not migrate to form divacancies which are the assumed defect giving rise to the short-range barrier to dislocation motion.

Again, if a modulus interaction is assumed, and then if a vacancy traps an oxygen interstitial, the modulus assigned to this complex cannot be zero. The result of a nonzero modulus is that the interaction energy decreases by an amount comparable with the increase from zero, for in calculations of Bullough and Newman (1962) a value of zero modulus was assigned to a vacancy.

The discussion, thus far, has been concerned primarily with changes in τ^*, i.e. the rate-controlling mechanism of deformation: the data are summarized in Fig. 64. The increase in τ_μ caused by an increase in the oxygen concentration is in agreement with the majority of the data obtained from bcc metals. However, the point to be made in this case is that oxygen atoms which act as srb's to dislocation motion are not the same ones which contribute to an increase in τ_μ, as some of the oxygen atoms are tied with dislocations and subgrain boundaries, and thereby increase the effectiveness of the dislocations and subgrain boundaries in terms of a contribution to τ_μ (Armstrong et al., 1963). Similarly, in the case of neutron irradiation, the defect which accounts for the srb is not the defect size which contributes to τ_μ. As a result of neutron irradiation there is, in general, a spectrum of defect sizes (Mastel and Brimhall, 1965) which partly affect the athermal and thermal components.

Hasson et al. (1973) have conducted an investigation of the effect of neutron irradiation on the RCM of slip in molybdenum. The reason for considering molybdenum was that the effect of impurity interstitials should be minimal due to the low solubility of interstitials in molybdenum.

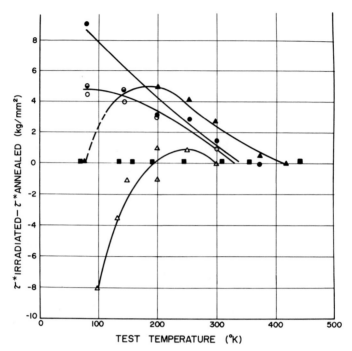

Fig. 64. A plot of the difference in the effective stress between irradiated and annealed single crystal and polycrystalline vanadium. The comparison is made between samples of the same interstitial concentration. Vanadium single crystal: ○, low oxygen, 1.6×10^{19} neutrons/cm^2; ○, low oxygen, 1.6×10^{20} neutrons/cm^2; △, intermediate oxygen, 1.6×10^{19} neutrons/cm^2. Vanadium polycrystal: ●, low oxygen, 1.6×10^{19} neutrons/cm^2; ▲, intermediate oxygen, 1.6×10^{19} neutrons/cm^2; ■, high oxygen, 1.6×10^{19} neutrons/cm^2.

The results obtained from neutron-irradiated single crystals indicate that radiation softening can occur in molybdenum. The magnitude of the softening was orientation dependent. The [100] oriented crystals showed a substantial decrease due to irradiation (Fig. 65); the [491] oriented crystals showed very little change; and the [110] oriented crystals actually showed an increase in the yield stress and effective stress below 150° K. The radiation softening is probably not due to a scavenging of interstitials as was proposed for radiation softening observed in vanadium containing an intermediate amount of oxygen for two reasons: (1) The solubility limit of impurity interstitials is low. (2) The amount of impurity interstitials present (< 5 wt ppm) is too low to account for a 5 kg/mm^2 reduction in yield stress reduction in τ^*. The decrease in τ^* is probably due to a mechanism that was initially proposed by Weertman (1958), whereby neutron-produced defects can aid in the formation of a double kink. The radiation hardening for the [110] orientation below 150° K is attributed to the interaction of disloca-

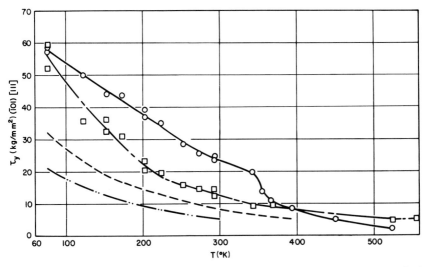

Fig. 65. Yield stress versus temperature for irradiated and nonirradiated molybdenum single crystals. [100] ○, nonirradiated; – – – –, nonirradiated proportional limit; □, irradiated (fluence = 1.6 × 10¹⁹ neutrons/cm²; — ·· —, unirradiated proportional limit (Stein).

tions with radiation-produced defect clusters and loops. The mechanism of weakening by Weertman (1958) does not invalidate the mechanism of Duesbery (1969). Arsenault and de Wit (1974) and Arsenault and Cadman (1973) have examined this weakening in great detail, and have shown that the weakening can be substantial. In the intermediate stress range, Duesbery predicts that the dislocation line is in the unextended mode during the formation of the double kink. Vanoni *et al.* (1972) have also observed radiation softening in high purity iron due to neutron irradiations at low temperatures.

Soo and Galligan (1969) and Oku and Galligan (1970) have reported different results, however, for another Group VI metal, tungsten. These investigators considered the differences in the proportional limit due to irradiation, and indicated that there was a significant increase in τ^*. If the present data from molybdenum are analyzed in the same manner, and if the proportional limit of the nonirradiated samples is compared to the yield stress of irradiated samples (i.e. in the irradiated samples the proportional limit is about equal to the yield stress), then there should also be a large increase in τ^* due to neutron irradiation for all orientations.

The radiation effects on bcc metals and alloys are discussed in greater detail elsewhere (Arsenault and Cadman, 1973b). However, one of the intriguing and confusing problems is how or why impurity interstitials have such a large effect on how the neutron damage affects τ^*.

X. The Superconducting State

The earliest indications that there might be a difference in dislocation behavior in the superconducting state, when compared to the normal state, came from the internal friction measurements of Tittmann and Bommel (1966). Kojima and Suzuki (1968) were the first to report the difference in flow in lead and niobium due to the change from the normal to the superconducting state. Figure 66 is a stress–strain curve of niobium at different strain rates and different magnetic fields (H). The flow stress is larger in the normal state; also the change in state, i.e. $\Delta\tau_{SN}$, is independent of strain rate. Suenaga and Galligan (1972) have written a review of the effect of the change in state from superconducting to normal in which they discuss the pertinent experimental results and the recent theories: only some aspects of the recent theories will be considered.

Granato (1971a,b) and Suenaga and Galligan (1970, 1971a–d) have proposed a vibrating string model. However, for the case of plastic deformation the title of the model might have to be changed. The dislocation is assumed to be pinned at two positions separated by a distance L, and initially only motion due to an applied shear stress is considered; $T = 0°$ K. The equation of motion of a dislocation segment is given by Koehler (1952, 1955):

$$A \, \partial^2\xi/\partial t^2 + B \, \partial\xi/\partial t - \Gamma \, \partial^2\xi/\partial x^2 = +b\tau \tag{85}$$

where ξ is the displacement of a dislocation, $A = \eta\rho b^2$ is the effective dislocation mass per unit length, B is the viscosity coefficient, and $\Gamma \simeq \mu b^2$ is the dislocation line tension. Note that ρ is the density of the material, and μ and b are the shear modulus and the Burgers vector, respectively. This equation is subjected to the following boundary conditions:

$$\partial\xi/\partial t = v_0(x) \quad \text{at} \quad t = 0 \tag{a}$$

$$\xi = \xi_0(x) \quad \text{at} \quad t = 0 \tag{b}$$

$$\xi = 0 \quad \text{at} \quad x = \pm L/2 \text{ for all } t \tag{c}$$

where $\xi_0(x)$ is a given function of x which describes the initial position of the dislocation segment, and $v_0(x)$ is the initial velocity. Equation (85) has several different types of solutions, but one of the interesting results of these solutions is that it is possible to define the underdamped case as follows:

$$B^2 < 4\pi^2 A\Gamma/L^2 \tag{86}$$

and for values of $L < 10^{-3}$ cm, the dislocation is underdamped. This means that the applied force at absolute zero does not have to be equal to the

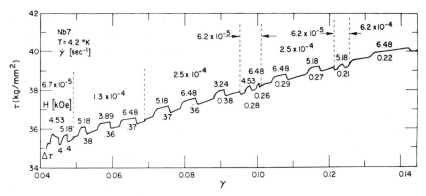

Fig. 66. Stress–strain curve of a niobium single crystal at different strain rates and magnetic fields *H*. *T* = 4.2° K (Kostorz, 1970).

maximum force of the opposing barrier in order for the dislocation to overcome the barrier. This added stress can be determined by the following equation:

$$\tau_{UD} = \tau^*[1 + \exp(-\beta t_{max})] \tag{87}$$

where τ^* is the effective stress acting on the dislocation in the unbowed position, $\beta = B/2A$, and t_{max} is required by the dislocation to reach the maximum bowout position. Inserting the appropriate values for a bcc metal, β_{max} ranges from 10 to 100, which means that for all practical purposes $\tau_{\mu D} = \tau^*$.

Returning now to superconductivity, it is possible to define the critical force for breakaway from the pinning points by Eq. (62). Suenaga and Galligan (1971a–d) obtained the following equation from these changes in flow stress due to a change in state:

$$\Delta\tau_{NS} \cong (\pi f_c/8b^3)(\pi\rho\mu)^{-1/2}B_N(1 - B_S/B_N) \tag{88}$$

where f_c is the critical breakaway force, B_N the drag coefficient in the normal state, and B_S is the drag coefficient in the superconducting state. The expression for $\Delta\tau_{NS}$ is independent of strain rate, which is in agreement with the experimental results.

However, it is necessary to consider the temperature dependency of the breakaway process. Suenaga and Galligan (1971a–d) took into account the temperature dependency by starting with the following equation:

$$\tau^* = [\Delta G_0 - kT \ln(\dot{\gamma}_0/\dot{\gamma})]/\bar{v}^* \tag{89}$$

where τ^* is the effective stress required to overcome the pinning points in the completely damped case, ΔG_0 is the total Gibbs force energy due to pinning, $\dot{\gamma}_0$ is a constant, and v^* is an average activation volume defined as follows:

$$\bar{v}^* = (1/\tau^*) \int_0^{\tau^*} v^*(\tau^*) \, d\tau^* \tag{90}$$

Equation (95) is slightly different from that of Suenaga and Galligan (1971a-d) for it takes into account the fact that v^* is stress dependent. Then for a dynamic interaction of the dislocation and the pinning point, Suenaga and Galligan (1971a-d) replaced τ^* with $\tau\mu_D$, and the following equation results:

$$\tau^* = [\Delta G_0 - kT \ln (\dot{\gamma}_0/\dot{\gamma})]/\bar{v}^*[1 + \exp (-\beta t_{max})] \tag{91}$$

Now it is possible to replace each τ^* by its respective value in the normal and superconducting state, i.e.

$$\Delta\tau_{NS} = \tfrac{1}{8}b^2[1/(AC)^{1/2}]\{[\Delta G_0 - kT \ln (\dot{\gamma}_0/\dot{\gamma})]/\bar{v}^*\}B_N(1 - B_S/B_N) \tag{92}$$

For the case where $\Delta G_0 \gg kT \ln (\dot{\gamma}_0/\dot{\gamma})$, i.e. in the low-temperature range, and $\Delta G/v \approx f_c$, then Eq. (92) becomes equivalent to Eq. (88). This means that at temperatures above $0°$ K, $\Delta\tau_{NS}$ is also predicted *not* to be strain rate dependent.

However, Eq. (92) is a *very* specific equation. It is only valid under quite stringent boundary conditions which can be seen by considering Figs. 67a and b. For Eq. (92) to be valid, the dislocation has to break away from the pinning point when the dislocation reaches its maximum bowout. In other words, the dislocation has to breakaway by a thermal fluctuation *during* the time interval Δt as is shown in Fig. 67b. The absolute value of Δt depends upon numerous parameters, but it is possible to make a conservative approximation of Δt: t_{max} can be obtained from the following equation assuming that maximum bowout is a semicircle

$$t_{max} = LB/2\tau^* b \tag{93}$$

Inserting representative values in the above equation, i.e. $B = 10^{-5}$ dyn sec/cm^2, $L = 10^{-4}$ cm, $\tau^* = 10^9$ dyn/cm^2, and $b = 2.5 \simeq 10^{-8}$ cm, the value of t_{max} is 10^{-10} sec. Therefore, Δt is some fraction of 10^{-10} sec. The thermal fluctuation breakaway time can be calculated using the following equation:

$$t_B = t_0 e^{+\Delta G/kT} \tag{94}$$

since it is assumed that there is a linear relationship between ΔG and T, and for most bcc metals ΔG_0 is $\simeq 1$ eV at $350°$ K and $t_0 = L/bv_D$, where v_D is the Debye frequency, 10^{+13}/sec; then $t_B \simeq 10^{+5}$ sec. From a consideration of t_B and Δt, it appears highly unlikely that thermal breakaway would occur in

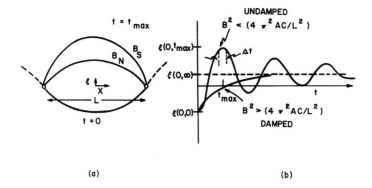

Fig. 67. (a) The various positions of the dislocation between pinning points. (b) The position of the center of the loop as a function of time for damped and undamped cases.

the time interval Δt. This criticism of this model may not have much meaning, for the thermal dependence is assumed to be insignificant by Suenaga and Galligan. However, a point which is much more significant is the fact that implicitly and explicitly it was assumed that the intrinsic lattice is not the srb to dislocation motion but some pinning defect, i.e. an impurity interstitial.

Another interpretation of the present observations has been given by J. C. M. Li (private communication, 1971) in terms of the kink model of deformation (Seeger and Schiller, 1966). In this model the mobility of the kinks is affected by the superconducting transition. As a result, the steady state density of kinks changes. In order to remove the excess density of kinks (relative to the steady state density), the sample deforms an increased amount quickly. After the density of kinks has been reduced to the steady state value, the sample deforms as in the normal state.

A transient increase of strain of the order of

$$\gamma_t = \rho_k^* b \chi_k (\alpha - 1)/\alpha \tag{95}$$

where ρ_k^* is the steady state kink density, χ_k is the distance traveled by a kink, and α is the increase in kink velocity due to the change in state from normal to superconducting. This expression would explain the abrupt decrease of stress during stress relaxation tests (Suenaga and Galligan, 1972).

The motion of the kinks may be resisted by electrons only, and as a result

$$v_k = \tau_k^* b/B \tag{96}$$

where τ_k^* is the effective stress exerted on the kinks. If, in the normal state, a steady state density of kinks has been reached and then the sample becomes superconducting, a *transient* change of flow stress would be predicted in a constant strain rate test:

$$\Delta\tau_{NS}^* = \tau_k^*[1 - (B_S/B_N)] \tag{97}$$

If B_S is small in comparison to B_N, the $\Delta\tau_{NS}$ is practically τ_k^* exerted on the dislocation in the normal state. This means that the predicted value of $\Delta\tau_{NS}$ could be large, i.e. greater than the experimentally observed values of $\Delta\tau_{NS}/\tau_k^* \sim 10\%$. However, this model depends only on the intrinsic lattice, which is more compatible with the vast majority of data obtained from bcc metals.

XI. Conclusions

The low-temperature deformation characteristics of bcc metals and their alloys are unusual when compared to fcc metals and their alloys. The probable cause of these unusual deformation characteristics of bcc metals is due to the nature of the core structure of the screw dislocations. This core structure can account for asymmetry of the yield stress and the predominance of screw dislocation which also means that the intrinsic lattice is the rate-controlling barrier to dislocation motion at low temperatures. However, there is evidence that impurity interstitials affect the low-temperature deformation characteristics of bcc metals and their alloys. The evidence is overwhelming that the impurity interstitials do not have a direct effect on dislocation kinetics. It is possible that impurity interstitials have an effect on the interatomic potential which controls the core structure of the screw dislocation.

The work-hardening characteristics of bcc single crystals are also unusual. The rate of working may be the same as that of fcc metals in stage II and there is a tendency to define or describe stage II hardening in bcc metals in terms of mechanisms that have been proposed for fcc metals. There are a few problems with this approach, i.e. the rate of work hardening in stage II of bcc is very temperature dependent, whereas in fcc it is not. A satisfactory explanation for the temperature dependence of stage II has not been developed.

The effect of irradiation damage on the low-temperature deformation characteristics of bcc metals is complicated by the fact that impurity interstitials have a very large effect on the magnitude of change in yield stress due to irradiation. Also it is possible for irradiation to result in a reduction of the yield stress provided that the irradiation is conducted at low temperatures.

Arsenault and co-workers have put forward a few mechanisms to account for this unusual behavior.

The electronic state, i.e. normal or superconducting, has an effect on the yield stress. A change from superconducting to the normal state results in an increase in the yield or flow stress. The reasons why this change in flow stress occurs have not been adequately explained; one of the intriguing aspects is the lack of strain rate sensitivity of the effect.

There have been a large number of investigations of bcc metals and their alloys in the past decade, and I am sure that I did not reference all of them. I apologize for all important and critical references that I may have missed.

ACKNOWLEDGMENTS

The author wishes to acknowledge the support of the United States Atomic Energy Commission under contract no. AT(40-1)-3612. The author also wishes to acknowledge a discussion with Professors J. C. M. Li and H. Conrad on various aspects of dislocation kinetics in bcc metals and their alloys and Professor B. Mordike concerning bcc solid solutions.

References

Argon, A. S. (1972). *Phil. Mag.* [8] **26**, 1053.

Armstrong, D. A., and Mordike, B. (1970). *J. Less-Common Metals* **22**, 265.

Armstrong, R. W., Bechtold, J. H., and Begley, R. T. (1963). "Refractory Metals and Alloys," Vol. 17, p. 159. Gordon & Breach, New York.

Arsenault, R. J. (1964). *Acta Met.* **12**, 547.

Arsenault, R. J. (1966a). *Acta Met.* **14**, 831.

Arsenault, R. J. (1966b). *Acta Met.* **14**, 1634.

Arsenault, R. J. (1967a). *Acta Met.* **15**, 501.

Arsenault, R. J. (1967b). *Acta Met.* **15**, 1853.

Arsenault, R. J. (1967c). "Experimental Methods of Materials Research," p. 215. Wiley, New York.

Arsenault, R. J. (1969). *Acta Met.* **17**, 1291.

Arsenault, R. J. (1971a). *Trans. Met.* **2**, 1472.

Arsenault, R. J. (1971b). *Phil. Mag.* [8] **24**, 259.

Arsenault, R. J. (1972a). *Trans. Met.* **3**, 1014.

Arsenault, R. J. (1972b). *Phil. Mag.* [8] **26**, 1481.

Arsenault, R. J. (1973). "The Microstructure and Design of Alloys," Vol. 1, p. 520. Institute of Metals.

Arsenault, R. J., and Cadman, T. (1973a). *Trans. Metal Soc. AIME* **18**, 41.

Arsenault, R. J., and Cadman, T. (1973b). *In* "Defects and Defect Clusters in B.C.C. Metals and Their Alloys" (R. J. Arsenault, ed.). AIME, New York.

Arsenault, R. J. and Cadman, T. (1973c). *Proc. Intern. Conf. Defects and Defect Clusters in bcc Metals and Their Alloys,* p. 41. Nucl. Metall. AIME.

Arsenault, R. J., and Cadman, T. (1975). *John Dorn Mem. Symp.* (to be published).

Arsenault, R. J., and de Wit, R. (1974). *Acta Met.* **22**, 819.

Arsenault, R. J., and Kuo, K. (1975). *Met. Trans.* To be published.

Arsenault, R. J., and Lawley, A. (1967). *Phil. Mag.* [8] **15**, 549.

Arsenault, R. J., and Li, J. C. M. (1967). *Phil. Mag.* [8] **12**, 1307.

Arsenault, R. J., and Pink, E. (1971). *Mater. Sci. Eng.* **8**, 141–151.
Arsenault, R. J., Crowe, C. R., and Carnahan, R. (1972). *In* "Keinststoffe in Wissenschaft und Technik," p. 345. Akadamie-Verlag, Berlin.
Attardo, M., and Galligan, J. M. (1966). *Phys. Status Solidi* **16**, 449.
Barnett, D., and Nix, W. (1973). *Acta Met.* **21**, 1157.
Basinski, Z. S., Duesbery, M. S., and Taylor, R. (1971). *Can. J. Phys.* **49**, 2160.
Blin, J. (1955). *Acta Met.* **3**, 199.
Bolton, D., and Taylor, G. (1972). *Phil. Mag.* [8] **26**, 1359.
Brailsford, A. D. (1961). *Phys. Rev.* **122**, Part I, 778.
Brailsford, A. D. (1962). *Phys. Rev.* **128**, Part II, 1033.
Brimhall, J. L. (1965). *Acta Met.* **13**, 1109.
Bullough, R., and Newman, R. C. (1962). *Phil. Mag.* [8] **7**, 529.
Cadman, T., and Arsenault, R. J. (1972). *Scr. Met.* **6**, 593.
Celli, V., Kabler, M., Ninomiya, T., and Thompson, R. (1963). *Phys. Rev.* **131**, 58.
Chen, C., and Arsenault, R. J. (1975). To be published.
Chou, Y. T. (1965). *Acta Met.* **13**, 251.
Chou, Y. T., and Sha, G. T. (1972). *Trans. Met.* **3**, 2857.
Christain, J. (1970). *In* "Second International Conference on Strength of Metals and Alloys," p. 29. Amer. Soc. Metals, Metals Park, Ohio.
Christain, J. W. (1968). *Scr. Met.* **2**, 369 and 677.
Conrad, H. (1963). *In* "Relation Between Structure and Mechanical Properties of Metals" (N. P. L. Teddington, ed.). HM Stationery Office, London.
Conrad, H., and Fredrick, S. (1962). *Acta Met.* **10**, 1013.
Conrad, H., and Wiedersich, H. (1960). *Acta Met.* **8**, 128.
Conrad, H., Hays, L., Schoeck, G., and Wiedersich, H. (1961). *Acta Met.* **9**, 367.
Cottrell, A. H. (1953). "Dislocations and Plastic Flow in Crystals," p. 53. Oxford Univ. Press, London and New York.
Cottrell, A. H., and Bilby, B. A. (1951). *Phil. Mag.* [7] **42**, 573.
Darken, L. S., and Gurry, R. W. (1953). "Physical Chemistry of Metals," pp. 213, 469. McGraw-Hill, New York.
Diehl, J., Seidel, G. P., and Weller, M. (1968). *Trans. Jap. Inst. Metals* **9**, 219.
Dorn, J., and Mukherjee, A. K. (1969). *Trans. Met. Soc. AIME* **245**, 1493.
Dorn, J., and Rajnak, S. (1964). *Trans. Met. Soc. AIME* **230**, 1052.
Duesberry, M. S. (1969). *Phil. Mag.* **19**, 501.
Duesberry, M. S., and Hirsch, P. (1968). "Dislocation Dynamics," p. 57. McGraw-Hill, New York.
Escaig, B. (1972). *3rd Int. Symp. Reinststoffe Wiss. Tech.*, Dresden.
Eustice, A. L., and Carlson, O. N. (1961). *Trans. AIME* **221**, 238.
Evans, J. H. (1970). *Acta Met.* **18**, 499.
Fischer, J. C. (1954). *Acta Met.* **2**, 9.
Fleischer, R. L. (1962a). *Acta Met.* **10**, 835.
Fleischer, R. L. (1962b). *J. Appl. Phys.* **33**, 3504.
Fleischer, R. L. (1964). *J. Appl. Phys.* **33**, 3504.
Fleischer, R. L. (1967). *Acta Met.* **15**, 1513.
Foreman, A. J. E., and Makin, M. J. (1966). *Phil. Mag.* **14**, 911.
Foxall, R. A., and Statham, C. D. (1970). *Acta Met.* **18**, 1147.
Foxall, R. A., Duesbery, M. S., and Hirsch, P. (1967). *Can. J. Phys.* **45**, 607.
Friedel, J. (1959). "Internal Stresses and Fatigue in Metals," p. 244. Elsevier, Amsterdam.
Friedel, J. (1964). "Dislocations," pp. 31 and 379. Addison & Wesley, Reading, Massachusetts.
Frost, H. J., and Ashby, M. F. (1971). *J. Appl. Phys.* **42**, 5273.

Gehlen, P. C., and Cohen, J. B. (1965). *Phys. Rev.* **139**, A844.

Gibbs, G. B. (1969). *Phil. Mag.* [8] **20**, 611.

Granato, A. V. (1971a). *Phys. Rev. Lett.* **27**, 660.

Granato, A. V. (1971b). *Phys. Rev. B* **4**, 2196.

Guiu, F. (1969). *Scr. Met.* **3**, 449.

Hahn, G. T., Gilbert, A., and Jaffee, R. I. (1963). *In* "Refractory Metals and Alloys II" (M. Semchyshen and I. Perlmutter, eds.), Vol. 17, p. 23. Wiley (Interscience), New York.

Hasson, D. F. (1973). *In* "Defects and Defect Clusters in B.C.C. Metals and Their Alloys" (R. J. Arsenault, ed.). AIME, New York.

Hasson, D., and Arsenault, R. J. (1971). *Scr. Met.* **5**, 75.

Hasson. D.. and Arsenault, R. J. (1972). This Series, Vol. 1, p. 179.

Hasson, D. F., and Arsenault, R. J. (1974). *Radiation Effects* **21**, 203.

Hasson, D., Huang, Y., Pink, E., and Arsenault, R. J. (1974). *Met. Trans.* **5**, 371.

Head, A. K. (1964). *Phys. Status Solidi* **6**, 461.

Hirsch, P. B. (1963). "The Relation Between the Structure and Mechanical Properties of Metals," p. 39. HM Stationery Office, London.

Hirsch, P. B. (1968). *Trans. Jap. Inst. Metals* **9**, Suppl., XXX.

Hirth, J. P. (1970). *In* "Elastic Behavior of Solids" (M. Kannien *et al.*, eds.), p. 281. McGraw-Hill, New York.

Hirth, J. P., and Nix, W. D. (1969). *Phys. Status Solidi* **35**, 177.

Huang, Y., and Arsenault, R. J. (1972). "Engineering Materials Group," Rep. No. C. University of Maryland.

Huang, Y., and Arsenault, R. J. (1973a). *Mater. Sci. Eng.* **12**, 111.

Huang, Y., and Arsenault, R. J. (1973b). *Radiation Effects* **17**, 3.

Johnson, R. A. (1964). *Phys. Rev.* **134**, 1329.

Johnson, R. A. (1966). *Phys. Rev.* **145**, 423.

Jonas, J. J., and Luton, M. (1971). *Met. Trans.* **2**, 3492.

Jonas, J. J., Lutton, M. J., and Surek, T. (1972). *Trans. Met.* **3**, 2295.

Jøssang, T., Lothe, J., and Skyland, K. (1965). *Acta Met.* **13**, 271.

Keh, A. S. (1965). *Phil. Mag.* [8] **12**, 9.

Kitujima, K. (1968). *Trans. Jap. Inst. Metals* **9**, 182.

Koehler, J. S. (1952). "Imperfections in Nearly Perfect Crystals," p. 197. Wiley, New York.

Koehler, J. S. (1955). *J. Phys. Soc. Jap.* **10**, 669.

Kojima, H., and Suzuki, T. (1968). *Phys. Rev. Lett.* **21**, 290.

Koppenaal, T. J., and Arsenault, R. J. (1965). *Phil. Mag.* [8] **12**, 951.

Koppenaal, T. J., and Arsenault, R. J. (1971). *Met. Rev.* **16**, 175.

Kostorz, G. (1970). *Scr. Met.* **9**, 95.

Kroner, E. (1958). "Kontinuumstheorie der Versetzungen und Eigen Spann ungen," p. 80. Springer-Verlag, Berlin and New York.

Kroupa, F., and Vitek, V. (1967). *Can. J. Phys.* **45**, 945.

Kubin, L. P., and Jouffrey, B. (1971). *Phil. Mag.* [8] **24**, 437.

Labusch, R. (1972). *Acta Met.* **20**, 917.

Lachenmann, R., and Schultz, H. (1970). *Scr. Met.* **4**, 709.

Lachenmann, R., and Schultz, H. (1972). *Scr. Met.* **6**, 731–736.

Lachenmann, R., and Schultz, H. (1973). *Scr. Met.* **7**, 155.

Li, J. C. M. (1965). *Trans. AIME* **233**, 219.

Lord, A. E., and Beshers, D. N. (1965). *J. Appl. Phys.* **36**, 1620.

Lothe, J. (1962). "Explicit Expressions for the Energy of Dislocation Configurations Made Up Piecewise of Straight Segments," Metals Res. Lab. Tech. Rep. Carnegie Institute of Technology.

McRickard, S. B. (1968). *Acta Met.* **16**, 969.

Mastel, B., and Brimhall, J. L. (1965). *Acta Met.* **13**, 1109.

Mitchell, T. E., and Raffo, P. L. (1967). *Can. J. Phys.* **45**, 1047.

Mitchell, T. E., Foxall, R. A., and Hirsch, P. B. (1963). *Phil. Mag.* [8] **8**, 1895.

Mordike, B., and Rogausch, K. P. (1970). *In* " Second International Conference of Strength of Metals and Alloys," p. 258. Amer. Soc. Metals, Metals Park, Ohio.

Ohr, S. M., and Beshers, D. N. (1963). *Phil. Mag.* [8] **8**, 1343.

Ohr, S. M., Wechsler, M. S., Chenand, C. W., and Hinkle, N. E. (1970). *In* " Second International Conference on the Strength of Metals and Alloys," p. 742. Am. Soc. Metals, Metals Park, Ohio.

Oku, T., and Galligan, J. M. (1970). *In* " Second International Conference on Strength of Metals and Alloys," p. 737. Amer. Soc. of Metals, Metals Park, Ohio.

Ono, K., and Sommer, A. W. (1970). *Met. Trans.* **1**, 877.

Owen, C. V., and Scott, T. E. (1970). *Met. Trans.* **1**, 1715.

Owen, C. V., and Scott, T. E. (1972). *Met. Trans.* **3**, 1915.

Pink, E., and Arsenault, R. J. (1972). *Metal Sci. J.* **6**, 1.

Predmore, R. (1975). Ph.D. Thesis, Univ. Maryland.

Ravi, K. V., and Gibala, R. (1969). *Scr. Met.* **3**, 547.

Ravi, K. V., and Gibala, R. (1970a). *Acta Met.* **18**, 623.

Ravi, K. V., and Gibala, R. (1970b). *In* " Second International Conference on the Strength of Metals and Alloys," p. 83. Amer. Soc. Metals, Metals Park, Ohio.

Ravi, K. V., and Gibala, R. (1971). *Met. Trans.* **2**, 1219.

Read, W. T. (1953). " Dislocation in Crystals," pp. 53 and 131. McGraw-Hill, New York.

Reed, R. E., and Arsenault, R. J. (1975). To be published.

Rogausch, K. D., and Mordike, B. (1970). *In* " Second International Conference on the Strength of Metals and Alloys," p. 168. Amer. Soc. Metals, Metals Park, Ohio.

Rudman, P. S. (1965). *Proc. Conf. Refractory Metals, 4th, 1965*.

Rudolph, G., and Mordike, B. L. (1967). *Z. Metallk.* **58**, 708.

Sato, A., and Meshii, M. (1973). *Acta Met.* **21**, 753.

Schoeck, G. (1965). *Phys. Status Solidi* **8**, 499.

Seeger, A. (1956). *Phil. Mag.* [8] **1**, 651.

Seeger, A., and Schiller, P. (1962). *Acta Met.* **10**, 348.

Seeger, A., and Schiller, P. (1966). *In* " Physical Acoustics." (W. P. Mason, ed.), Vol. 3, Part A, p. 361. Academic Press, New York.

Sherwood, P. J., Guiu, F., Kim, H. C., and Pratt, P. L. (1967). *Can. J. Phys.* **45**, 1045.

Slining, J. R., and Koss, D. A. (1973). *Met. Trans.* **4**, 1261.

Smialek, R. L., and Mitchell, T. (1970). *In* " Second Internal Conference on the Strength of Metals and Alloys," p. 73. Amer. Soc. Metals, Metals Park, Ohio.

Smidt, F. (1965). *J. Appl. Phys.* **36**, 2317.

Soo, P. (1969). *Trans. AIME* **245**, 985.

Soo, P., and Galligan, J. M. (1969). *Scr. Met.* **3**, 153.

Statham, C. D., and Christain, J. (1970). *Scr. Met.* **5**, 399.

Statham, C. D., Koss, D., and Christain, J. (1972). *Phil. Mag.* [8] **26**, 1089.

Stein, D. F. (1967). *Can. J. Phys.* **45**, 1063.

Stein, D. F. (1968). *In* " Dislocation Dynamics," p. 353. McGraw-Hill, New York.

Stein, D. F., and Low, J. R. (1966). *Acta Met.* **14**, 1183.

Stein, D. F., Low, J. R., and Seybolt, A. E. (1963). *Acta Met.* **11**, 1253.

Suenaga, M., and Galligan, J. M. (1970). *Scr. Met.* **4**, 697.

Suenaga, M., and Galligan, J. M. (1971a). *Scr. Met.* **5**, 63.

Suenaga, M., and Galligan, J. M. (1971b). *Phys. Rev. Lett.* **27**, 721.

Suenaga, M., and Galligan, J. M. (1971c). *Scr. Met.* (to be published).
Suenaga, M., and Galligan, J. M. (1971d). *Spring Meet. AIME, 1971.*
Suenaga, M., and Galligan, J. (1972). *In* " Physical Accoustics " (W. P. Mason, ed.), Vol. 9, p. 1. Academic Press, New York.
Surek, T., Lutton, M. J., and Jonas, J. J. (1973). *Phil. Mag.* [8] **27**, 425.
Suzuki, H. (1969). " Dislocation Dynamics," p. 679. McGraw-Hill, New York.
Suzuki, H. (1971). *Nachr. Akad. Wiss. Göettingen, Math.-Phys. Kl.,* 2 No. 6, p. 1–66.
Taylor, G., and Bowen, K. (1970). *In* " Second International Conference on Strength of Metals and Alloys," p. 78. Amer. Soc. Metals, Metals Park, Ohio.
Tittmann, B. R., and Bommel, H. E. (1966). *Phys. Rev.* **151**, 178.
Urakami, A., and Fine, M. E. (1970). *Scr. Met.* **4**, 667.
Vanoni, F., Groh, P., and Moser, P., (1972). *Scr. Met.* **6**, 777.
Vitek, V. (1966). *Phys. Status Solidi* **18**, 687.
Vitek, V., and Christain, J. W. (1973). " The Microstructure and Design of Alloys." Vol. 1, p. 534. Inst. of Metals.
Vitek, V., and Kroupa, F. (1966). *Phys. Status Solidi* **18**, 703.
Wechsler, M. S., Tucker, R. B., and Bode, R. (1969). *Acta Met.* **17**, 541.
Weertman, J. (1956). *Phys. Rev.* **101**, 1429.
Weertman, J. (1958). *J. Appl. Phys.* **29**, 1685.
Weertman, J., and Weertman, J. (1964). " Elementary Dislocation Theory." Macmillan, New York.
Westlake, D. G. (1967). *Trans. AIME* **239**, 1341.
Westlake, D. G. (1969a). *Trans. AIME* **245**, 287.
Westlake, D. G. (1969b). *Trans. AIME* **245**, 1969.
Westlake, D. G. (1969c). *Trans. Amer. Soc. Metals* **62**, 1000.
Zenner, C. (1942). *Trans. AIME* **147**, 361.

Cyclic Deformation of Metals and Alloys

CAMPBELL LAIRD

School of Metallurgy and Materials Science
University of Pennsylvania
Philadelphia, Pennsylvania

I. Introduction

A. The Development of Cyclic Stress–Strain Response

Cyclic deformation universally implies the idea of fatigue, the failure of materials by cyclic loading at stresses either below or above the unidirectional yield stress. Reference to any of the standard books or review articles on fatigue shows that the bulk of the work on fatigue has covered the fracture aspects (Thompson and Wadsworth, 1958; Kennedy, 1963; Manson, 1966; Forsyth, 1969; Plumbridge and Ryder, 1969; Grosskreutz,

1971, 1973; Laird and Duquette, 1972). The reason for this lies partly in the old-fashioned stress-based approach of design against fatigue, exemplified in the book by Forrest (1962), partly in the fact that the high strain fatigue problem had not been properly recognized, so that research was carried out at low stresses where strains were difficult to measure, and partly in the lack of suitable measuring devices. In the stress-based design approach, an elastic analysis is carried out on a structure or a component, and the fatigue life is designed on the basis of the S-N curve, the relationship between life and applied stress. This approach, although notable for some spectacular catastrophes, was largely successful on the basis of the testing which also accompanied the design. Sometimes, as many as four iterations of testing and redesign were required before a satisfactory component was achieved (Forrest, 1962). The materials problems associated with this approach were relatively simple. All that was needed was basic S-N data, and provision for mean stresses, notch effects, variable loading, and environment. Although it must be admitted that all these problems are very complex, the design approach was simple, and some of the problems, such as environment, remain largely ignored. A major difficulty with this approach lay in the definition of fatigue "damage," which is vague in relation to stress and gave problems in cumulative fatigue.

During the 1950's, the importance of high strain fatigue in many applications, mainly in aerospace and energy conversion, was recognized by Coffin and Manson. Their work led not only to the well-known Coffin–Manson law of failure $\Delta\varepsilon_p N_f^n = $ const, where $\Delta\varepsilon_p$ is the plastic strain range of the cycle and N_f the number of cycles to failure, but also emphasized the considerable hardness changes which developed in the material during cycling. Although these hardening phenomena were discovered decades earlier by Bauschinger, they appeared largely irrelevant to real problems and had been ignored. Unfortunately, cyclic deformation does not always cause hardening. Kenyon (1950) rediscovered that cold-worked copper used for power cables underwent serious softening due to high frequency cycling by wind loading. The recognition that the unidirectional properties used for static design could be seriously altered by cyclic loading also directed more attention to cyclic deformation.

Out of this work, mainly carried out at high plastic strains which were reasonably easy to measure, a new design approach has emerged. In this approach, fully described by Manson (1966), some attempt is made to take plasticity into account. Designers make use either of conventional unidirectional plasticity computations available in the standard texts of mechanical engineering, or of cyclic plasticity. In the conventional approach, only one cycle of loading is considered, and the virgin unidirectional stress–strain curve of the material is used to tie the equilibrium and compatibility equations together. The strain range developed during this single cycle is then assumed to be the strain range after a large number of cycles as well. If the strain ranges at the critical points in a component can be calculated by such techniques, then the life can be determined by the Coffin–Manson law.

In the approach of cyclic plasticity, the plasticity which occurs after cyclic hardening or softening has developed is used instead of the unidirectional stress–strain curve. This method is invariably superior to the unidirectional approach, which can give disastrous predictions in the case of cyclic soften-ing material (e.g. tempered ferrous martensites and cold-worked metals), or overconservative predictions in those materials which cyclically harden. This approach is intuitively more satisfactory than a stress-based approach since the connection between fatigue damage and plastic strain is concrete.

It is clear that the field of cyclic deformation attempts to define the stress response of a material to cyclic straining. It is fundamental to the failure problem in two ways: (1) the fatigue fracture phenomena are closely asso-ciated with cyclic plasticity in that cracks nucleate in active slip bands and subsequently propagate along them, and (2) a proper understanding of stress and strain in a structure is prerequisite to making accurate fatigue predic-tions. The field has developed to the extent that a simple textbook has been written on the subject (Sandor, 1972). However, much of this work has been done from an engineering viewpoint and less attention has been given to the role of microstructure in controlling cyclic stress–strain response, and to the underlying mechanisms of cyclic deformation. The present review, in keep-ing with the rest of this volume, is devoted to these aspects of cyclic deforma-tion and no attempt is made here to treat the associated failure phenomena, which have already been reviewed recently in great detail (Grosskreutz, 1971, 1973; Plumbridge and Ryder, 1969).

The phenomena to be covered here therefore include the rapid hardening or softening which is caused by cyclic straining in the early stages of fatigue life. However, a stage is eventually reached, usually at a relatively small fraction of the total life, when these hardness changes are exhausted and a saturation flow stress is established for a particular applied plastic strain. It is this saturated state which is of most interest to engineers and scientists because the failure processes usually occur only after it has been attained (Laird and Duquette, 1972). In more complex materials, however, saturation need never be attained. For example, a gradual softening may persist during the whole fatigue life of a material. Such problems will also be covered. For materials which display a steady saturation stress–strain response, the *cyclic stress–strain curve*, a plot of saturation flow stress versus applied plastic or total strain has been defined. This important curve is usually quite different from the unidirectional tensile or compressive curve, but is represented em-pirically by a similar kind of expression, $\sigma = K'(\Delta \varepsilon_\mathrm{p}/2)^{n'}$ where σ is the saturation value of the stress, $\Delta\varepsilon_p$ the applied plastic strain range, and K' and n' are material constants. Industries which are interested in using cyclic plasticity for design purposes are beginning to compile K' and n' for various materials (e.g. R. W. Landgraf, Ford Motor Co., Dearborn, Michigan), just as endurance limits were compiled in the older design philosophy.

In order to emphasize the close association of cyclic deformation and failure, a recent application of cyclic plasticity to the problem of fatigue prediction under variable loading will be briefly described (Wetzel, 1971).

B. Application of Cyclic Stress–Strain Response

Wetzel (1971) has developed a procedure for predicting the fatigue life of parts and components subjected to complex load histories of the kind that occur in service. One of the most difficult problems encountered in summing such damage, by older stress-based methods, is making allowance for mean stress effects (i.e. when a fluctuation occurs entirely within the tensile or compressive stress regime). Wetzel accounts for this problem in the following way. He plots life data as $2\sigma_{max}\Delta\varepsilon_t$ versus N_f, where σ_{max} is the maximum stress in a cycle and $\Delta\varepsilon_t$ is the total strain range, rather than as $\sigma_{max}-N_f$, or $\Delta\varepsilon_t-N_f$. The usefulness of this parameter lies in the fact that measurements from constant stress amplitude tests in which a mean stress is present tend to fall on the same curve as measurements from constant amplitude tests in which no mean stress is present (Smith et al., 1970). Thus Wetzel counts the damage for any cycle by determining $2\sigma_{max}\Delta\varepsilon_t$ and taking the reciprocal of the life associated with that parameter as the damage. [This procedure is essentially one of Miner summation (Miner, 1945).] Damage determined in this way includes the effect of any mean stress, thus avoiding the problems of mean stress correction by traditional methods.

Now consider a variable loading sequence. From knowledge of cyclic stress–strain response, it is possible to plot for each fluctuation in the sequence a stress–strain hysteresis loop which Wetzel assumed to be the *saturation* hysteresis loop. Typical response of this kind, actually determined by experiment, is shown in Fig. 1. The problem essentially consists of dividing the hysteresis loops into segments, assigning an increment of damage to each segment and then summing the damage for a prediction of life under such a loading sequence. For example, consider the segment BF in Fig. 2. Wetzel shows, again from cyclic stress–strain response, that the points O and O' can be constructed in Fig. 2, in which the dotted lines represent fictitious hysteresis loops associated with each end of the segment BF. Damage is calculated for the loop O'–B–O' by entering $2\sigma_j^b\varepsilon_{tj}$ on a plot of this parameter versus N_f and taking the reciprocal of the life which results. This unit

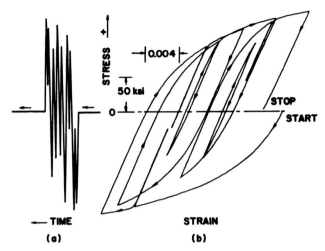

Fig. 1. The cyclic stress–strain response of SAE 4340 steel subjected to the random stress history shown (a) on the left of the hysteresis loops (b). (Courtesy of Dolan, 1965).

of damage is called D_b. In a similar way, D_f the damage for O′–F–O′ is calculated by entering $2\sigma_j^f(\varepsilon_{tj} + \Delta\varepsilon_j)$ on the life curve. Wetzel defines the damage for the element BF as $\Delta D_j = D_f - D_b$. Wetzel uses computer techniques, of course, to simulate the cyclic stress–strain response from a loading sequence, again taking saturation behavior in each cycle, and then to sum the damage of the elements of all the cycles, i.e. $\sum_1^n \Delta D_j$, fracture being predicted when that damage sum reaches unity.

Figures 3 to 5 are taken from Wetzel's paper and compare the accuracy of his cyclic stress–strain approach with two older stress-based methods of handling cumulative damage, the range count method and the histogram

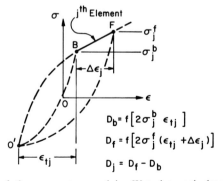

Fig. 2. Definition of the parameters used by Wetzel to calculate the fatigue damage associated with the *j*th element in the cycling history. — — —, Extended curve assuming all elements were used to their fullest extent. (Courtesy of Wetzel, 1971).

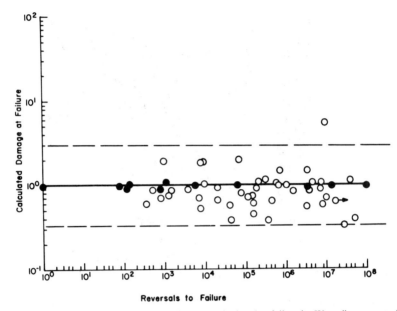

Fig. 3. Observed life to failure versus damage calculated at failure by Wetzel's computerized cyclic plasticity method. ●, Data from Wetzel's Fig. 30; ○, fatigue failure data; ○→, runout. (Courtesy of Wetzel, 1971).

method. Figure 3 is a plot of experimentally observed lives versus damage calculated at failure. For example, if the cyclic plasticity method predicted a life of 20,000 reversals† for a specimen which actually lasted 10,000, then the calculated damage at failure is 0.5, and a point is plotted at 10,000 reversals and 0.5 damage. The scatter band depicted in Fig. 2 is typical of the scatter exhibited by structural fatigue tests. Wetzel notes that only one point determined by the plasticity method lies outside this band.

In Figs. 4 and 5, Wetzel plots the same measurements for the older methods. More than one third of the points in each of these figures lies outside the scatterband and is less accurate than the least accurate points of Fig. 3. These differences are extremely important, as Wetzel points out, since analytical techniques should be compared on the *ranges* of their predictions, and range of applicability. Any technique may be accurate in special cases. Furthermore, some of the histories analyzed by Wetzel came from real components and since stress-based range count and histogram analyses remain typical of current engineering practice, the large errors evident in

---

† It has been found more convenient in recent work to use the concept of the "reversal" rather than cycles. A reversal is counted whenever the direction of stressing is reversed; for example, there are two reversals to a cycle.

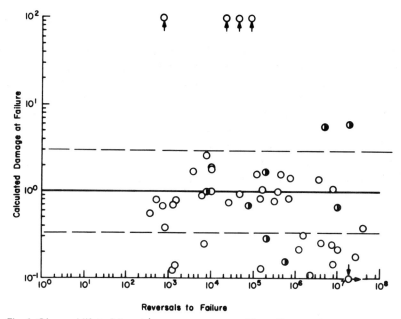

Fig. 4. Observed life to failure (same measurements as Figure 3) versus damage calculated at failure by the range count method. ○, Fatigue failure data; ◑, indicates two points; ○→, runout; ⸙, data off scale. (Courtesy of Wetzel, 1971).

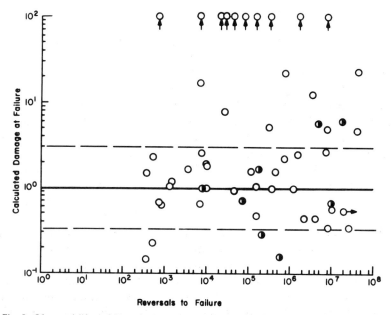

Fig. 5. Observed life to failure (same measurements as Fig. 3) versus damage calculated at failure by the histogram method. ○, Fatigue failure data; ◑, indicates two points; ○→, runout; ⸙, data off scale. (Courtesy of Wetzel, 1971).

Figs. 4 and 5 are representative of inaccuracies that occur in practice. One further point should be noted. Wetzel found that analyses based on saturation behavior, i.e. the cyclic stress–strain curve, were as accurate as the more complicated type where he included transient changes in hardness. As will be shown later, this result must be closely associated with his specific choices of materials—various steels and aluminum alloys.

The successes achieved by Wetzel and his colleagues, and earlier by Manson, in the application of cyclic plasticity to design should stimulate further development in this area. At the same time, they underscore the need for more material oriented studies, which are the subject of the present review.

II. Cyclic Stress–Strain Response of Single Phase Metals and Alloys

A. Rapid Hardening or Softening

1. HIGH STRAIN

Although many authors have made important contributions to cyclic response over two decades, for example, Kenyon (1950), Polakowski (1952), Polakowski and Palchoudhuri (1954), Dugdale (1959), Wood and Segall (1958), Coffin and various co-workers (Coffin and Read, 1956; Baldwin et al., 1957; Coffin and Tavernelli, 1958), Manson and his co-workers (see Manson, 1966), and Morrow (1965), to mention a few, no attempt will be made to document these in detail. Instead of using a catalogic approach, it seems preferable to focus on the main phenomena of cyclic response, citing recent work which synthesizes to some extent the older contributions, emphasizes underlying mechanisms, and to judge from recent reviews (Grosskreutz, 1971, 1973; Coffin, 1972), appears to represent a consensus. Current problems in understanding will also be focused.

The techniques chosen for studying cyclic stress–strain response have been essentially of two kinds. In one, a material is cycled under a constant load amplitude. Periodically, the cycling is interrupted and the hardness changes are measured either by a hardness test (Kemsley, 1958) or by subjecting the material to a conventional tension or compression test (Broom and Ham, 1957). In the other kind, the plastic strain amplitude may be controlled to be constant during tension–compression testing, and the peak stress required to enforce that strain is observed as the measure of hardening. Alternatively, if stress is controlled as constant, the plastic strain will reflect hardness changes. With the introduction and development of electrohydraulic testing equipment, controlled plastic strain tests are the kind most frequently used to study cyclic response. Of course, all the techniques used in the study of unidirectional deformation, such as transmission electron microscopy, single crystal techniques, and the interplay of variables such as temperature and alloy contents, have been employed in cyclic deformation.

The general response of both cold-worked and annealed metals to a constant plastic strain range, $\Delta \varepsilon_p$, is shown schematically by the hysteresis loops in Fig. 6 (Feltner and Laird, 1967a). For cold-worked materials, the stress

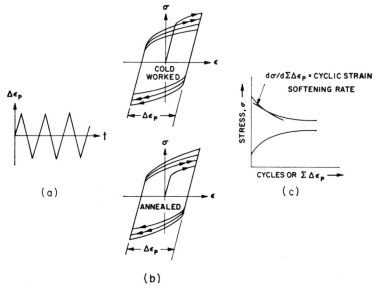

Fig. 6. Schematic of cyclic stress–strain response of annealed and cold-worked metals. (a) Control condition. (b) Hysteresis loops. (c) Cyclic strain softening and hardening curves. (Courtesy of Feltner and Laird, 1967a).

needed to enforce the strain range usually decreases with increasing cycles, indicating softening. Conversely, annealed metals show an increase in stress with cycles indicating hardening. Plots of such stress values versus the cumulative plastic strain (or reversals or cycles) experienced by a specimen are termed cyclic strain softening or hardening curves, as the case may be, and generally show that rapid hardening or softening occurs in the first few percent of the life (rapid hardening or transient stage). After the transient stage, a steady state or saturation condition is usually established during which the hysteresis loops maintain an essentially constant shape until just prior to complete fracture. The curve drawn through the tips of the steady state loops obtained from specimens tested at different amplitudes is called the cyclic stress–strain curve, as mentioned in the Introduction.

One of the variables which has been most useful in helping to elucidate the mechanisms of cyclic response has been " slip mode." Slip mode or character is defined in terms of two phenomena which are easy to observe, namely, the nature of surface slip bands and the dislocation structure of the material when deformed. This concept, arising from the work of Johnston, McEvily, and their co-workers (see McEvily and Johnston, 1965) varies from the largely two-dimensional or planar glide of materials like Cu–7.5% Al and Fe–3% Si, which show narrow straight surface slip bands and planar, intersecting, dislocation structures, to the largely three-dimensional or wavy glide of materials like pure copper or iron, which show wavy slip bands and dislocation cell structures. The materials exemplified represent extremes of planar and wavy slip. The criteria for the transition of wavy to planar slip are not well-defined but, certainly, one of the chief factors which controls the spreading of glide from two to three dimensions is the cross slip of screw dislocations. The problem of slip mode, however, is a great deal more complicated than this. For example, Fe–3% Si, although a planar slip material, is nevertheless capable of rather easy cross-slip (Low and Turkalo, 1962). Alloying effects, concentrations of point defects, and precipitate structures all can play a role in determining how wavy or planar the slip behaves, and it is difficult to predict from first principles. Clearly more research is needed in this area. In practice, most materials can be assigned to the wavy or planar class fairly definitely, with only certain alloys, generally of low solute content and intermediate stacking fault energy, being somewhat indefinite. Fortunately, in cyclic deformation studies, the alloys chosen for study were generally well defined.

Typical changes in hysteresis loops with increasing cycles are shown in Fig.7 for both annealed and cold-worked copper and Cu–7.5% Al (polycrystalline material) (Feltner and Laird, 1967a). For copper, the hysteresis loops are symmetrical in tension and compression after only a few cycles but this is only true of polycrystals. In single crystals, the loops can be asymmetrical by a few percent. The hysteresis loops for Cu–7.5% Al are asymmetrical in tension and compression even in polycrystals, with the flow stress in compression always being lower by 5–7% than the flow stress in tension when the test is started in tension. The reverse is true when the tests are started in compression. This kind of effect is more marked at low temperature and for alloys, especially those with planar slip character.

Feltner and Laird (1967a) studied the shapes of the loops as a function of cycles by a normalizing procedure due to Wadsworth (1963). The stress value at any point in the loop was divided by the maximum stress of that cycle and plastic strain values were divided by the width of the hysteresis

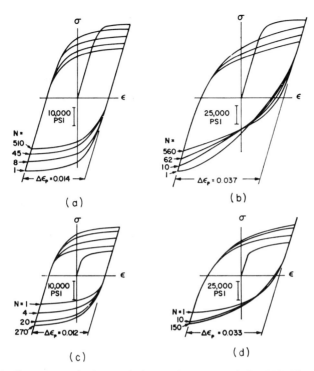

Fig. 7. Cyclic stress–strain hysteresis loops of copper and Cu–7.5% Al with different histories. (a) Cu, 23% reduction in diameter, 300 K. (b) Cu–7.5% Al, 5% reduction in diameter, 78 K. (c) Annealed copper, 300 K. (d) Annealed Cu–7.5% Al, 78 K. (Courtesy of Feltner and Laird, 1967a).

loop. As seen from Fig. 8, for copper, after the first cycle, all subsequent hysteresis loops had the same shape for a given plastic strain range, *when the material was annealed.* Cu–7.5% Al showed equivalent behavior. By contrast, the loops changed continually for both metals in the cold-worked condition. A typical example is shown for Cu–7.5% Al in Fig. 8b.

The general shapes of the softening and hardening curves for the two metals, again taken from the work of Feltner and Laird (1967a), are shown in Fig. 9. Results for a range of amplitudes are shown, and it is clear that the softening rate of the cold-worked material (as measured from the slopes of the softening curves) increases as the amplitude decreases while the hardening rate of the annealed materials decreases with decreasing amplitude. It is also apparent from these curves that Cu–7.5% Al initially softens at greater rates than does copper.

These results may be regarded as typical of wavy and planar slip materials. However, their applicability must be regarded with care. For example,

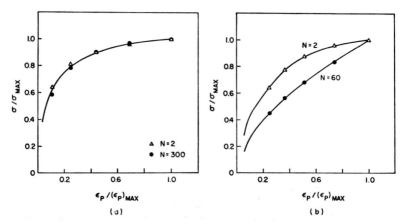

Fig. 8. Normalized hysteresis loops (tensile half only): (a) Annealed copper, 300 K, $\Delta\varepsilon_p = 0.013$. (b) Cu–7.5% Al, 5% reduction in diameter, 300 K, $\Delta\varepsilon_p = 0.032$ (Courtesy of Feltner and Laird, 1967a).

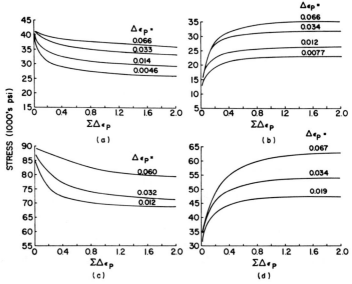

Fig. 9. (a), (c) Softening curves and (b), (d) hardening curves for copper and Cu–7.5% Al at room temperature and various strain amplitudes. (a) Cu, 23% reduction in diameter. (b) Cu, annealed. (c) Cu–7.5% Al, 5% reduction in diameter. (d) Cu–7.5% Al, annealed. (Courtesy of Feltner and Laird, 1967a).

although cold-worked metals usually soften when cyclically strained, as pointed out by Feltner and Laird (1967a), they do not always soften. Clearly, if a metal is cycled at a plastic strain range equal to that of the prestrain, cyclic hardening will occur (Coffin and Tavernelli, 1958). Feltner and Laird (1967a) have observed cyclic hardening to occur even when the plastic strain range was less than one tenth of the prestrain. This occurred in copper which had been cold-worked at room temperature and then cycled at 78° K—see Fig. 10.

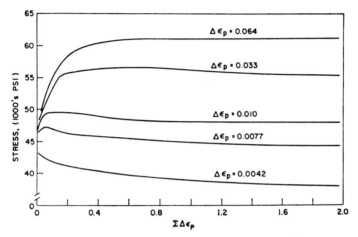

Fig. 10. Cyclic hardening at strains > 0.01, and softening at strains < 0.01, at 78 K, of copper which had been cold-worked, 5% reduction in diameter at 300 K. (Courtesy of Feltner and Laird, 1967a).

Figures 11 and 12 illustrate two other interesting results (Feltner and Laird, 1967a) for copper and Cu–7.5% Al. In these figures comparison is made between the hardening rates of these materials, in the annealed condition, and for Cu–7.5% Al, as a function of temperature. Clearly, slip character has no significant effect on the cyclic strain hardening rate of polycrystals. Further, although the steady state flow stress increases with decreasing temperature, the rate of hardening of Cu–7.5% Al is virtually insensitive to temperature. For copper, only the hardening rate in the earliest stages of cycling is insensitive to temperature.

2. LOW STRAIN

At short fatigue lives, plastic strain obviously contributes more to the total strain than does elastic strain, and the reverse is true at long lives (Sandor, 1972). Clearly, there is a life where the two strain components are equal; this life is called the transition fatigue life. It now appears conventional to define

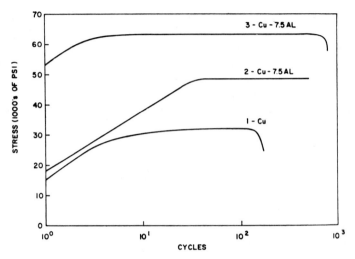

Fig. 11. Comparison of cyclic hardening behavior of copper and Cu–7.5% Al, showing that slip character has no significant effect on the cyclic hardening rate. $\Delta\varepsilon_p = 0.032$, 300 K. Annealing treatment: curve 1, 1 hr at 450 C; curve 2, 1 hr at 800 C; curve 3, 1 hr at 500 C. (Courtesy of Feltner and Laird, 1967a).

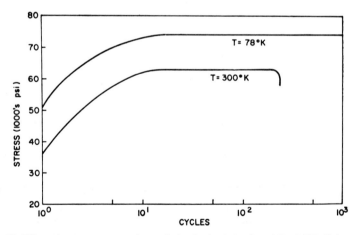

Fig. 12. Effect of temperature on the cyclic hardening behavior of Cu–7.5% Al, $\Delta\varepsilon_p = 0.066$. (Courtesy of Feltner and Laird, 1967a).

the high strain short-life regime as that where lives are shorter than the transition life. The low strain regime, naturally, is for lives longer than the transition life. As might be expected, the transition life is highly structure sensitive (see Fig. 13 taken from the work of Landgraf, 1970), being quite short for materials of great hardness and in the range 10^4 to 10^5 for soft pure metals of the kind being considered here.

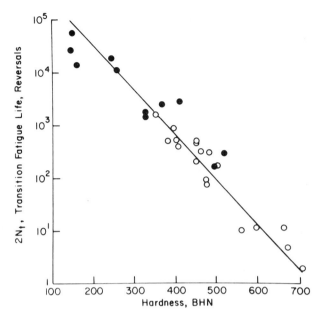

Fig. 13. Transition fatigue life as a function of hardness for steel. Steels: ○, data from Landgraf, 1970; ●, data from Smith, Hirschberg, and Manson, 1963. (Courtesy of Landgraf, 1970).

Hardening and softening behavior is different in many respects at the low strains which give lives greater than the transition life (Avery and Backofen, 1963). As shown in Fig. 14, the amount of softening which occurs reaches a maximum at strains in the region of the transition life but, at longer lives, is quite limited, and in some cases (e.g., Laird and Krause, 1970a,b) cannot be detected by techniques such as tensile testing, although the softening may be noticeable in sensitive stress–strain monitoring equipment (Laird and Krause, 1970a).† Although hardening at low strains occurs in a relatively short fraction of the total life, as in high strain fatigue, an important difference exists in the hardening rates of wavy and planar slip materials (Avery and Backofen, 1963; Lukás and Klesnil, 1973). The cyclic strain hardening stage is greater in Cu–7.5% Al alloy than in pure copper crystals. For example, Lukás and Klesnil (1973) report that the transient stage for copper represents less than 1% of the total number of cycles to failure, whereas in Cu–31% Zn, roughly 20% of the total number of cycles to failure is necessary to reach complete saturation of the stress amplitude.

Further differences with respect to high strain are shown in the effects of other variables such as temperature. Feltner (1965) has shown, for example,

† This would indicate that the older methods of measuring cyclic response at long lives are not reliable.

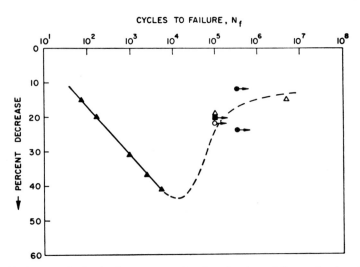

Fig. 14. Amount of softening in copper as a function of cycles to failure; taken from Feltner and Laird, 1967a. The references are as follows: ●, Polakowski and Palchoudhuri (1954); △, Kemsley (1958); ■, Hein and Dodd (1964); ○, Broom and Ham (1957); ▲, this paper; →, did not fail.

that with increasing temperature, the cyclic strain hardening rate increases and the extent of the rapid hardening stage decreases (Fig. 15). To a reader familiar with the results of unidirectional deformation, it may not be obvious how a comparison can even be considered between the results of *polycrystals* in high strain and the results obtained with *single crystals* at low strains. However, fatigue is unique in the respect that many of the observations made on polycrystals are similar to those in single crystals. For example, in the work by Avery and Backofen (1963) and Lukás and Klesnil (1973) cited above, similar results were obtained although the former used single crystals and the latter polycrystals.

Of course, not all results are similar. Polycrystals tend to show unique rapid hardening behavior, but, as in the unidirectional deformation of single crystals, strong orientation effects are shown in the cyclic deformation of single crystals. Typical hardening curves taken from the work of Kemsley and Paterson (1960) for three orientations are shown in Fig. 16. Clearly, cyclic hardening is more rapid for orientations which permit multiple slip. The impression also conveyed by this figure that cyclic hardening is less than unidirectional hardening is of course correct for the relatively low values of cumulative strain considered, but it may be misleading. At large values of cumulative strain, the hardening rate for unidirectional deformation falls off more rapidly than in cyclic deformation, and ultimately more hardening is accomplished by cycling.

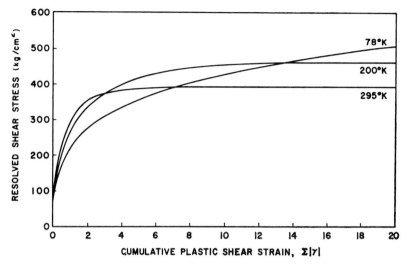

Fig. 15. Cyclic hardening curves for copper crystals [110] subjected to a plastic strain amplitude $\gamma = \pm 0.008$ at 78, 200, and 295 K. (Courtesy of Feltner, 1965).

B. Saturation Behavior

1. High Strain

It will be clear from the figures of the previous section that, after relatively few cycles, the slopes of the cyclic strain softening and hardening curves approach zero and the saturation condition is achieved. It persists until just prior to fracture, at which point the specimen contains a large crack and the tensile load required to enforce the strain falls off rapidly.

The most important result regarding saturation behavior is shown schematically in Fig. 17, "which delineates the effect of slip character and material history" (Feltner and Laird, 1967a). For a given temperature of testing, a material with a wavy slip character (Cu) has a unique cyclic stress–strain curve *independent of history*, whereas a planar slip material (Cu–7.5% Al) has a different cyclic stress–strain curve for each different initial condition of the material. The evidence supporting this general result is shown in Figs. 18 and 19. The uniqueness of wavy slip saturation behavior is widely supported. This is probably the reason, for example, why Wetzel (1971) has achieved such success in his cumulative damage procedure described in the introduction of the present paper. If, after a load change, the material rapidly attains the steady state condition appropriate to the new loading, then use of the cyclic stress–strain curve to calculate the stresses and strains in the material will be accurate. It happened that the structural steels

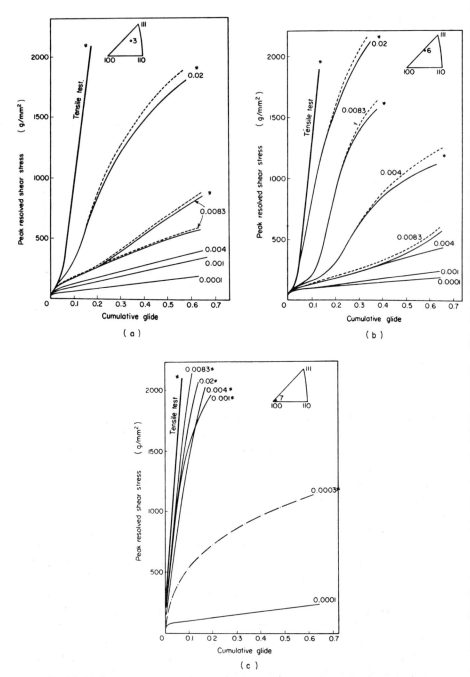

Fig. 16. Cyclic hardening curves for copper single crystals of different orientation. (a) Single slip, (b) Double slip. (c) Multiple slip. – – –, Compressive half-cycles; ———, tensile half-cycles. (Courtesy of Kemsley and Paterson, 1960).

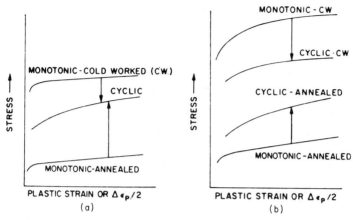

Fig. 17. The effect of the initial condition of a material on cyclic stress–strain curves of (a) wavy slip material and (b) planar slip material. (Courtesy of Feltner and Laird, 1967a).

and aluminum alloys of interest to Wetzel were indeed materials of a wavy slip character. Further support can be obtained by analysis of results on aluminum and low carbon steels reported by Coffin and Tavernelli (1958) and Klesnil *et al.* (1965). Laird (1962) demonstrated the same result for nickel and mild steel, and it can also be deduced from the work of Oldroyd *et al.* (1965) on copper and various steels. Finally Bauschinger claimed a similar result. The only work known to me which is in conflict with this result was reported by Tuler and Morrow (1963). These authors took "as

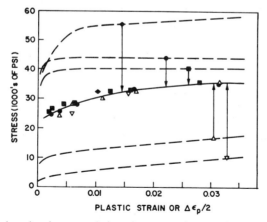

Fig. 18. Cyclic (——) and monotonic (– – –) stress–strain curves for copper at 300 K, initial condition. △, Annealed 1 hr at 450 C; ▽, annealed 1 hr at 750 C; ▲, annealed 24 hr at 450 C; ■, 5% reduction in diameter; ●, 23% reduction in diameter; ◆, annealed (△) then prestrained 50% in tension at 78 K. (Courtesy of Feltner and Laird, 1967a).

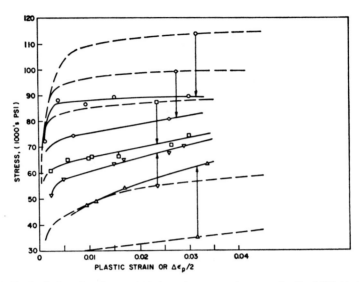

Fig. 19. Cyclic (——) and monotonic (- - -) stress–strain curves for Cu–7.5% Al at 300 K, initial condition. △, Annealed 24 hr at 550 C; ▽, annealed 1 hr at 500 C; □, 5% reduction in diameter; ○, 23% reduction in diameter. (Courtesy of Feltner and Laird, 1967a).

received" (cold-drawn) copper and cycled specimens in that condition, and also in conditions which they described as "partially annealed" and "fully annealed." No attempts were made to check the microstructures by metallurgical techniques and no details were given of the annealing procedures. The authors reported different cyclic stress–strain curves for the three conditions. Since this work is widely referenced in the engineering literature (e.g. Sandor, 1972), it is necessary to discuss its significance. In the present author's view, such a result should be discounted in the face of overwhelming evidence to the contrary. In speculation as to why it occurred, I note the possibility that the chemistry of the copper was altered during the annealing procedures, perhaps by oxygen contamination. This, of course, would radically alter the cyclic response.

Feltner and Laird's (1967a) claim that planar slip materials, in contrast to wavy materials, do show a strong history dependence was later substantiated by them (Feltner and Laird, 1969) when they cycled to saturation the following wavy slip materials: iron, disordered Ni_3Mn, Cu–3% Co (precipitation-hardened), aluminum, disordered Fe–Co–V, and cadmium; and the following planar slip materials: Fe–3% Si, ordered Ni_3Mn, ordered Fe–Co–V, and zinc in both annealed and cold-worked conditions. All the wavy materials showed unique cyclic stress–strain response (CSSR) and all the planar materials showed a history dependence. It should be noted that representative bcc, fcc, and hcp metals were cycled, as well as materials with slip characteristics varied by factors such as order.

A further point for emphasis is that the *nature* of the history does not affect the conclusion regarding the uniqueness of a wavy slip materials' CSSR. As shown in Fig. 20a, a material cycled at different strains and temperatures quickly adopts the response unique to those conditions. This result is responsible for a practice now widely adopted by engineers, that of determining CSSR by the " incremental step test." Instead of using different

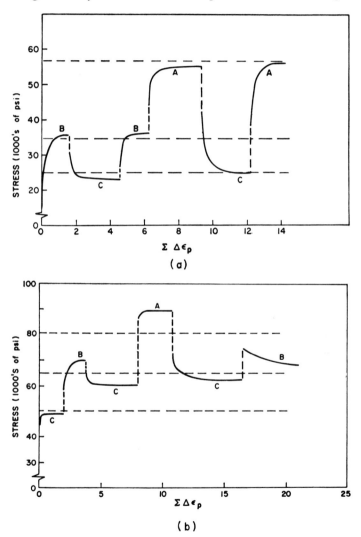

Fig. 20. Effect of change in amplitude and/or temperature on the saturation stress of (a) copper and (b) Cu–7.5% Al. – – –, Average saturation stress for conditions shown. A, $\Delta\varepsilon_p = 0.05$, $T = 78$ K; B, $\Delta\varepsilon_p = 0.05$, $T = 300$ K; C, $\Delta\varepsilon_p = 0.005$, $T = 300$ K. (Courtesy of Feltner and Laird, 1967a).

specimens to establish the points on the CSS curve, the following procedure can be adopted. A single specimen is first cycled at a low strain and the saturation stress noted; the strain is then increased, and the stress appropriate to that condition noted, and so forth. As will be clear from Fig. 20b, such a practice would give an erroneous result for a planar slip material.

Temperature has no effect on the role of slip character in determining the uniqueness of the cyclic stress–strain curve, provided the temperatures are low. However, as might be expected, temperature does cause the unique curve of a wavy slip material to rise to higher flow stresses as the temperature is decreased (Feltner and Laird, 1967a). Further, since temperature has such a large influence on slip mode when temperatures are high, by promoting cross-slip, we might expect that all materials will show a unique curve when the temperature of testing is a sufficiently large fraction of the melting temperature. This expectation is borne out in the work of Dawson *et al.* (1967) and Dawson (1967), who cycled austenitic steels at 600 K. Although, at room temperature, we should expect such steels to have a planar slip character, the results of these workers indicate that unique behavior (and thus wavy slip character) is the rule at high temperatures.

The shapes of hysteresis loops have also been of interest during saturation. Although the stable loop shapes are independent of history in wavy slip materials (Feltner and Laird, 1967a), some authors have reported them to depend strongly on amplitude and temperature, e.g. Fig. 21 showing loops described by the normalizing procedure of Wadsworth (1963). Feltner and Laird (1967a) report that the expression which describes the CSS curve does not provide a good fit for the shape of the stable hysteresis loop. On the other hand, Halford and Morrow (1962) have pointed out that the curves which connect consecutive reversal points (i.e. one side of a hysteresis loop) in an incremental step test can be made to coincide with appropriate translations of axes. Further, according to Wetzel (1971), these curves are congruent with the CSS curve if they are reduced in size by one-half. Figure 22 shows this in the case of 4340 steel. It would be interesting to know why these differences occur.

2. LOW STRAIN

Evidence relating to cyclic response in long life fatigue is relatively rare or difficult to interpret (Ebner and Backofen, 1959) because of the complex stressing employed. Recently, however, Lukás and his co-workers at Brno have ingeniously modified and instrumented an Amsler vibrophore, so that they are capable of measuring extensions of microinches across the shoulders of a push–pull specimen. They are among the few who have effectively overcome the difficult problems of measuring such low strains on a day to day basis. Lukás and Klesnil (1973) have measured the CSS curves of Cu

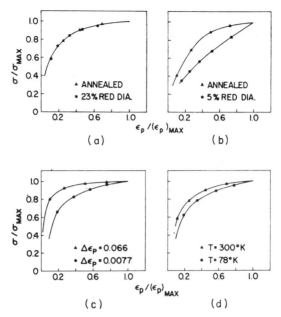

Fig. 21. Effect of different variables on the shapes of stable hysteresis loops. (a) Cu, $\Delta\varepsilon_p = 0.013$, $T = 300$ K. (b) Cu–7.5% Al, $\Delta\varepsilon_p = 0.032$, $T = 300$ K. (c) Annealed copper, $T = 300$ K. (d) Annealed copper, $\Delta\varepsilon_p = 0.013$. (Courtesy of Feltner and Laird, 1967a).

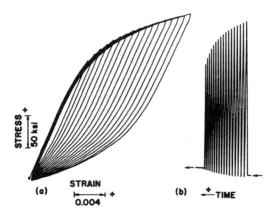

Fig. 22. (a) Hysteresis loops for 4340 steel cycled at increasing strain with the minimum strain held constant, (b) The "outsides" of the hysteresis loops are nearly the same size and shape. (Courtesy of Dolan, 1965).

and Cu–31% Zn for different histories at amplitudes giving lives greater than 10^5 cycles to fracture. Unlike the high strain result, the CSS curves for copper do depend on the degree of prior monotonic deformation, but the dependence is considerably weaker for copper than for Cu–31% Zn (see Fig.23). However, with regard to changes in *cyclic strain* levels, the saturation stress amplitude does not depend on the preceding cyclic strain levels, but adopts a value characteristic of a virgin specimen (of appropriate monotonic history). Lukás and Klesnil therefore conclude that copper very nearly obeys a cyclic mechanical equation of state relating stress amplitude uniquely with plastic strain amplitude. However, Cu–Zn is far removed from this.

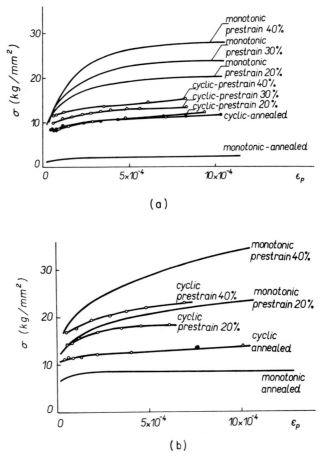

Fig. 23. Low strain cyclic and monotonic stress–strain curves for (a) copper and (b) Cu–31% Zn. (Courtesy of Lukás and Klesnil, 1973).

In another investigation, Laird and Krause (1970a) studied the low strain response of a commercial Al alloy (5083, essentially Al–4% Mg) which they had reason to believe was a planar slip material. Using an LVDT extensometer, they were not as accomplished as Lukás and Klesnil in overcoming measurement difficulties and their results show much greater scatter. However, within a rather broad band, they found a unique CSS curve (Fig. 24a). A complication revealed by the cyclic curves (Fig. 24b) was connected with the capacity of the alloy for dynamic strain aging. Note that, in specimens cycled under constant stress conditions, the plastic strain range is measured as an index of hardness changes. An *increase* of plastic strain indicates softening. However, as shown in Fig. 24b, a peak strain was attained and was then followed by slight hardening as the plastic strain decreased. The latter hardening was interpreted as due to strain aging, and it seems possible that the uniqueness of the CSS curves may be associated with this strain aging. An interesting result of that investigation was that the plastic strain was appreciable below the fatigue limit.† Lukás and M. Klesnil have observed the same result in steels (private communication, 1973).

More research in CSSR at low strains is clearly necessary, especially since Lukás and Klesnil (1973) have demonstrated beyond any doubt that the Coffin–Manson law of failure extends over the whole range of lives. This means that the cyclic plasticity approach of fatigue design should be accurate in any cyclic stress regime.

C. Dislocation Structures

1. RAPID HARDENING

The most complete study of the dislocation structures which develop during the rapid, cyclic hardening of an annealed metal has been made by Hancock and Grosskreutz (1969) on copper single crystals. Their results were later confirmed in essentials by Canning (1971). They oriented their crystals for single slip and sectioned them after cycling to cumulative strains in the rapid hardening regime. Sections were taken for examination by transmission electron microscopy both parallel to, and normal to, the primary slip plane, and the work is notable for its carefulness. A typical electronmicrograph, taken after a few cycles, is shown in Fig. 25. The dominant features are bundles of dislocations (multipoles) separated by regions relatively free of dislocations. The bundles tend to be oriented perpendicular to the direction of the primary Burgers vector, and along the traces of intersection of the primary glide plane with the critical and conjugate glide planes.

---

† It does not appear widely appreciated that materials other than steels can show a fatigue limit. Stubbington (1958) found that Al–Mg alloy, under certain conditions, is one of them.

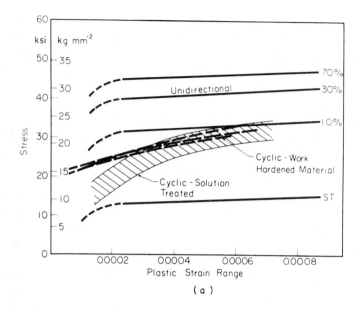

(a)

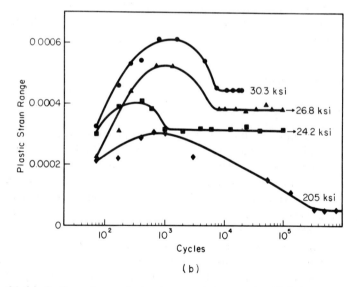

(b)

Fig. 24. (a) Cyclic and monotonic stress–strain curves for 5083Al–Mg alloy in the solution-treated and cold-worked (10%, 30%, and 70%) states. Note that both the cyclic and monotonic results are plotted against plastic strain range; for direct comparison of the cyclic and monotonic curves, the latter should be translated to the right once with respect to strain. (b) Softening and hardening curves for 5083Al–Mg alloy cold-worked 70% and cycled at the constant stresses indicated. (Courtesy of Laird and Krause, 1970b).

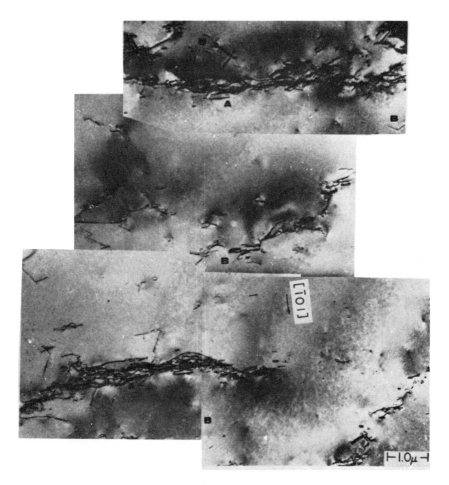

Fig. 25. Dislocation structure after $4\frac{1}{4}$ cycles (9 reversals) at $\gamma_t = \pm 0.0075$, 300 K, in a copper crystal oriented for single slip. Foil normal = [111]; $g = \bar{2}02$. The primary Burgers vector is $a/2$ [$\bar{1}01$]. Edge bundles at A; faulted dipoles at B. (Courtesy of Hancock and Grosskreutz, 1969).

The majority of the dislocations within the bundles were found to be primary edge dislocations on parallel slip planes, which had trapped each other over portions of their lengths to form dipoles and multipoles. Hancock and Grosskreutz observed very few primary screw dislocations at any stage of the hardening and concluded that they were mutually annihilated by cross-slip during cyclic straining.

Figure 26, also taken from Hancock and Grosskreutz (1969), shows the dislocation arrangement after 100 cycles. Here, an indefinite cell structure is

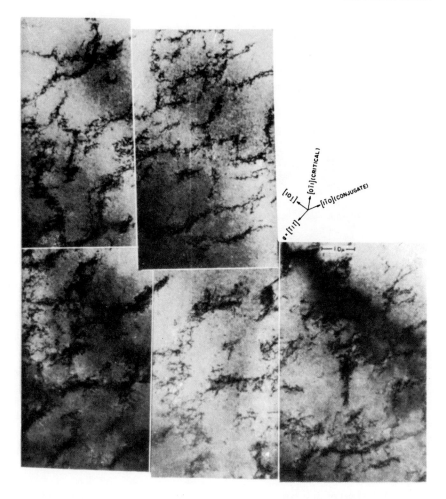

Fig. 26. Dislocation structure after 100 cycles under same conditions as Fig. 25. $g = \bar{1}1\bar{1}$. Dislocations with the primary Burgers vector are not in contrast except near the lower left-hand corner. (Courtesy of Hancock and Grosskreutz, 1969).

present, and predominantly planar arrays of dislocations extend along the primary slip plane. Many dislocations with other than the primary Burgers vectors are also present.

Hancock and Grosskreutz (1969) concluded from observations such as these, as had several of their predecessors, that the principal dislocation interaction during rapid hardening is the mutual trapping of primary edge dislocations, which develops bundling. The bundles provide obstacles to further dislocation motion and reduce the mean free path progressively. As hardening proceeds, the bundles are continuously fragmented into dense tangles of loops and dipoles, and the dislocation structures subsequently extend normal to the primary slip plane to build up a cell structure. Of course, the precise nature of the final structure depends to a large extent on the strain amplitude. As the reader realizes, once saturation has been attained, no further continuous changes take place within the dislocation structures. Feltner and Laird (1968a) have documented the factors which influence the saturation dislocation structures in fatigued metals. As shown in Fig. 27, the structures range from rather complete cells at high strains and short lives, through incomplete cells of the type shown developing in Fig. 26 at intermediate lives, to dense rafts (or veins) of dislocation dipoles and multipoles, which decrease in density with decreasing strain, at long lives. Depending on the strain amplitude, dislocation multiplication via the

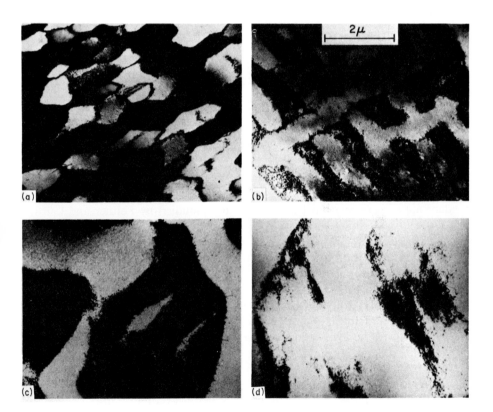

Fig. 27. Dislocation structures in copper at room temperature after cyclic straining to 50% N_f at amplitudes resulting in lives of 10^4 to 10^7 cycles: (a) $N_f = 10^4$; (b) $N_f = 10^5$; (c) $N_f = 10^6$; (d) $N_f = 10^7$. (Courtesy of Feltner and Laird, 1968a).

processes described by Hancock and Grosskreutz stops short of cells at the appropriate saturation level, or proceeds to cells at the highest strains. The higher the strain, the smaller is the cell size (Feltner and Laird, 1967b, 1968a; Pratt, 1965). A schematic summary of dislocation structures in fcc metals, after saturation has been obtained, as a function of amplitude, temperature, and slip character, is shown in Fig. 28 (Feltner and Laird, 1968a).

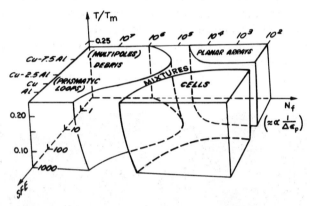

Fig. 28. Schematic summary of dislocation structures in fcc metals as a function of amplitude, temperature, and slip mode (SFE). (Courtesy of Feltner and Laird, 1968a).

2. SATURATION

The saturation structures of fatigued metals (polycrystals), and their dependence on specimen history, have been studied by Feltner and Laird (1967b). Wavy slip materials, whatever their structures before strain cycling, develop cell structures if cycled at high strains. A typical example of the structure before and after cycling in copper, for both an annealed and a preworked condition, is shown in Fig. 29. The annealed copper before cycling contains only a few stray dislocations, but the initially cold-worked copper before cycling shows a ragged cell structure regularly expected of unidirectional deformation. After cycling, both the annealed (Fig. 29c) and cold-worked (Fig. 29d) copper have a well-developed cell structure with the interiors of the cells almost completely free of dislocations. The cell structures induced by cycling are different from these formed in unidirectional deformation, being tighter and showing virtually no misorientation across cell walls. This indicates that the signs of the dislocations within the walls balance nicely.

Feltner and Laird (1967b) observed that the size of the cells depended on both amplitude and temperature; it increased with decreasing strain and

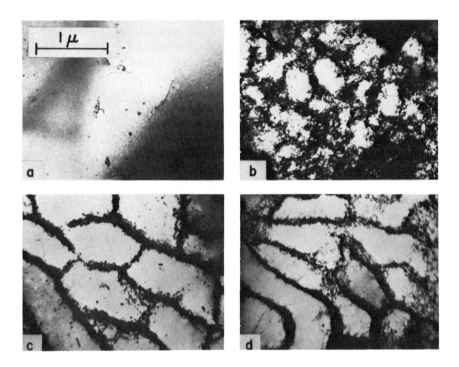

Fig. 29. Dislocation structures in copper before, (a) and (b), and after cycling, (c) and (d), at high strain and room temperature. (a) $g = \bar{1}\bar{1}1$; (b) $g = 002$; (c) $g = \bar{1}\bar{1}1$; (d) $g = 1\bar{1}1$. (Courtesy of Feltner and Laird, 1967b).

increasing temperature, but did not depend on history. This result was strongly supported by a statistical evaluation of the cell size in the different specimens.

In contrast to the structures seen in copper, a cell structure was never observed in the example of a planar slip material, Cu–7.5% Al, studied by Feltner and Laird (1967b). Instead the dislocations were observed to be dissociated and were generally arranged in planar arrays. A typical example of the structure before and after cycling, as a function of history, is shown in Figure 30. Clearly, after cycling, many planar arrays have formed in the annealed alloy (Fig. 30c) but they are considerably less numerous than those in the cold-worked alloy (Fig. 30d), which is difficult to distinguish from its initial condition (Fig. 30b). Burgers vector analysis of the dislocations within the bands by the regular techniques of electron microscopy showed that all the partial dislocations which lay in the plane of the band were represented. Within any grain, partials making up the primary slip direction were dominant, but slip traces associated with secondary slip planes were

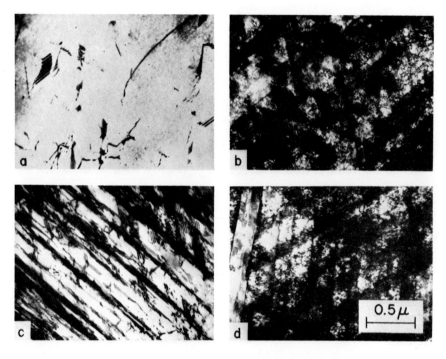

Fig. 30. Dislocation structures in Cu–7.5% Al before, (a) and (b), and after cycling, (c) and (d), at high strain and room temperature. (a) $g = 002$; (b) $g = 11\bar{1}$; (c) $g = 111$; (d) $g = \bar{1}\bar{1}1$. (Courtesy of Feltner and Laird, 1967b).

also frequent. Observations of obvious cross slip out of the primary bands in the cycled alloy were infrequent.

The studies of Feltner and Laird thus described were largely confined to the high strain regime, except that they did additional work at low strains, yielding the summary shown in Fig. 28. More recently, Lukás and Klesnil (1973) have taken full advantage of their accurate low strain machine to study the dislocation structures associated with history effects in *long life fatigue*. They found that, even in copper, cell substructure initially introduced by unidirectional deformation or high strain cycling could not be removed by subsequent cycling at low strains (i.e. well beyond the transition life). To some extent the unidirectional cell substructure could be "tidied up" and the cell size slightly enlarged. Interestingly enough, it was possible for copper to have the same saturation stress for a modified initial cell structure as for the virgin dislocation structure (veins of dislocation dipoles and multipoles). The substructures which formed in Cu–Zn (planar arrays and multipoles) were resistant to change when cyclic amplitude changes at low strain were made (Lukás and Klesnil, 1973). This, of course, is to be expected of a planar slip material.

An even more important series of papers by the Czechoslovak workers consists of their documentation of near surface dislocation structures. It has been recognized for many years (see Thompson and Wadsworth, 1958) that fatigue cracks nucleate in marked slip bands, called "persistent slip bands" connected to the free surface of the specimen. Lukás and his co-workers concentrated on a detailed examination by transmission microscopy of these bands. It appears widely believed that, in low strain, most of the strain is concentrated in these bands (Laird and Duquette, 1972) and they should therefore play a central role in cyclic deformation at long life. The work of the Czechoslovak group was summarized in Lukás and Klesnil (1972). Briefly, their conclusions are as follows: In wavy slip materials at the highest strains, cell structure was observed in the whole crystal, and no differences could be discerned between the interior and near surface structures. At longer lives, where persistent slip bands are most prominent, differences between slip band and interior structures were observed. Typically, in a bulk structure comprising veins of dipoles, the persistent slip band, observed "in elevation," appears as a ladder structure (Fig. 31). Observed "in plan view," it appears as an array of cells (Fig. 32). These slip bands widen slightly close

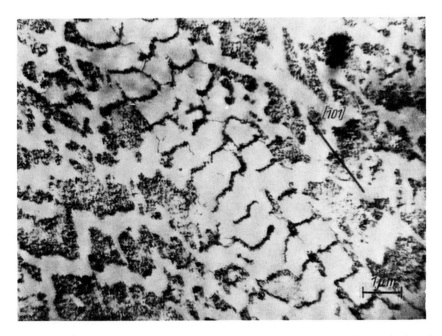

Fig. 31. The dislocation structure of a persistent slip band in a copper single crystal oriented for single slip. The plane of the foil was ($1\bar{2}1$) and the section was taken 50 μ below the surface. Stress amplitude ± 6.5 kg/mm^2; $N_f \sim 10^6$; specimen given 20% of N_f. (Courtesy of Lukás and Klesnil, 1968).

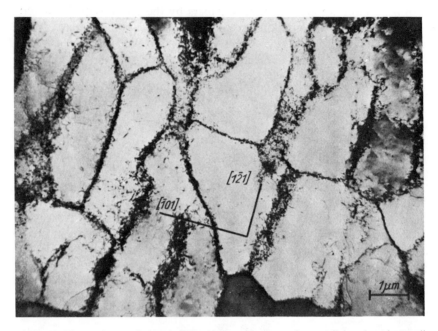

Fig. 32. The persistent slip band of Fig. 31 is seen in the section parallel to the primary slip plane (111). (Courtesy of Lukás and Klesnil, 1968).

to their surface and are directly associated with notch and peak surface topography. They persist to considerable depths in the crystal and, in some cases, probably run right through. A schematic diagram of a persistent slip band, deduced from sections such as those shown in Fig. 31 and 32, is given in Fig. 33, along with a diagram of the single crystals used by Lukás and Klesnil (1968). This diagram indicates the orientations of the sections shown in Figs. 31 and 32. The persistent slip band clearly consists of cylindrical cells packed along the slip band, with the cylinder axes normal to the plane of the band.

In planar slip materials, the situation is different (Lukás and Klesnil, 1970). For all the strain amplitudes used by these workers, the general structure was the "planar arrays" type. The only difference between the bulk structure and the near surface layer was a lower dislocation density in the surface layer.

D. Mechanisms of Cyclic Deformation

An adequate theory of cyclic deformation must account for the phenomena in rapid hardening and in saturation and all the details of history

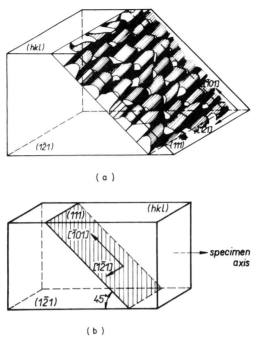

(a)

(b)

Fig. 33. (a) A space model of the persistent slip bands observed by Lukás and Klesnil (1968). (b) The orientation of the copper single crystals used by them in obtaining the sections shown in Figs. 31 and 32.

dependence, temperature effects, variable load cycling, and slip mode. In particular, it must explain how, in saturation, hundreds of percent of strain can be accumulated while the dislocation structures remain essentially static from cycle to cycle. An attempt to do this was made by Feltner and Laird (1967b) on the basis of a suggestion by Avery and Backofen (1963) that a distinction must be drawn between high and low strain cyclic deformation. Feltner and Laird further pointed out that clear parallelisms exist between low strain deformation and the stage I unidirectional deformation of a single crystal, and between high strain cyclic deformation and stage II unidirectional deformation, as follows:

1. Stage I and low strain deformation (Feltner and Laird, 1967b): In stage I deformation, the hardening rate with respect to strain is low, and increases with increasing temperature. The extent of stage I decreases with increasing temperature, increasing tendency to multiple slip (due to orientation effects), and with increasing tendency to cross-slip. During this stage, dislocation dipoles are prominent and they tend to cluster. Theories of hardening in stage I rely on the assumption that dislocations act singly. The phenomena

for low strain cyclic deformation described in the previous sections are encompassed nicely by these generalities. For example, when one considers the effect of slip mode, one finds that, for unidirectional deformation, the hardening rate for brass is lower than for copper and the extent of stage I in brass is greater than in copper. By comparison, low amplitude fatigue studies (see Sections II,A,2 and II,C,1) have shown that the cyclic strain hardening rate is lower and the extent of rapid hardening is greater in Cu–7.5% Al alloy (both polycrystals and single crystals) than in copper.

2. Stage II and high strain deformation (Feltner and Laird, 1967b): The properties of stage II deformation are as follows: (a) the transition from stage I to II is marked by dislocation tangling, (b) the rate of hardening is virtually independent of temperature and slip mode, (c) multiple slip orientations harden more rapidly than do those of single slip, (d) cell formation is normal in wavy materials, (e) the dislocation density from secondary slip systems increases with increasing strain, and (f) the onset of stage III is dependent on slip mode, occurring at lower strains/stresses as cross-slip becomes easier. When we compare the properties of cyclic deformation at high strains described in the previous sections, we note essentially equivalent behavior.

As a consequence of these parallelisms, Feltner and Laird (1967b) expected that the mechanism of rapid hardening in low and high strain deformation are essentially similar to those which occur in stage I and II unidirectional deformation respectively. For example, Feltner (1965) formulated a dipole hardening mechanism for low strain fatigue in which the increase in the flow stress in the rapid hardening stage is determined by the increase in density of the dislocation dipole obstacles. The changes in dislocation structure which Hancock and Grosskreutz (1969) observed during rapid hardening, Section II,C,1, are essentially consistent with this view. Further, in high strain fatigue, the rapid hardening should be accomplished by the same kinds of dislocation tangling and cell formation which occur in stage II unidirectional deformation. Hancock and Grosskreutz appear to disagree with this view in that they require the hardening process to pass through stage I first. Of course, since the strains even in high strain cyclic deformation are low relative to unidirectional deformation, this will be true for single slip orientations. However, the view of Feltner and Laird (1967b) is more general in that it accounts for behavior in polycrystals. Here, because of multiple slip orientations, stage II types of deformation begin essentially with the first cycle.

Since the saturation stress of wavy slip materials is also a unique function of the plastic strain range, temperature, and strain rate (Raouf and Plumtree, 1971; Raouf et al., 1971), there is a cyclic mechanical equation of state. Consequently, metals initially cold-worked will have to soften to reach this state. Feltner and Laird (1967b) have demonstrated that *reversed plastic strain* is required for softening to occur. This is consonant with the way the degree of softening falls off in low strain fatigue (Section II,A,2) and the observation of Ham and Broom (1962) who showed that, in prestrained copper crystals tested in completely reversed stress, there is a cyclic stress below which softening will not occur. Feltner and Laird (1967b) believed that this stress corresponds to that which will produce the measurable reversed plastic strain (8×10^{-5}) which they observed. At such low strains softening will not be complete, as Lukás and Klesnil (1973) have also observed. This is because the strains are insufficient to change the type of dislocation structure, from cells to veins. On the other hand, softening at high amplitudes requires only a change in the degree of the dislocation structure, not the kind. Thus initially introduced ragged unidirectional cell structures require only slight rearrangement and changes in dislocation density to achieve complete softening. Of course, the three-dimensional slip capacity of wavy slip materials is necessary for such rearrangements. The situation in planar slip materials, where cross-slip is infrequent and slip is essentially two-dimensional, will not allow complete softening to take place even at high strain amplitudes. However, due to shuffles in the planar arrays, partial softening does occur.

It has to be admitted that the precise dislocation mechanisms which produce softening are not understood and little electron microscopy has been carried out on the dislocation structures formed in the transient condition. It is believed, however, (see, for example, Feltner and Laird, 1967b) that dislocation interactions are mainly responsible for softening, while the point defects which are created in large concentrations during cycling straining (Feltner, 1965; Broom and Ham, 1959; Hull, 1958; Wadsworth and Hutchings, 1964; Johnson and Johnson, 1965; Polak, 1969) play a lesser part. Presumably the annihilation of these defects helps to complete the softening in wavy slip mode materials.

In general, the mechanisms used to explain rapid hardening or softening appear to be in wide agreement in that most authors employ a débris mechanism involving mutual dislocation trapping and related models; quantitative aspects of the problem are not understood. It is the saturation state, comprising cell structures at high strains and multi-pole veins at low strains, where controversy mainly exists. Different dislocation models are required for each of these structures. At high strains, Feltner and Laird (1967b) suggested that the saturation stress is controlled by the reaction of free mesh lengths of dislocations against bowing out of the cell walls. This idea should apply either to a cell or to a tangled pileup of the kind which occurs in planar slip materials. According to these workers, once a link has gone critical, it moves freely through the volume between cell walls. A loop emitted from such a link was imagined as growing to a quite large size before it contacted the walls of the containing cell and became tangled and partly annihilated (enough to prevent accumulations of dislocations from cycle to cycle). This process repeats itself when the stress is reversed, new links bowing out of the walls or tangles and roughly retracing the paths in the forward direction. This model is known as the "shuttling" model and it adequately accounted for both stresses and strains (Feltner and Laird, 1967b).

However, Basinski et al. (1969) have criticized the shuttling model on the grounds that it would produce new dislocations at too large a rate. To work, at least 99% of the shuttling dislocations would have to be annihilated each cycle. In view of the large densities of dislocations present at high strains, and the equal numbers of dislocations that exist with opposite sign, it does not appear unreasonable to this author.

An alternative explanation, which has been championed by Grosskreutz (1971), is that mostly the *same* dislocations (about 5% of the total number) shuttle to and fro to produce the saturation strain, and that the flow stress is provided by a friction stress encountered by the dislocations as they traverse the cell volume. This model invokes the presence of point defects and clusters to produce the friction stress. In order to show this, Piqueras et al. (1972) measured the yield stress for unidirectional deformation as a function of temperature between 78 and 300 K on copper single crystals which had been fatigue-hardened into saturation at room temperature. They also used weak beam transmission electron microscopy to observe point defect clusters, which were found to be preferentially located in dislocation bundles and walls, but also existed in fair density between the walls. Piqueras et al. concluded that the saturation stress in fatigued copper single crystals is determined by the thermally activated surmounting of defect clusters and the *athermal* overcoming of long-range stresses by mobile dislocations passing through the regions of low dislocation and defect cluster density. J. C. Grosskreutz and H. Mughrabi (private communication, 1972) later carried out a very elaborate and difficult experiment in which they contrived to remove a fatigue specimen *under load* from a fatigue machine (the specimen had been cycled to saturation), neutron irradiated it to anchor the dislocations in their as-loaded configuration, and thinned the specimen for observation by electron microscopy. They showed that many dislocations traversed the cells (Hancock and Grosskreutz, 1969) and claimed support for the friction stress model. While there is not much doubt that point defect clusters can help to increase the saturation stress, Grosskreutz' model suffers from the following serious drawback. It is well known that, once a cell structure has been established by cycling, reduction of the strain amplitude to a lower value quickly causes the cells to enlarge. If Grosskreutz' model—

the to-and-from motion of dislocations without significant interaction with dislocations in cell walls—is valid, then a reduction in strain should only reduce the amplitude of the to-and-from motion, without causing changes in the cell walls. In this respect, the shuttling model of Feltner and Laird (1967b) shows up in a much better light. Since this model requires the shuttling dislocations to annihilate in the walls, the opportunity for a rapid rearrangement of the walls does exist when the strain range is reduced. This softening observation implies that the balance between the mobile dislocations and the cell walls is quite fragile, and it also implies that the walls are easily penetrable. Clearly, much more work is needed to sort out these difficult problems.

At low strains, the situation is even more confused. Feltner (1965) proposed that the cyclic strain could be accommodated without an increase in dislocation density from cycle to cycle if the dislocation dipoles, which comprise the bulk of the structures at low strains, could undergo a "flip-flop" motion with the cyclic stress. That is, at equilibrium, the dislocations making up a dipole lie at 45° degrees with respect to each other and their slip plane. On application of the stress, the dislocations in the dipole can flip past one another so as to adopt the 45° stance in the opposite sense. Reversal of the stress flops the dipole back to its original position. This model has been criticized (Grosskreutz, 1971) for not being capable of predicting the correct flow stress; however, there are enough adjustable parameters to permit a reasonable calculation of the flow stress: for example, the dipole width and the distribution of widths, the lengths of the dipoles (involving end effects) and the distribution of lengths. It has also been criticized on the grounds that there are not enough dipoles to accommodate the cyclic strain. Since Feltner has studied dipoles most thoroughly (1965, 1966), and he claims an adequate number, the present author is inclined to defer to his view. However, the flip-flop model is subject to more serious criticism in that the only temperature-dependent quantity it contains is the elastic modulus, and the saturation flow stress is much more temperature dependent than the modulus. Moreover, Watt and Ham (1966) took issue with the model on the following grounds. A copper single crystal was cycled into saturation at a given low strain, at which point the cyclic hysteresis loop was measured. Then the strain amplitude was lowered and the hysteresis loop shrank from the one marked ABCD in Fig. 34 to that marked EFGH. Now, if the flip-flop model is valid, it should be possible to construct loop EFGH from ABCD, because the dipoles which begin to flip on passing from tension to compression at C on the curve BCD should be the same dipoles that flip on passing G on the curve FGH. Here Watt and Ham assumed that the flipping of these dipoles (the widest ones) would not be influenced by the state of all the other dipoles. Therefore, by translating BCD to the left,

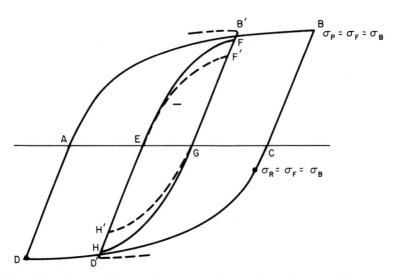

Fig. 34. Actual hysteresis loops, ABCD and EFGH, at high and low amplitudes (both in the low strain regime), as compared to the Feltnerian loop EF'GH'. (Courtesy of Watt and Ham, 1966).

B′GH′ is obtained; similarly DAB yields D′EF′. The "Feltnerian" loop so constructed, EF′GH′, is too soft compared to the measured loop, EFGH. This means that there must exist back stresses at the end of stroke DAB which assists the reverse plastic flow. Although this result is in conflict with that shown in Fig. 22, it raises serious doubts about the flip-flop model.

Of course, the real problem of low strain fatigue is that we do not understand how much of the plastic strain is in fact carried by the persistent slip bands shown in Figs. 31–33. The possibility exists that most of the strain is carried by these bands and a totally new approach to interpreting cyclic deformation at low strains is required.

E. A Surface Hardening Interpretation of Cyclic Response

In 1969, Kramer published a paper (1969a) in which he criticized the studies described in Sections II,A–D of the present review, on the grounds that they have not incorporated observations relating to the role of the surface. Kramer and his associates have made numerous studies on this subject over the last fifteen years (Kramer, 1961, 1963a,b, 1965a,b,c, 1967a,b,c, 1969a,b; Kramer and Haehner, 1967; Kramer and Demer, 1961; Feng and Kramer, 1965; Shen et al., 1966; Shen and Kramer, 1967; Kramer and Balasubramanian, 1973; Kramer and Podlaseck, 1963). In effect, Kramer (1969a) asserted that cyclic hardening and softening are accounted for *entirely* in terms of "the surface layer stress." Kramer defines that concept by the following. If a specimen is hardened, then unloaded, and immediately reloaded, the resulting stress–strain curve will be as schematically shown in Fig. 35a, that is, the stress–strain curve will immediately rejoin the continuation of curve CDA at F. However, if, after unloading, the specimen is permitted to rest or else a thin layer is removed from the surface, the stress–strain curve will be as shown in Fig. 35b; that is, the stress departs from the elastic curve at a low value, *G*, and the quantity *S* is the surface

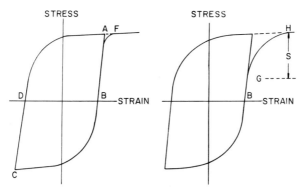

Fig. 35. Kramer's definition of the surface layer stress: (a) the immediately reloaded response BF, following unloading from A; (b) the reloading response BGH which follows relaxation or surface removal after unloading from A. The surface layer stress is the quantity S.

layer stress. It is this quantity which Kramer claims as being responsible for hardening or softening in distinction from the bulk hardening described above. It should be pointed out that Kramer's work is in conflict with a list of publications even longer than his own, obtained by numerous investigators, for example, Fourie (1967, 1968, 1970). However, Kramer explains away such difficulties by arguing that the interior dislocation structures modify the surface layer stress.

Since the present author firmly believes in the validity of bulk hardening mechanisms, Kramer's work presented a fascinating puzzle until, recently, J. M. Finney had an opportunity to independently test Kramer's basic definition shown in Fig. 35. J. M. Finney (private communication, University of Pennsylvania, 1973) cycled a copper single crystal to saturation under constant plastic strain, and then unloaded at zero strain. The hardening accomplished in cycling to saturation represented an increase in flow stress of two orders of magnitude. Finney next reloaded the specimen immediately and cycled between the same strain limits as used for saturating the crystal. Once again he unloaded the specimen, rested it overnight, and removed 0.006 in. from the surfaces at which the active slip steps emerged. Finally, he reloaded and recycled the specimen. Instead of observing a Kramerian curve (Fig. 35b), he could find no significant difference between this curve and the one he previously obtained by immediately reloading.† Since Finney's work was done with great care and accuracy, the present author is unable to accept Kramer's assertion that the surface layer is responsible for cyclic hardening. It would be more satisfactory, however, if independent observers would check Kramer's results (which are very interesting) and help to resolve fully the problems which he has raised. There does seem to be a surface effect, in certain circumstances, and it should not be obscured by the overstatement which Kramer made (1969a) in an attempt to focus deserving attention on this problem.

F. Cyclic Response under Variable Loading

Some of the most interesting observations in cyclic deformation in the last six years have been obtained on specimens not cycled under the constant strains/stresses of interest in the sections above. For example, Neumann (1967, 1968, 1969) cycled single crystals oriented for single slip in push–pull, so that the stress amplitude increased linearly with time to a preselected final value, at which point the cyclic stress was held constant. During the stress buildup, the plastic strain amplitude occurred almost entirely as strong bursts (Fig. 36). Between bursts, which lasted for about 50 cycles, the plastic strain was quite small. After the stress amplitude became constant, the plastic strain decreased to a small value. Neumann observed that each burst originated from slip all over the specimen and not from localized slip. Such bursts were observed in a wide variety of metals (Neumann and Neumann, 1970), but not in polycrystals.

Neumann (1969) explains the instability which gives rise to bursts in the following way [Basinski et al. (1969) have also observed the instability of fatigue structures]. He imagines the structures more or less as Hancock and Grosskreutz (1969) observed them during rapid hardening, namely, multipole tangles, with unpaired dislocations moving to and fro. As the free dislocations move around, they find partners with which to annihilate or to form stable dipoles, so that the strain is low. After more cycles elapse, and the stress is increased, the widest of the dipoles dissociate producing free dislocations. These dislocations are able to push stable dipoles into clumps counterbalanced by dislocations of opposite sign. Finally, by local stress intensification due to pileups, the stable dipoles in the volumes between the

---

† In fairness to Dr. Kramer who was privately shown these curves it should be pointed out that he does claim a difference. Even if Dr. Kramer is correct, his own evaluation is at variance with the claims of his (1969a) paper.

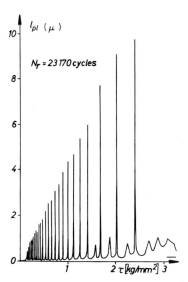

Fig. 36. Strain bursts in a copper single crystal which occur during the gradual increase of the stress amplitude from zero Strain $= l_{pl}/8.5$ mm. (Courtesy of Neumann, 1969).

pileups (which are stable only to the applied stress) are forced to dissociate and multiply free dislocations. Thus a strain avalanche develops which disturbs the nearly unstable equilibrium of neighboring dislocations until the burst spreads over the whole specimen. Neumann has used these observations in interesting applications to crack nucleation (Neumann, 1969) and to crack propagation (Neumann, 1973). Clearly such phenomena will have a big effect in variable loading conditions and the subject looks one of the most interesting currently being explored.

Another equally fascinating development is a result reported by Landgraf *et al.* (1969) in which specimens were subjected to blocks of gradually increasing and then decreasing *strain* (not stress, as Neumann explored). After a few of these blocks, the metal cyclically stabilizes and, surprisingly enough, the flow stress corresponding to any of the peak amplitudes in the blocks turns out to be the saturation stress that the specimen would show in a constant strain test at such a value of strain. A typical test record of controlled strain amplitude blocks and the stress response of the specimen to this strain program, both as functions of time, are shown in Fig. 37; of course, this result should only apply for wavy slip materials such as the steels, SAE 1045, quenched and tempered, which Landgraf *et al.* used. The reasons for this behavior appear not to have been investigated. It would be most interesting to understand it and variations on this type of experiment could probably yield very fruitful contributions to our general understanding of cyclic deformation.

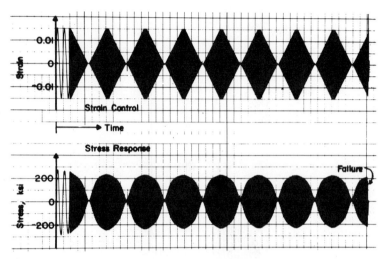

Fig. 37. The strain–time and stress–time records of triangular block testing on quenched and tempered SAE 1045 steel, 450 BHN. Since no hardening changes are shown, these results were recorded only after saturation was attained. (Courtesy of Landgraf *et al.*, 1969).

III. Cyclic Stress–Strain Response of Multiphase Materials

A. *Microstructures Containing Particles Penetrable by Dislocations*

In spite of the commercial importance of multiphase materials, fundamental studies of the cyclic response of these materials are relatively rare. Of course, the cyclic stress–strain curves of conventional structural and tool steels, aluminum alloys, titanium alloys, and other useful materials have been established (e.g., Manson, 1966; Wetzel, 1971). Simple interpretation of the cyclic stress–strain response of such materials is virtually impossible at present because of their microstructural complexity and the lack of evidence. However, a number of studies of simpler materials has now been published and the underlying phenomena are becoming clearer.

In many microstructures where the particles are very small and closely spaced, the bowing stress for dislocations to pass between the particles is very large. Consequently, the dislocations must cut the particles when cyclic strains are applied to such materials. Two examples of the hardening curves which result for such microstructures are shown in Figs. 38a,b. In one, Fig.38a, the Al–4% Cu alloy of interest was aged to contain θ'' plate precipitates which are on the order of 20 Å thick and a few hundred angstroms in diameter (Calabrese and Laird, 1974a). In the other, due to McGrath and

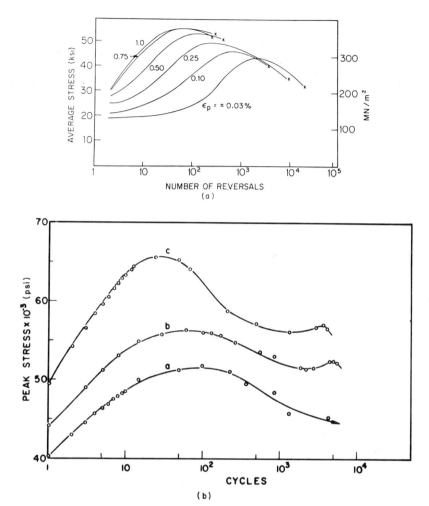

Fig. 38. (a) Cyclic hardening and softening curves for Al–4% Cu alloy aged at 160 C for 5 hr to produce a microstructure containing the metastable θ'' precipitate. The crosses at the ends of the curves denote fracture. (Courtesy of Calabrese and Laird, 1974a). (b) Cyclic hardening and softening curves for Fe–0.04% C containing fine precipitates. Quenched from 700 C. Aged at 25 C, 10 days. a, $\Delta\varepsilon_p = 0.005$; b, $\Delta\varepsilon_p = 0.007$; c, $\Delta\varepsilon_p = 0.010$. (Courtesy of McGrath and Bratina, 1969).

Bratina (1969), an Fe–0.04% C alloy was quenched and aged at room temperature to produce a fine dispersion of carbides. In both these materials, hardening initially occurred until a peak stress was attained, and then softening occurred. Softening is most marked at intermediate strains. This kind of behavior is not observed in single phase materials, except that Feltner and

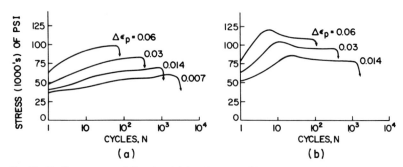

Fig. 39. Cyclic response of disordered (a) and ordered (b) Ni_3Mn alloy. (Courtesy of Feltner and Laird, 1970).

Laird (1970) found an alloy of Ni_3Mn, in the *ordered* condition, to give such a curve. This curve is shown in Fig. 39 along with the hardening response of the Ni_3Mn in the disordered condition.

Before the study of cyclic deformation in these kind of alloys was undertaken, considerable interest centered round their fatigue fracture behavior, which was generally regarded as disappointing when the endurance stress at 10^8 cycles to failure was compared to the monotonic ultimate strength attainable. Typical observed ratios lay in the range 0.25 to 0.35. However, it should be noted that *all* high strength materials, *when hardened to their ultimate capacities by any means whatever*, show such ratios. This fact can easily be checked by reference to standard compilations of fatigue properties. Although the disappointment was therefore inappropriate, it stimulated much research by electron microscopy and optical microscopy into the nature of the slip bands which generated the fatigue cracks. This work has been reviewed at length by Laird and Thomas (1967) and Krause and Laird (1967/1968). Several suggestions have been made to explain why fatigue cracks nucleated and propagated easily in the slip bands:

1. The precipitates within the bands undergo overaging, thus weakening the microstructure locally.

2. The precipitates within the bands are repeatedly cut by the dislocations and are thereby chopped into a size too small to be stable; consequently they redissolve, and locally the structure is weakened.

3. The materials of interest contain aging inhomogeneities along slip bands which are initially weak and thus concentrate the cyclic deformation.

Explanation 3 cannot be general, although aging inhomogeneities, if present, will clearly act as suggested, and no evidence of a definitive kind has been found for 1. Explanation 2, the "resolution hypothesis" is widely popular and has been taken over, e.g. by Abel and Ham (1966) and by Kralik and Schneiderhan (1972), to explain softening of the kind shown in Fig. 38. The present author has taken issue with the resolution hypothesis on numerous grounds and finds it to be inadequate. For example, Krause and Laird (1967/1968) prepared microstructures in Al–4% Cu alloy of widely different kinds but similar in the scale of the structure. On testing these structures, they could find no significant differences in their fatigue properties although the structures differed widely in their propensity for resolution. Furthermore, Calabrese and Laird (1974a) have shown that the electron microscope evidence for so-called precipitate free bands can actually be a misinterpretation of a contrast effect. In their work, they found that the intense bands within the grains were actually kink bands packed with both dislocations and precipitates, which differed slightly in orientation from the remainder of the grain. Thus, if a band were appropriately oriented with respect to the contrast reflection in the electron microscope, it could be made to appear free of any structure. Much of the old work criticized by Laird and Thomas (1967) suffered from the lack of the good goniometer stage which was available to Calabrese and Laird (1973a). For further criticisms, the reader is referred to the reviews/investigations cited above.

Instead of the resolution hypothesis, Calabrese and Laird (1974a) have advanced a new explanation for softening. From a study of the occurrence of the intense bands with respect to the stages of the hardening curve, they agree that softening is localized in the intense bands. Typical bands are shown in Fig. 40, and the occurrence of the bands is shown schematically in Fig.41. During the initial hardening only a few of the grains, those most favorably oriented for slip, are marked with general slip and intense bands. As the "soft" grains harden, more of the strain is transferred to adjacent grains, and at the peak stress (for moderately high strains) all the grains show evidence of slip. In the subsequent softening, the number and intensity of the intense bands increase. Calabrese and Laird point out that the turbulent to and fro motion of the dislocations within the bands can lead not to resolution, but to a disordering of the ordered structure of the precipitate, thus eliminating that component of the hardening due to the creation of anti-phase boundaries (APB's) when dislocations cut the precipitates. According to this explanation, the "disordering hypothesis" should also operate in single phase materials which are ordered. The fact that this is so is shown by Fig. 39. The amount of softening is also consistent with the APB contribution to hardening (Calabrese and Laird, 1974a).

The disordering hypothesis so offered is useful for interpreting generally the cyclic response of this class of materials. For example, Udimet 700, a nickel-based superalloy hardened by means of the ordered precipitate Ni_3Al, has been observed to harden and soften in the manner shown in Fig. 38 (Wells and Sullivan, 1964). Since the precipitate is ordered, this is exactly what we would expect to happen. However, in Cu–3% Co (Feltner and

Fig. 40. Intense deformation bands in Al–4% Cu alloy aged at 160 C for 5 hr so as to contain θ'' precipitates and cycled at $\varepsilon_p = \pm 0.0003$ to fracture. $\bar{g} = \langle 111 \rangle$; ZA $= \langle 112 \rangle$. (Courtesy of Calabrese and Laird, 1973a).

Laird, 1968b) and in Fe–1.5% Cu alloy (McGrath and Bratina, 1970), where the hardening precipitates are essentially pure Co and Cu respectively, no softening was observed. Clearly, in a precipitate composed of a pure solute, no disordering could occur. Typical results from McGrath and Bratina's paper, which are essentially similar to those of a single phase material, are shown in Fig. 42.

On the basis of the above, the most reasonable interpretation of cyclic response in this class of materials is the following. As cycling begins the

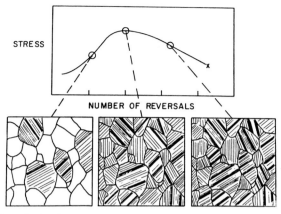

STRESS

NUMBER OF REVERSALS

Fig. 41. Surface slip band structure at various stages of the cyclic response of Al–4% Cu alloy when containing small precipitates penetrable by dislocations. (Courtesy of Calabrese and Laird, 1974a).

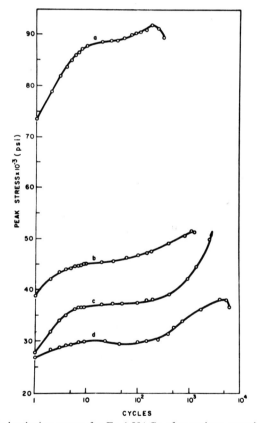

Fig. 42. Cyclic hardening curves for Fe–1.5% Cu after various quench–aging treatments, $\Delta\varepsilon_p = 0.015$. a, Fe–1.5% Cu, quenched from 840 C, aged at 500 C for 18 hr; b, Fe–1.5% Cu, quenched from 840 C, aged at 650 C for 5 hr; c, Fe–1.5% Cu, quenched from 840 C; d, Fe–0.04% C, annealed at 700 C, furnace cooled. (Courtesy of McGrath and Bratina, 1970).

grains in softest orientation and with least constraint from adjacent grains deform and harden by dislocation accumulation and by dislocation–precipitate interaction. Either because the structure is not homogeneous due to aging inhomogeneities, or because stress concentrations are associated with grain boundary triple points, some of the deformation becomes concentrated into intense slip bands. All the types of hardening contributions familiar in unidirectional deformation may be expected to operate in appropriate circumstances. By the time the peak stress is attained, all the grains capable of deformation will be hardened, and softening, if any, will subsequently occur by a disordering mechanism. Of course, atomic rearrangements may affect not only the APB contribution to hardening. The elastic properties may be affected as well and softening could be expected on that account also.

Although the resolution hypothesis is generally unacceptable, there is one situation where it can operate with plausibility, namely in Fe–C alloys. Here the evidence, by electron microscopy, for resolution is more convincing (McGrath and Bratina, 1969); the very strong interaction between carbon atoms and the dislocations introduced by the cycling could presumably lead to dissolution of the carbides. However, electron microscopic evidence on iron specimens is not satisfactory because of astigmatism problems, and the investigation of a disordering (or related) hypothesis with respect to this system may be worthwhile in future studies.

B. Microstructures Containing Particles Circumvented by Dislocations

When the particles which cause hardening are either well separated or so strong as to be impenetrable by dislocations, the dislocations circumvent the particles and a different cyclic response occurs. Most of the investigations in this area involve materials containing oxide dispersions (Leverant and Sullivan, 1968; Kupcis et al., 1973; Woo et al., 1974; Stobbs et al., 1971), but plate precipitates have also been studied (Calabrese and Laird, 1974b; Park et al., 1970).

When the microstructure contains an oxide dispersion, such as the TD nickel of Leverant and Sullivan (1968), the saturation behavior is most interesting in that it is closely similar to that of the pure matrix material. This result is shown in Fig. 43 where the cyclic stress–strain curves of the TD nickel and a commercially pure Ni are compared. Leverant and Sullivan found that, for a given applied strain, the dislocation cell structures in the two materials were different in size but equivalent in cell wall structure. Thus, in the TD nickel, the thoria particles provided a framework on which the cells formed and which determined the cell size. However, since the cell

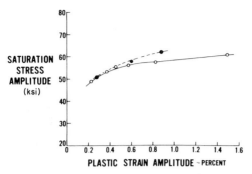

Fig. 43. Cyclic stress–strain curves for TD nickel, ○, and Ni-200, ●, which is commercially pure nickel. (Courtesy of Leverant and Sullivan, 1968).

wall structures were equivalent in Ni and TDNi, the distributions of "free links" were also equivalent. Consequently, by application of the shuttling model, the stress–strain behavior of the two materials could be reconciled.

This simple picture of the fatigue hardened (saturated) state has been elaborated, on a qualitative basis, by Sastry and Ramaswami (1973). They have carefully documented, by electron microscopy, the dislocation structures in copper single crystals containing coherent alumina particles. As shown in the composite Fig. 44, the dislocation arrangement consists of plates of dislocations enclosing a three-dimensional interconnected network. Sastry and Ramaswami point out that this arrangement is different from that observed in pure copper crystals (e.g., Hancock and Grosskreutz, 1969), and they thus appear to be at variance with the results of Leverant and Sullivan (1968). In spite of the fact that Sastry and Ramaswami's crystals were oriented for single slip, they found a significant amount of slip on the critical and conjugate systems. The latter slip greatly increased the forest dislocation density, leading the authors to ascribe to these dislocations a major role in the hardening process. On account of the complexity of the dislocation interactions they were not able to justify their conclusions quantitatively. It is interesting that the dislocation structures can vary greatly with particle distribution (Woo et al., 1974), but that again (see Leverant and Sullivan, 1968), the saturation flow stress is not significantly affected by the presence of the particles.

The development of the dislocation structure prior to saturation, when oxide particles are present, has been studied by Woo et al. (1974). These authors have also studied (along with Stobbs et al., 1971) the effects of particle size and strain amplitude in matrix single crystals, for a range of particle size up to several hundred angstroms. The rapid hardening rate falls off with increasing particle size partly because the saturation stress is determined only by the matrix but the first cycle flow stress is determined

Fig. 44. The arrangement of dislocations, at saturation, in a copper single crystal containing coherent αAl_2O_3 particles, as observed on the (111), ($\bar{1}$01), and (1$\bar{2}$1) planes. $\gamma_p = \pm 0.0105$. (Courtesy of Sastry and Ramaswami, 1973).

both by the yield stress and the dislocation multiplying capacity of the particles (Fig. 45). Within the range of particle size studied both of these quantities increased with particle size.

While well dispersed, essentially equiaxed particles are disappointingly ineffective in increasing the cyclic flow stress, plate precipitates perform rather better (Calabrese and Laird, 1974b). As shown in Fig. 46, the saturation flow stress does depend on the interparticle spacing, being larger for finer distributions of plates. Note that the precipitate employed, the metastable θ' plate in Al–4% Cu alloy, was too large to be penetrated by the

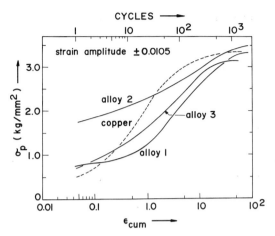

Fig. 45. The effect of particle size (Al_2O_3) on the cyclic hardening curves of copper single crystals containing: alloy 1, mean particle diameter (m.p.d.) = 86 Å; alloy 2, m.p.d. = 565 Å; alloy 3, m.p.d. = 135 Å. (Courtesy of Woo *et al.*, 1974).

dislocations. Consequently Calabrese and Laird were able to employ Ashby's (1970) idea of "geometrically-necessary" dislocations for interpreting the flow stress. In this idea, an attempt to shear a matrix which contains impenetrable plates leads to curvature of the matrix. The curvature is accommodated by dislocations stored at the plate–matrix interfaces. The cyclic hardening is then a function of the interaction between the mobile dislocations and the geometrically necessary ones on the precipitate interfaces. The hardening is not as dependent on the interplate spacing as Ashby's model requires, however, because in coarse plate structures dislocations are also

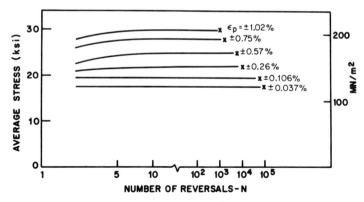

Fig. 46. Cyclic hardening curves for Al–4% Cu alloy aged at 250 C for 5 hr to produce a uniform dispersion of θ' plate precipitates. (Courtesy of Calabrese and Laird, 1974b).

stored in the matrix between plates. At extremely large strains, the matrix will eventually determine the flow stress uniquely, just as it does in the case of equiaxed particles. However, for the kinds of strains normally encountered in service, the cyclic flow stress is appreciably greater than that of the matrix alone and will be slightly greater than the monotonic flow stress of the material. Note that, as shown in Fig. 46, the rapid hardening stage is very short in plate-containing microstructures. At the lowest strains, all the hardening is accomplished in the first reversal and the geometrically necessary dislocations subsequently suffice to accommodate the cyclic plastic strain. At the highest strains, only a few cycles are required for hardening, mainly by further accumulation of dislocations, as cells, between the plates.

It will be observed that the similarities pointed out in Section II,D between the unidirectional and cyclic deformation mechanisms of single phase materials also apply to duplex phase materials. It is customary in studies of precipitation hardening by unidirectional deformation to divide the field into a regime where precipitates are small and penetrable and into another regime, the Orowan, where dislocations bypass the precipitates. This natural division also works well in cyclic deformation and, as shown above, distinct classes of behavior operate in the two regimes.

IV. The Value of Microstructural Studies in Predicting Cyclic Stress–Strain Response

If cyclic stress–strain curves are to be determined, the most definitive method is to employ plastic strain cycling. Such testing is far from simple and, as currently performed, requires rather expensive equipment. The data from such tests are necessary, however, for design purposes. Especially in the early stages of design, it is desirable to make estimates based on available data of the most limited type without having to resort to expensive testing. Thus, workers with engineering interests, such as Manson (1966), have developed approximate relations involving readily available data. These approximations can serve a very useful purpose provided their limitations are recognized. In this section, typical approximations are described and then examined in the light of the fundamental results of the preceding sections. For the final analysis of a significant design, it would appear reasonable to expect that the actual properties of the material involved will be determined. It would of course be economically beneficial if the correct choice of material could be made at the first try, and it is in this respect (as one example) that fundamental research is valuable.

The most widely available data are the results of standard unidirectional tension tests. As is well known, great differences exist in these data, even for a single material, depending on history. As can be seen from the cyclic response studies described in Section II,B, strain cycling causes the cyclic stress–strain curves for different conditions to become much more alike. For wavy slip materials, the cyclic stress–strain curves are exactly alike; even for planar slip materials, the differences between different histories are less for cyclic deformation than for unidirectional deformation. The first essential information a designer might require is therefore: where will the cyclic stress–strain curve lie in relation to the unidirectional curve, at higher or lower stresses, and by how much? An approximate comparison of materials with respect to this problem was made by Smith et al. (1965). For a wide range of commercial materials, they evaluated the ratios of the saturation stress at a given cyclic strain range to the virgin tensile stress at a corresponding value of strain. This ratio is less than unity for materials which cyclically soften and greater than unity for materials which harden. Smith et al. (1965) found a fair correlation between the degree of softening or hardening and the parameter S_u/S_{ys} (the ratio of the ultimate tensile strength to the conventional 0.2% yield strength). As is shown in Fig. 47, taken from their paper, all materials for which $S_u/S_{ys} = 1.2$ or less softened under cyclic straining and those for which $S_u/S_{ys} = 1.4$ or greater underwent cyclic hardening. At intermediate values, $1.2 < S_u/S_{ys} < 1.4$, both hardening and softening were observed. Smith et al. (1965) offered this correlation as a guide for estimating from tensile data the degree of hardening or softening that may be expected in the absence of the requisite fatigue/cyclic deformation data.

In order to check this correlation, data available from the fundamental papers referenced in the present review are tabulated in Table I. The question of whether or not the prediction of Smith et al. is upheld was answered for a cyclic plastic strain of 1%, which was the strain used by those authors. Of course, at lower strains, softening is likely to be greater and hardening less in annealed materials, and the particular value of strain of concern to the engineer will be important in predicting behavior (see Sections II and III). From Table I, it is clear that Smith et al.'s prediction works well in simple cases, but is frequently in error. For example, pure Cd undergoes both hardening and softening; cold-worked planar slip material (Cu–7.5% Al) may soften at large values of S_u/S_{ys}; a θ' microstructure in Al–4% Cu should be expected to harden considerably because of its high value of S_u/S_{ys}, but in fact, because "geometrically necessary" dislocations can move efficiently in a reversible manner, the cyclic stress–strain curve is nearly coincident with the monotonic stress–strain curve (Calabrese and Laird, 1974b); many other examples can be cited. It is possible that commer-

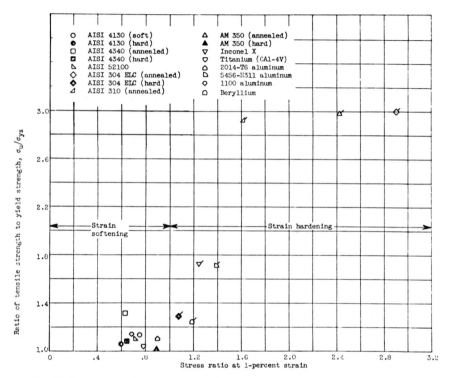

Fig. 47. Comparison of the ratio of cyclic saturation stress to monotonic flow stress (at $\varepsilon_p = 1\%$) with virgin tensile properties. Nontailed symbols denote strain-softening materials. Tailed symbols denote strain-hardening materials. (Courtesy of Smith *et al.*, 1965).

cial materials will match the prediction of Smith *et al.* better than the model materials of interest here. However, commercial materials are of such variety that it would be dangerous to rely too firmly on the prediction. This author therefore suggests that a much more reliable method of predicting cyclic response will be to use Smith *et al.*'s prediction *in conjunction with a qualitative prediction of response from microstructural knowledge of the material.* The relevant questions to ask in this context are:

(a) Is the material planar or wavy in its slip character? If it is wavy, and therefore possessed of a unique cyclic stress–strain curve, is there a material sufficiently close to it in chemistry with a known cyclic stress–strain curve? If the material is planar in slip, is its history, as shown in the S_u/S_{ys} ratio, determined by cold work or by other methods of hardening.

(b) If the material is precipitation hardened, are the precipitates ordered or disordered, of a pure solute or an intermetallic compound? If the precipitates are ordered, both hardening and softening can confidently be expected; if the precipitates are a "pure" phase, e.g. Cu in Fe–Cu alloys, then no softening will occur.

TABLE I: Test of the Correlation between Tensile and Cyclic Deformation Suggested by Smith et al. (1963)

Material	Condition[a]	S_u/S_{ys}[b]	Test temperature (K)	σ_s/σ_m[c]	Is the prediction upheld?	Reference
Cu	Annealed	3.43	300	2.50	Yes	Feltner and
(pure)	5% CW	1.02	300	0.77	Yes	Laird (1967a)
	23% CW	1.01	300	0.70	Yes	
	Annealed	5.19	78	4.00	Yes	
	5% CW	1.45	78	1.14	Yes	
	23% CW	1.40	78	1.04	Yes	
Cu–7% Al	Annealed	2.28	300	1.26	Yes	Feltner and
(pure)	5% CW	1.17	300	0.77	Yes	Laird (1967a)
	23% CW	1.39	300	0.80	No	
	Annealed	2.18	78	1.48	Yes	
	5% CW	1.24	78	0.66	No	
	23% CW	1.33	78	0.67	No	
Fe	Annealed	> 2.0	300	1.68	Yes	Feltner and
	CW	> 1.3	300	1.03	No	Laird (1969)
Fe–3% Si	Annealed	> 1.4	300	1.42	Yes	Feltner and
	CW	~ 1.07	300	0.84	Yes	Laird (1969)
Ni₃Mn	Annealed	> 1.4	300	—	Yes	Feltner and
(disordered)	CW	~ 1.0	300	—	Yes	Laird (1969)
Ni₃Mn	Annealed	> 1.4	300	Not constant	No	Feltner and
(ordered)	CW	~ 1.0	300	Not constant	No	Laird (1969)
Cu–3% Co	As aged	> 1.4	300	—	Yes	Feltner and
	Aged and CW	~ 1.1	300	—	Yes	Laird (1969)
Cd	Annealed	> 1.4	300	Not constant	No	Feltner and Laird (1969)
Ni-200	Annealed	> 1.4	300	—	Yes	Leverant and
TD-Ni	—	> 1.4	300	—	Yes	Sullivan (1968)
Fe–1.5% Cu	Quenched	> 1.4	300	—	Yes	McGrath and
	Quenched and aged (500 C)	~ 1.0	300	—	No	Bratina (1970)
	Quenched and aged (650 C)	> 1.4	300	—	Yes	
Fe–0.04 C	Quenched	> 1.4	300	—	Yes	McGrath and
	Quenched and aged (25 C)	> 1.4	300	—	No	Bratina (1969)
	Quenched and aged (200 C)	> 1.4	300	—	Yes	
Al–4% Cu	GPI zones	2.22	300	Not constant	No	Calabrese and
	θ''	2.16	300	Not constant	No	Laird (1974a,b)
	θ'	2.50	300	1.12	No	

[a] CW = cold worked.

[b] In many cases, these data were not explicitly given, but were estimated by the present author on the basis of partial and indirect information.

[c] σ_s/σ_m = ratio of cyclic saturation stress at $\varepsilon_p = \pm 1\%$ to the unidirectional flow stress at $\varepsilon_p = 1\%$. These data were frequently unavailable, mainly for lack of the unidirectional response.

(c) What is the morphology of the precipitates? Microstructures which contain platelike precipitates, or a large volume fraction of impenetrable equiaxed precipitates, can be expected to harden by "geometrically necessary" dislocations and therefore will have cyclic stress–strain curves little different from their monotonic curves? If the precipitates are present in low volume fraction, and are equiaxed and impenetrable, then behavior typical of TD nickel will be observed, i.e. the cyclic stress–strain response of the matrix alone will then be appropriate for design.

(d) Are there combinations of hardening mechanisms involved? For example, maraging steels contain dense masses of dislocations introduced by the martensite reaction involved in their processing, along with ordered precipitates of the type Ni_3Al. Consequently, initial cyclic softening will be associated with the dislocation structures followed by a more gradual softening due to the disordering of the ordered precipitates.

Careful attention to the metallurgical specifications of the material with, if necessary, simple metallographic observations can easily provide the answers to these questions. Primary metal producers document the microstructural nature of their materials on a routine basis. Reference to the contents of the present review should then permit the cyclic stress–strain response to be predicted with accuracy. It should of course be noted that this method can be extended to other aspects of fatigue, for example to crack propagation (Laird, 1966) and to life (Feltner and Beardmore, 1970). However, one must always bear in mind, in applying this method, at least two other factors which impinge on the choice of a material: (1) economics, prescribed by manufacturing methods and the consequences of failure, and (2) whether or not other mechanisms of fracture are relevant to the design problem, for example stress corrosion. It may be that the microstructural material most suitable for fatigue resistance may not have adequate resistance to some other failure mode.

V. Summary

Although the author has been charged to review the cyclic deformation of metals, an adequate attempt could not be made of such a large field in the limited space available. He has therefore proceeded to review on a selective basis, drawing only from a few papers where agreement appears to be general, but expanding to a larger literature where controversies exist, where problems still remain to be solved, or where the phenomena are so complicated that more detailed treatment is appropriate. Further, an attempt has been made to illustrate explicitly both the value of cyclic stress–strain response data for fatigue design and the value of microstructural studies in cyclic stress–strain response.

Thus the development of cyclic deformation from fatigue fracture studies is described and an example is provided of a new method of designing against cumulative damage, based on cyclic plasticity. This method is shown to be greatly superior to older stress-based methods.

The main part of the review is concerned with the phenomena of rapid hardening (or softening) and saturation under constant applied cyclic strains and for single phase materials. The factors which control these phenomena are described and also the dislocation structures which are associated with them. In so far as the mechanisms of cyclic deformation are understood these too are described and a number of parallels between cyclic and unidirectional deformation are pointed out. A similar approach is then taken for materials which contain second phases for strength. Note that these studies apply to cycling at ambient temperatures or below. High-temperature cyclic deformation is beyond the scope of this review.

Finally, an engineering method of predicting cyclic stress–strain response from tensile testing data is examined in the light of the fundamental knowledge described, and is shown to be severely limited. A method of improving such prediction by introducing additional microstructural information which is readily available is therefore suggested.

ACKNOWLEDGMENTS

I am indebted to J. M. Finney for his detailed criticism of the manuscript, to R. Wetzel for permission to describe his unpublished plasticity method of predicting cumulative damage, and to O. T. Woo, B. Ramaswami, O. A. Kupcis, and J. T. McGrath for their kindness in showing me their recent manuscripts before publication. The support of the Army Research Office (Grant No. DA-ARO-D-31-124-72-G29) and of the National Science Foundation (Grant No. NSF-GH-33633) is also acknowledged and deeply appreciated.

References

Abel, A., and Ham, R. K. (1966). *Acta Met.* **14**, 1495–1504.
Ashby, M. F. (1970). *Phil. Mag.* [8] **21**, 399–424.
Avery, D. H., and Backofen, W. A. (1963). *Acta Met.* **9**, 352–361.
Baldwin, E. E., Sokol, G. J., and Coffin, L. F. (1957). *Proc. Amer. Soc. Test. Mater.* **57**, 567–586.
Basinski, S. J., Basinski, Z. S., and Howie, A. (1969). *Phil. Mag.* [8] **19**, 899–924.
Broom, T., and Ham, R. K. (1957). *Proc. Roy. Soc., Ser. A* **242**, 166–179.
Broom, T., and Ham, R. K. (1959). *Proc. Roy. Soc., Ser. A* **251**, 186–199.
Calabrese, C., and Laird, C. (1974a). *Mater. Sci. Eng.* **13**, 141–157.
Calabrese, C., and Laird, C. (1974b). *Mater. Sci. Eng.* **13**, 159–174.

Canning, E. J. (1971). Ph.D. Thesis, University of Pennsylvania, Philadelphia.

Coffin, L. F. (1972). *Annu. Rev. Mater. Sci.* **2**, 313–348.

Coffin, L. F., and Read, J. H. (1956). *Proc. Int. Conf. Fatigue Metals, 1956.*

Coffin, L. F., and Tavernelli, J. (1958). "Cyclic Straining and Fatigue of Metals," Rep. No. 58-RL-2100. G. E. Res. Lab., Schenectady, N.Y.

Dawson, R. A. T. (1967). *In* "Thermal and High Strain Fatigue," pp. 40–54. Metals and Metallurgy Trust, London.

Dawson, R. A. T., Elder, W. J., Hill, G. J., and Price A. T. (1967). *In* "Thermal and High Strain Fatigue," pp. 239–269. Metals and Metallurgy Trust, London.

Dolan, T. J. (1965). *Midwest. Mech. Conf., Proc., 9th, 1964,* pp. 3–21.

Dugdale, D. S. (1959). *J. Mech. Phys. Solids* **7**, 135–142.

Ebner, M. L., and Backofen, W. A. (1959). *Trans. AIME* **215**, 510–520.

Feltner, C. E. (1965). *Phil. Mag.* [8] **12**, 1229–1248.

Feltner, C. E. (1966). *Phil. Mag.* [8] **14**, 1219–1231.

Feltner, C. E., and Beardmore, P. (1970). *Amer. Soc. Test. Mater., Spec. Tech. Publ.* **467**, 77–112.

Feltner, C. E., and Laird, C. (1967a). *Acta Met.* **15**, 1621–1632.

Feltner, C. E., and Laird, C. (1967b). *Acta Met.* **15**, 1633–1653.

Feltner, C. E., and Laird, C. (1968a). *Trans. AIME* **242**, 1253–1257.

Feltner, C. E., and Laird, C. (1968b). "The Role of Slip Behavior in Steady State Cyclic Stress-Strain Behavior," Ford Motor Co., Dearborn, Mich.

Feltner, C. E., and Laird, C. (1969). *Trans. AIME* **245**, 1372–1373.

Feltner, C. E., and Laird, C. (1970). Reported in Feltner and Beardmore (1970).

Feng, C., and Kramer, I. R. (1965). *Trans AIME* **233**, 1467–1473.

Forrest, P. G. (1962). "Fatigue of Metals." Pergamon, Oxford.

Forsyth, P. J. E. (1969). "The Physical Basis of Metal Fatigue." Amer. Elsevier, New York.

Fourie, J. T. (1967). *Can. J. Phys.* **45**, 777–786.

Fourie, J. T. (1968). *Phil. Mag.* [8] **15**, 187–198.

Fourie, J. T. (1970). *Phil. Mag.* [8] **21**, 977–985.

Grosskreutz, J. C. (1971). *Phys. Status Solidi* **47**, 11–31 and 359–396.

Grosskreutz, J. C. (1973). *Proc. Int. Conf. Fract. 3rd, 1973* PLV-212.

Halford, G. R., and Morrow, J. (1962). *Proc. Amer. Soc. Test. Mater.* **62**, 695–707.

Ham, R. K., and Broom, T. (1962). *Phil. Mag.* [8] **7**, 95–103.

Hancock, J. R., and Grosskreutz, J. C. (1969). *Acta Met.* **17**, 77–97.

Hein, E., and Dodd, R. A. (1964). *Trans. AIME* **221**, 1095–1101.

Hull, D. (1958). *Phil. Mag.* [8] **3**, 513–518.

Johnson, E. W., and Johnson, H. H. (1965). *Trans. AIME* **233**, 1333–1339.

Kemsley, D. S. (1958). *J. Inst. Metals* **87**, 10–16.

Kemsley, D. S., and Paterson, M. S. (1960). *Acta Met.* **8**, 453–467.

Kennedy, A. J. (1963). "Processes of Creep and Fatigue in Metals." McGraw-Hill, New York.

Kenyon, J. N. (1950). *Proc. Amer. Soc. Test. Mater.* **50**, 1073–1084.

Klesnil, M., Holzmann, M., Lukás, P., and Rys, P. (1965). *J. Iron Steel Inst.* **203**, 47–53.

Kralik, G., and Schneiderhan, H. (1972). *Scripta Met.* **6**, 843–849.

Kramer, I. R. (1961). *Trans. AIME* **221**, 989–993.

Kramer, I. R. (1963a). *Trans. AIME* **227**, 529–533.

Kramer, I. R. (1963b). *Trans. AIME* **227**, 1003–1010.

Kramer, I. R. (1965a). *Trans. AIME* **233**, 246–247.

Kramer, I. R. (1965b). *Trans. AIME* **233**, 1462–1467.

Kramer, I. R. (1965c). "Environment-Sensitive Mechanical Behavior," pp. 127–146. Gordon & Breach, New York.

Kramer, I. R. (1967a). *Trans. Amer. Soc. Metals* **60**, 310–317.

Kramer, I. R. (1967b). *Trans. AIME* **239**, 520–528.
Kramer, I. R. (1967c). *Trans. AIME* **239**, 1754–1758.
Kramer, I. R. (1969a). *Trans. Amer. Soc. Metals* **62**, 521–535.
Kramer, I. R. (1969b). *Proc. Air Force Conf. Fatigue Fract., 1968*, pp. 271–284.
Kramer, I. R., and Balasubramanian, N. (1973). *Acta Met.* **21**, 695–700.
Kramer, I. R., and Demer, L. J. (1961). *Trans. AIME* **221**, 780–786.
Kramer, I. R., and Haehner, C. L. (1967). *Acta Met.* **15**, 199–202.
Kramer, I. R., and Podlaseck, S. (1963). *Acta Met.* **11**, 70–71.
Krause, A. R., and Laird, C. (1967/1968). *Mater. Sci. Eng.* **2**, 331–347.
Kupcis, O. A., Ramaswami, B., and Woo, O. T. (1973). *Acta Met.* **21**, 1131–1138.
Laird, C. (1962). Ph.D. Thesis, University of Cambridge.
Laird, C. (1966). *Amer. Soc. Test. Mater., Spec. Tech. Publ.* **415**, 130–180.
Laird, C., and Duquette, D. J. (1972). *In* "Corrosion Fatigue: Chemistry, Mechanics and Microstructure" (A. J. McEvily and R. W. Staehle, eds.), pp. 88–117. Nat. Ass. Corrosion Eng., Houston, Texas.
Laird, C., and Krause, A. R. (1970a). *In* "Inelastic Behavior of Solids" (M. F. Kanninen *et al.*, eds.), pp. 691–715. McGraw-Hill, New York.
Laird, C., and Krause, A. R. (1970b). *Metal Sci. J.* **4**, 212–214.
Laird, C., and Thomas, G. (1967). *Int. J. Fract. Mech.* **3**, 81–97.
Landgraf, R. W. (1970). *Amer. Soc. Test. Mater., Spec. Tech. Publ.* **467**, 3–36.
Landgraf, R. W., Morrow, J., and Endo, T. (1969). *J. Mater.* **4**, 176–188.
Leverant, G. R., and Sullivan, C. P. (1968). *Trans. AIME* **242**, 2347–2353.
Low, J. R., and Turkalo, A. M. (1962). *Acta Met.* **10**, 215–227.
Lukás, P., and Klesnil, M. (1968). *Phys. Status Solidi* **27**, 545–558.
Lukás, P., and Klesnil, M. (1970). *Phys. Status Solidi* **37**, 833–842.
Lukás, P., and Klesnil, M. (1972). *In* "Corrosion Fatigue-Chemistry, Mechanics and Microstructure" (A. J. McEvily and R. W. Staehle, eds.), pp. 118–132.
Lukás, P., and Klesnil, M. (1973). *Mater. Sci. Eng.* **11**, 345–356. Nat. Ass. Corrosion Eng., Houston, Texas.
McEvily, A. J., and Johnston, T. L. (1965). Sendai Fracture Conf., Sendai, Japan.
McGrath, J. T., and Bratina, W. J. (1969). *Czech. J. Phys.*, *B* **19**, 284–293.
McGrath, J. T., and Bratina, W. J. (1970). *Phil. Mag.* [8] **21**, 1087–1091.
Manson, S. S. (1966). "Thermal Stress and Low Cycle Fatigue." McGraw-Hill, New York.
Miner, M. A. (1945). *J. Appl. Mech.* **12**, 159–164.
Morrow, J. (1965). *Amer. Soc. Test. Mater., Spec. Tech. Publ.* **378**, 45–63.
Neumann, P. (1967). *Z. Metallk.* **58**, 780–789.
Neumann, P. (1968). *Z. Metallk.* **59**, 927–934.
Neumann, P. (1969). *Acta Met.* **17**, 1219–1225.
Neumann, P. (1973). *Proc. Int. Conf. Fract., 3rd, 1973*, I 232 and III 233.
Neumann, R., and Neumann, P. (1970). *Scr. Met.* **4**, 645–650.
Oldroyd, P. W. J., Burns, D. J., and Benham, P. P. (1965). "Strain Hardening and Softening of Metals Produced by Cycles of Plastic Deformation," Report. Imperial College, London.
Park, B. K., Greenhut, V., Luetjering, G., and Weissman, S. (1970). Tech. Rep. AFML-TR-70-195, Air Force Mater. Lab., Dayton, Ohio.
Piqueras, J., Grosskreutz, J. C., and Frank, W. (1972). *Phys. Status Solidi* **11**, 567–580.
Plumbridge, J. H., and Ryder, D. A. (1969). *Met. Mat. Rev.* **3**, No. 136.
Polak, J. (1969). *Czech. J. Phys.*, *B* **19**, 315–321.
Polakowski, H. N. (1952). *Proc. Amer. Soc. Test. Mater.* **52**, 1086–1097.
Polakowski, H. N., and Palchoudhuri, A. (1954). *Proc. Amer. Soc. Test. Mater.* **54**, 701–712.
Pratt, J. (1965). *Theoret. Appl. Mech. Rep.* No. 652. University of Illinois, Urbana.

Raouf, H. A., and Plumtree, A. (1971). *Met. Trans.* **2**, 1863–1867.

Raouf, H. A., Benham, P. P., and Plumtree, A. (1971). *Can. Met. Quart.* **10**, 87–92.

Sandor, B. I. (1972). "Fundamentals of Cyclic Stress and Strain." Univ. of Wisconsin Press, Madison.

Sastry, S. M. L., and Ramaswami, B. (1973). *Phil. Mag.* (Submitted for publication).

Shen, H., and Kramer, I. R. (1967). *Trans. Int. Vac. Met. Conf., 1966*, pp. 263–283.

Shen, H., Podlaseck, S. E., and Kramer, I. R. (1966). *Acta Met.* **14**, 341–346.

Smith, K. N., Watson, P., and Topper, T. H. (1970). *J. Mater.* **5**, 767–776.

Smith, R. W., Hirschberg, M. H., and Manson, S. S. (1963). *NASA Tech. Note* **NASA TN D-1574.**

Stobbs, W. M., Watt, D. F., and Brown, L. M. (1971). *Phil. Mag.* [8] **23**, 1169–1184.

Stubington, C. A. (1958). *Metallurgia* **58**, 165–171.

Thompson, N., and Wadsworth, N. J. (1958). *Advan. Phys.* **7**, 72–169.

Tuler, F. R., and Morrow, J. (1963). "Cycle Dependent Stress-Strain Behavior of Metals," Theor. Appl. Mech. Rep. No. 239. University of Illinois, Urbana.

Wadsworth, N. J. (1963). *Acta Met.* **11**, 663–673.

Wadsworth, N. J., and Hutchings, J. (1964). *Phil. Mag.* [8] **10**, 195–217.

Watt, D. F., and Ham, R. K. (1966). *Nature (London)*, **211**, 734–735.

Wells, C. H., and Sullivan, C. P. (1964). *Trans. Amer. Soc. Metals* **57**, 841–855.

Wetzel, R. M. (1971). "A Method of Fatigue Damage Analysis," Report. Ford Motor Co., Dearborn, Mich.

Woo, O. T., Kupcis, D. A., Ramaswami, B., and McGrath, J. T. (1970). *Int. Conf. Strength Alloys, 2nd, 1969.* Paper 12.5.

Woo, O. T., Ramaswami, B., Kupcis, O. A., and McGrath, J. T. (1974). *Acta Met.* **22**, 385–398.

Wood, W. A., and Segall, R. L. (1958). *J. Inst. Metals* **86**, 225–228.

High-Temperature Creep

Department of Mechanical Engineering
University of California
Davis, California

I. Introduction

Creep is the continuing plastic deformation of materials subjected to a constant load or constant stress. Slightly more than twenty years ago, based principally on the marriage of dislocation theory with the theory for kinetics of reactions, an atomistic approach to unraveling the complicated nature of creep was initiated. Considerable progress has been made over the recent past in uncovering the role of many of the important factors that determine the physical mechanisms of high-temperature creep.

It has now been firmly established that creep of crystalline solids occurs as the result of thermally-activated migration of dislocations, grain boundary shearing, and diffusion of vacancies. It can take place at all temperatures above the absolute zero. Because of the great versatility of dislocations and the numerous interactions they may undertake with one another, additional lattice defects, and various substructural details, a number of different dislocation mechanisms can control creep. At low temperatures, high intensity thermal fluctuations in energy are so infrequent that dislocations can surmount only the lowest energy barriers that impede their motion. As they pass over these they arrive at new and often higher energy barriers which they now surmount with an ever decreasing frequency. Consequently low-temperature creep is generally characterized by an ever diminishing creep rate.

At higher temperatures the thermal fluctuations needed to stimulate the low energy mechanisms such as the cutting of repulsive trees, the nucleation of kink pairs, cross-slip, etc. becomes so frequent that such barriers to dislocation motion are no longer effective. A larger "apparent initial creep strain" is therefore obtained at higher temperatures for the same applied stress. When these conditions prevail, it might be expected that the creep would rapidly approach immeasurably small rates as the gliding dislocations become arrested at higher energy barriers that might result from long-range stress fields, attractive junctions, etc. If, however, the temperature is above about $T_m/2$, one-half the melting temperature, creep nevertheless continues to take place despite the fact that glide dislocations now become arrested at quite high energy barriers. In contrast, however, it is just at these high temperatures that creep continues ad infinitum until it is modified by such auxiliary phenomena as grain boundary fissuring, necking, and grain boundary cavitation which lead to rupture. The significant observation to be made here is that at $T_m/2$, diffusion first becomes reasonably rapid. Above this temperature, dislocations are known to acquire a new degree of freedom in their motion; edge components of dislocations are no longer confined to glide exclusively on their original slip planes and may climb to new planes. Although other suggestions have been made from time to time, it seems inevitable that climb itself must necessarily play a role in any model for high-temperature creep; it need not, however, be the rate-controlling mechanism. For example, if climb were extremely rapid and some glide processes much slower, creep might then take place in an almost fully climb-recovered substructure as controlled by a slow glide mechanism.

It is the objective of this article to evaluate the present status of experimental knowledge on the high-temperature diffusion-controlled creep of some metals and alloys with particular reference to the various creep mechanisms. The work is presented in three major sections. The first will be concerned with the effects of the independent variables of stress and temperature on the creep rate and the possible influence of crystal structure, modulus of elasticity, stacking fault energy and grain size on these data. In the second section emphasis will be given to a review of the major substructural changes that attend high-temperature creep. And thirdly a summary will be presented on the theoretical implications of the known experimental facts relative to the validity of various proposed high-temperature creep mechanisms.

A number of reviews are now available on the broad aspects of creep deformation, i.e. Weertman (1968, 1972), Mukherjee *et al.* (1969), Sherby and Burke (1967), Bird *et al.* (1969), Dorn and Mote (1963), McLean (1966), Garafalo (1965), and Lagneborg (1972a). No attempt will be made in this presentation to cover all aspects of creep literature. Instead, attention will be directed toward areas where a reasonable understanding has been achieved in uncovering the underlying mechanisms of creep, thereby leading to a satisfactory correlation between the experimental data and theoretical predictions. As will be illustrated, a number of different diffusion-controlled mechanisms are known to determine high-temperature creep rates and each is influenced in its own unique way by substructural and microstructural modifications.

II. Typical Creep Curves

Several common types of high-temperature creep curves are illustrated schematically in Fig. 1. Most metals and alloys exhibit a reasonably extensive range of steady state creep, stage II, followed by an increasing creep rate, stage III, characterized by microfissuring, local plastic deformation, and creep rupture. Major differences between the creep curves of various materials, however, are noted over the initial creep period, stage I. Type B, which enters the steady state almost immediately, is typical of materials in which the substructure pertinent to creep remains substantially constant. In

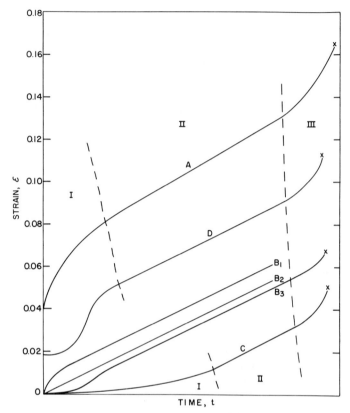

Fig. 1. Common types of creep curves.

contrast, the usual creep curves of Type A, obtained for annealed metals by Hazlett and Hansen (1954) and some alloys, exhibit a decelerating primary creep rate, illustrating the continued formation of a more creep-resistant substructure during the transient stage. These same metals and alloys give creep curves of Type C when they have been previously crept at a higher applied stress (Raymond and Dorn, 1964) or cold-worked (Hazlett and Hansen, 1954). The increasing creep rate over stage I denotes the recovery of the pertinent substructure to a steady state condition. The same steady state creep rate is obtained in these metals and alloys regardless of their previous mechanical treatment (Sherby *et al.*, 1956). It is clear, therefore, that the secondary creep rate is obtained when a balance is reached between the rate of generation of the creep-resistant substructure and the rate of its thermal recovery under the applied stress. The sigmoidal type of creep curve over stage I shown by Type D suggests (G. A. Webster, personal communication,

1968) the nucleation and spread of slip zones until the steady state is achieved. It has been observed in certain special dispersed phase alloys (Webster and Piearcey, 1967). More complicated types of creep curves than those shown in Fig. 1, however, are obtained as a result of periodic recrystallization (Gilbert and Munson, 1965), or as a result of microstructural changes attending precipitation hardening, overaging, etc. (Glen, 1958). In order to limit this section to tractable dimensions, the latter types of creep behavior will not be covered here.

III. Mechanical Behavior

The dependence of the creep rate, $\dot{\varepsilon}$, on the stress, σ, and temperature, T, is basic to an understanding of creep. Any acceptable theory of creep must account well for all such reliable data. Several factors, however, serve to complicate the picture. Frequently test materials are not adequately characterized relative to purity, grain size, and other pertinent quantities. Good data require excellent control of the temperature and stress. In order to cover a wide range of stresses at a given constant temperature, some tests must be conducted over very long periods of time. As will be described later, the subgrain size developed at the secondary stage of creep, although independent of the test temperature, depends on the stress. Consequently, the effect of stress may not easily be separated from the effects of coincident substructural changes. Furthermore the proportion of grain boundary shearing to the total creep strain increases as the stress is lowered. Other factors such as crystal structure, modulus of elasticity, stacking fault energy, diffusivity, etc. also require detailed consideration.

A. Stress Law

The effect of stress on the secondary creep rate will be emphasized here; first because there are more data in this area than others and secondly because most existing creep theories are concerned principally with the secondary stage of creep. Although other empirical functions have been suggested, the secondary creep rate can usually be represented as a power of the stress according to

$$\dot{\varepsilon}_s = \sigma^n f\{T\} \tag{1}$$

In general the exponent n is substantially constant over a wide range of stress and temperature. At high stresses the secondary creep rate appears to increase more rapidly than that given by a constant value of n. Occasionally, at quite low stress levels (Harper and Dorn, 1957), the secondary creep rate is somewhat higher than the value obtained by extrapolation of Eq. (1). In some metals this is believed to arise (Harper *et al.*, 1958) from the greater percentage of grain boundary shearing accompanying grain straining at such low stresses.

The values of n vary from about 3 to about 7 for various solid-solution alloys and pure metals. Most frequently pure metals give values of $4.2 < n < 6.9$ which suggests (Sherby, 1962) an average value of $n = 5$. Some alloys, e.g. Ni–Au (Sellers and Quarrell, 1961–1962), exhibit values of $n \simeq 3$ whereas others, e.g. Fe–Si and Fe–Al, have values as high as about $6 < n < 7$ (Davies, 1963; Lawley *et al.*, 1960). The observed variations in n for metals and alloys will be discussed later in conjunction with creep mechanisms.

B. *Apparent Activation Energy*

Since creep is a thermally-activated process, its temperature dependence is described by an Arrhenious type of expression. Early investigations (Dorn, 1962) revealed that the temperature dependence was frequently given by the expression

$$\dot{\varepsilon}_s = f\{\sigma\}e^{Q_{c'}/RT} \tag{2}$$

with rather good accuracy. The apparent activation energy, Q'_c, was shown to be insensitive to the stress; it was in very close agreement with the activation energy for self-diffusion, H_D. More recent investigations (Sherby and Burke, 1967) have invariably served to confirm the validity of Eq. (2), and to reaffirm the agreement of Q'_c with H_D.

C. *Modulus of Elasticity*

When Eqs. (1) and (2) are combined, the empirical relationship

$$\dot{\varepsilon}_s e^{Q_{c'}/RT} = A'\sigma^n \tag{3}$$

results. A', however, differs greatly from metal to metal. McLean and Hale (1961) made the important observation that variation in A' among the various metals and alloys seems to depend systematically on their moduli of elasticity. They have shown that the correlation of secondary creep rates among the various metals and alloys is considerably improved when Eq. (3) is rewritten as

$$\dot{\varepsilon}_s e^{Q_c/RT} = A''(\sigma/G)^n \tag{4}$$

where G is the shear modulus of elasticity. Similarly Sherby (1962; Sherby and Young, 1967) suggested the empirical relationship

$$\dot{\varepsilon}_s/D = A'''(\sigma/E)^n \tag{5}$$

where D is the self-diffusivity and E is Young's modulus of elasticity, as a basis for correlating the secondary creep rates of metals. When the temperature variation of E was included, slightly improved correlations were obtained.

D. True Enthalpy of Activation

Mukherjee *et al.* (1969) have shown that when creep is diffusion-controlled, the steady state creep rate can be represented somewhat better by

$$\dot{\varepsilon}_s = A(DGb/RT)(\sigma/G)^n \tag{6}$$

where A is a parameter which is constant at steady state and b is the Burger's vector of the order of interatomic distance. Because of the inclusion of G and T in Eq. (6), the apparent activation energy for creep

$$Q_c = [\partial \ln \dot{\varepsilon}/\partial(-1/RT)]_{\sigma=\text{const}} \simeq [\Delta \ln \dot{\varepsilon}/\Delta(-1/RT)]_{\sigma=\text{const}} \tag{7a}$$

TABLE I

COMPARISON OF ACTIVATION ENTHALPIES FOR CREEP AND DIFFUSION[a]

Metal	n	H_c (kcal/mole)	H_D (kcal/mole)
Al	4.4	34.0	34.0
Cu	4.8	48.4	47.1
Au	5.5	48 ± 5	41.7
Ni	4.6	66.5	66.8
Ag	~ 5.3	43	44.1
Pb	~ 4.2	24.2 ± 2.5	24.2
α–Fe	6.9	63.0	60–64
Fe-4% Si	5.75	71.4	54.4
Ta	4.2	114 ± 4	110.0
Mo	4.3	~ 93	92.2
β–Tl	5.8	21	20.0
α–Tl	5.3	21.0	22.7
Cd	4.3	19 ± 2	19.1
Zn	6.1	21.6 ± 1	24.3

[a] From Mukherjee *et al.* (1969).

is always slightly different from the activation enthalpy of diffusion

$$H_d = \partial \ln D/\partial(-1/RT) \tag{7b}$$

It is convenient therefore to define an activation enthalpy for creep, H_c, which does not contain terms arising from either G or T:

$$H_c = Q_c - RT[(n - 1)(T/G)(-\partial G/\partial T) - 1] \tag{8}$$

When creep is exclusively diffusion-controlled it should be found that $H_c = H_d$, as shown in Table I where H_c was calculated from Eq. (8). As shown in Table I, the true enthalpy of activation of creep, H_c, is in better agreement with the activation energies H_D of self-diffusion of metals than was the apparent activation energy of creep (Garafalo, 1965). The Fe–4% Si alloy data must be discounted in this context because of the presence of complexities arising from short-range ordering which results in a higher activation energy for creep than that for the diffusivity of Fe.

E. Activation Volume

The close correlation between the activation enthalpies of creep and diffusion strongly suggests that self-diffusion is the ultimate rate-controlling process in the steady state creep of pure metals. Weertman (1968) has shown that studies of the effect of hydrostatic pressure on the diffusion coefficient and on the steady state creep rate have strengthened the conviction that self-diffusion is the ultimate rate-controlling step in the high-temperature creep of pure metals and alloys. The activation volume of self-diffusion and

TABLE II

ACTIVATION VOLUMES, V^*, OF HIGH-TEMPERATURE
CREEP AND SELF-DIFFUSION (in $cm^3/mole$)[a]

Material	V^* creep	V^* diffusion
Sn	5.12	5.3
Zn	5.95	$5\perp$, $16\parallel$ to c axis
Na	9.8	9.6
Pb	13.9	13–15
P	30.1	30
AgBr	38	42

[a] After Weertman (1968).

that for high-temperature creep can be determined from hydrostatic pressure experiments. Experimental data on such activation volumes are not extensive. Table II shows that for the available data (Weertman, 1968), these two activation volumes are approximately equal.

F. Effects of Phase Change

The significance of diffusion to high-temperature creep is further emphasized by two striking examples: As illustrated in Fig. 2, Sherby (1962) has shown that the secondary creep rates of austenite under a constant stress

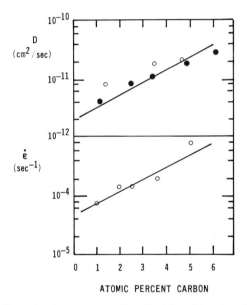

Fig. 2. Parallel effects of carbon content on the diffusivity of Fe and on the steady state creep rate in austenite at 1000° C (after Sherby, 1962). Upper half: ○, Gruzin, Korner, and Kurdinov; ●, Mead and Birchenall. Lower half: ○, P. Feltham; $\sigma = 3640$ psi (256 kg/cm^2).

and temperature increase with carbon content in a manner that closely parallels that for the effect of C on the self-diffusivity of Fe in austenite. Such an effect of alloy softening cannot be easily explained on any other basis. As shown in Figs. 3 and 4 (Sherby and Burke, 1967), the secondary creep rate for α–Fe at a constant stress shows almost the same variation through the magnetic-transformation range of temperature as does the self-diffusivity of iron in α–Fe. This remarkable correlation again suggests that the self-diffusion process ultimately controls the high-temperature creep rate.

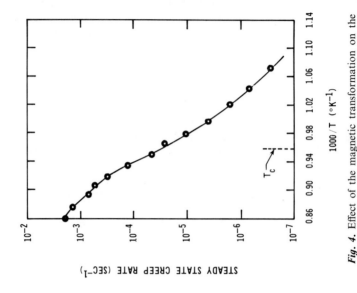

Fig. 4. Effect of the magnetic transformation on the steady state creep rate in α-Fe (after Sherby and Burke, 1967). Fe–0.02% C; 2550 psi.

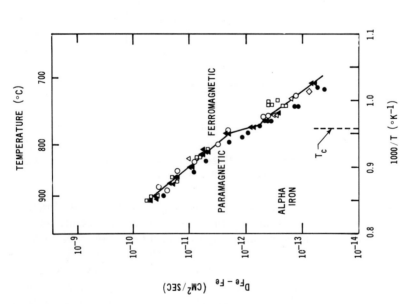

Fig. 3. Effect of the magnetic transformation on the self-diffusivity in α-Fe (after Sherby and Burke, 1967). □, Buffington, Bakalar, and Cohen; ○, △, Birchenall and Mehl; ●, Birchenall and Borg; ◇, Birchenall; ▲, Buffington, Hirano, and Cohen.

IV. Comparison with Experimental Results

Since high-temperature creep is diffusion-controlled, comparisons of the secondary creep rates might be based on the following form of Eq. (6)

$$\dot{\varepsilon}_s kT/DGb = A(\sigma/G)^n \tag{9}$$

where the Burgers vector, b, is tentatively taken to be the significant dimension in either diffusion or deformation to provide the required dimensionless expression.

The transition from Eq. (5) to Eq. (9) does not alter the significance of n, and changes only slightly the temperature effects. Nevertheless the change is not trivial when comparisons are made between various metals in an attempt to uncover what other factors might be involved in creep. Equation (9) is still largely empirical and attempts only to systematize the effects of D, T, G, and b on the secondary creep rate. If other factors are also pertinent, Eq. (9) will have to be reformulated so as to take these factors into appropriate consideration.

A detailed evaluation of the nominal correlation of experimental data with the trends suggested by Eq. (9) suffers a number of severe handicaps:

1. Few constant stress creep data are available because most tests have been conducted under constant load.

2. Often creep data are obtained at temperatures below or above the temperature range where diffusion-controlled creep occurs.

3. Data on the unrelaxed shear modulus of elasticity at high temperatures are often not available.

4. Data on self-diffusivities are not always accurate.

5. In some tests grain growth and even recrystallization took place at the higher test temperatures. These changes are known to produce abnormally high creep rates in instances where they have been observed. Occasionally investigators neglected to comment or even check on these possibilities when reporting their data.

6. Precautions to limit possible errors due to high temperature oxidation were exercised in only a few cases.

As stated in the Introduction, we shall restrict our discussion to high-temperature creep mechanisms where the rate-controlling step is due to the process of self-diffusion. We shall analyze first the experimental data on creep where dislocation motion is not involved, i.e. Nabarro–Herring creep. We shall then extend our correlation to dislocation climb-controlled creep in pure metals and alloys and finally to creep controlled by viscous glide of dislocations in some alloys. Later in this article we shall compare these various mechanisms over extended ranges of temperature, stress and creep rate in order to reveal the ranges of operation of the various individual mechanisms.

A. Nabarro–Herring Creep

This type of creep (Nabarro, 1948) results from the diffusion of vacancies from regions of high chemical potential at grain boundaries subjected to normal tensile stresses to regions of lower chemical potential where the average tensile stresses across the grain boundaries are zero. Atoms migrating in the opposite direction account for the creep strain. When volume diffusion controls, the tensile creep rate, $\dot{\varepsilon}_s$, is given by

$$\dot{\varepsilon}_s = A_n(Db^3\sigma/d^2kT) \tag{10}$$

where d is the mean grain diameter and kT is the Boltzmann constant times the absolute temperature. The diffusivity, D, is obtained from the tracer diffusivity, D^*. The dimensionless constant A_n depends insensitively on the geometry of grains, but is generally estimated to have a value of 8 (Nabarro, 1948) to 5 (Gibbs, 1966).

Nabarro–Herring creep does not involve the motion of dislocations. It predominates over high-temperature dislocation-dependent mechanisms only at low stress levels, and then only for fine-grained materials. Since it is unaffected by substructural changes, Nabarro–Herring creep gives curves of Type B_2 (see Fig. 1), in which small transients are occasionaly observed. Nabarro–Herring creep is characterized by creep rates that increase linearly with the stress and inversely with the square of the grain diameter, and have an activation energy that is nearly equal to that for self-diffusion.

In order to provide a ready comparison of Nabarro creep with other mechanisms to be discussed later, the data have been plotted in accordance with a reformulation of Eq. (9), namely

$$\dot{\varepsilon}_s kT/DGb = A_n(b/d)^2(\sigma/G)^1 \tag{11}$$

where G is the shear modulus of elasticity. The accuracy of the theoretical expression for Nabarro–Herring creep can be judged directly from Fig. 5, due to Bird et al. (1969). The agreement between theory and experiment is generally quite good in view of the difficulties of accurately characterizing grain size and geometry, and in estimating A_n.

The evidence for Nabarro–Herring creep, at low stresses and high temperature, is often convincing. Limitations of Nabarro–Herring creep will be discussed in more detail later on. Henceforth, experimental information on secondary creep rates by other mechanisms will be summarized in figures identical to Fig. 5. Creep data obtained from tests performed with single crystals are distinguished from the more common polycrystalline test data by the symbol [S]. Alloy compositions are stated in terms of atomic percent unless otherwise noted.

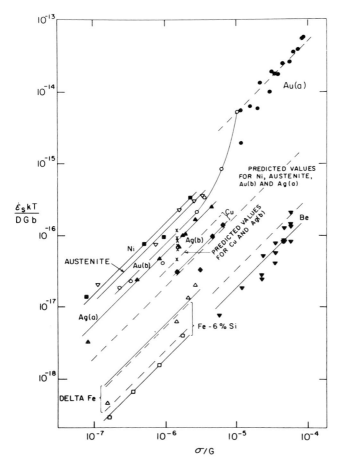

Fig. 5. Correlation between Nabarro–Herring creep [Eq. (11)] and experimental results. ——, Experimental data; – – –, predicted from Nabarro equation. Code for experimental data: ●, Au(a); ○, Au(b); ▲, Ag(a); ×, Ag(b); ◆, Cu; ■, Ni; ▼, Be; ∇, austenite; △, δ–Fe; □, Fe–6% Si.

B. Climb-Controlled Creep in Pure Metals

At high temperatures the rate of creep of most metals and many solid solution alloys is controlled by the recovery of the dislocation substructure. When the temperature is high enough, in excess of from 0.55 to 0.7 of the melting point T_m, dislocation climb appears to be the rate-controlling recovery process and consequently the rate-controlling step of creep.

The rate of dislocation-climb creep depends sensitively upon the dislocation substructure that is present in the metal. Eventually, a steady state is achieved in which both the creep rate and the rate-controlling substructural details remain constant.

For climb-controlled creep of pure metals and solid solution alloys, n of Eq. (9) generally ranges in value from 4.2 to 7. Since Eq. (9) is based upon the assumption that steady state creep is diffusion controlled, its use is restricted to temperatures where the activation enthalpy for creep, H_c, is equal to that for self-diffusion, H_D.

All available data on the high-temperature steady state creep rates of nominally pure fcc metals are correlatable in terms of Eq. (9), as documented in Fig. 6. The trends shown in Fig. 6 confirm the validity of Eq. (9) for individual fcc metals. In fact, for a given value of σ/G, the values of

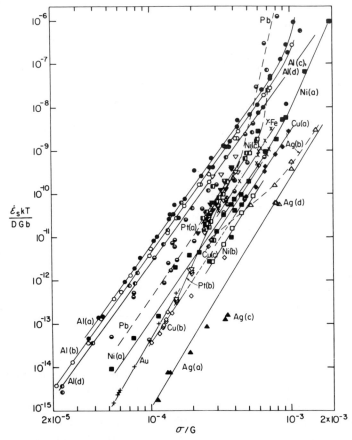

Fig. 6. Steady state creep rates of fcc metals correlated by Eq. (12). ●, Al(a); ○, Al(b); ⊕, Al(c); ◐, Al(d) single crystal; ■, Ni(a); □, Ni(b); ◧, Ni(c); ◆, Cu(a); ◇, Cu(b); ◈, Cu(c); ▲, Ag(a); △, Ag(b); ⊿, Ag(c); ▲, Ag(d); ▼, Pt(a); ▽, Pt(b); +, Au; ◓, Pb; ×, γ-Fe.

$\dot{\varepsilon}_s kT/DGb$ for the different metals do not deviate more than about 60 times from the mean of the band.

If all factors pertinent to steady state creep of fcc metals at high temperatures had been correctly incorporated into Eq. (9), all reliable data given in Fig. 6 should have clustered about a single straight line within experimental accuracy of determining $\dot{\varepsilon}_s$, T, D, G, and σ. The fact that such a unique correlation is not obtained reveals that additional factors, not included in Eq. (9), are also pertinent for correlations of secondary creep rates.

In the case of fcc metals it appears that the major factor arises from effects of stacking fault width on the dislocation climb mechanism. The creep rates of fcc metals seem to increase as the stacking fault energy increases. Barrett and Sherby (1964) interpreted this as an effect of stacking fault energy on A of Eq. (9). This concept was extended recently (Mukherjee et al., 1969) with emphasis on the fact that this approach assumes n to be the same constant for all fcc metals. The alternative possibility, however, is that A for climb-controlled creep is a universal constant and that the remaining parameter, n, of Eq. (9) increases with the dimensionless quantity Gb/Γ where Γ is the stacking fault energy.

The nominal validity of the concept that n depends on the stacking fault energy is shown in Fig. 7, where the datum points refer to average reported values of n and Gb/Γ and the vertical and horizontal bars give the range of error estimated for these values. Excluding the data of Pt(b), Ni(b) which are obviously inconsistent with other investigations on the same metals, n increases systematically with Gb/Γ. Deviations from the "best-fit" solid line are within the range of scatter of the experimental data. All fcc metals exhibiting normal creep trends have about the same values of $A = 2.5 \times 10^6$ when the "best-fit" value of n is adopted. Excluding the metal Pt, for which Gb/Γ is not well established, the secondary creep rates of all fcc metals are given by Bird et al. (1969),

$$\dot{\varepsilon}_s kT/DGb = 2.5 \times 10^6 (\sigma/G)^n \qquad (12)$$

within a factor of about 2 when n is selected from the "best-fit" curve of Fig. 7.

Similar analysis (Bird et al., 1969) of high-temperature steady state creep rates of bcc and hcp metals reveal that the correlation suggested by Eq. (12) can be used for these metals as well. In contrast to the fcc metals, the values of n for bcc metals seem to vary much more widely, from 4 to 7. This might suggest that some additional factors, not documented in Eqs. (9) and (12), influence the steady state creep rates of bcc metals. Although there is no firm evidence at present as to what factors might be responsible for the variance in n of the steady state creep rates of bcc metals, it appears likely that purity might have some effect. The creep of hcp metals at temperatures greater than

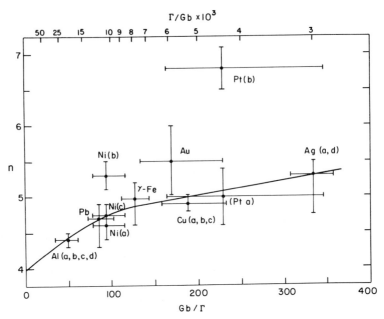

Fig. 7. Relation between the stress dependence of the creep rate, n, and stacking fault energy, Γ. Brackets indicate the estimated accuracy of n and Γ/Gb. Solid line shows the "best-fit" value of n.

0.6 to 0.7 T_m appears (Flinn and Gilbert, 1968) to be governed by a separate mechanism which is characterized by a distinctly higher activation energy. The mechanism is as yet unidentified and will not be discussed here.

C. Climb-Controlled Creep in Solid-Solution Alloys

At sufficiently high temperatures some solid-solution alloys creep in accord with Weertman's (1957b) viscous drag mechanism and others creep in nominal agreement with Weertman's (1957a, 1968) dislocation climb mechanism. Only alloys in the second group will be considered in this section. As shown in Fig. 8, the steady state creep rates of the fcc alloys that obey the dislocation climb mechanism fall into a wide band similar to that of the fcc metals. Bird *et al.* (1969) have shown that the value of $\dot{\varepsilon}_s \, kT/DGb$ was reduced by solid-solution alloying, in some cases only slightly but in others by over an order of magnitude.

It seems likely that the stacking fault energy, expressed as Gb/Γ, has the same effect on the steady state creep rates of fcc alloys which undertake the

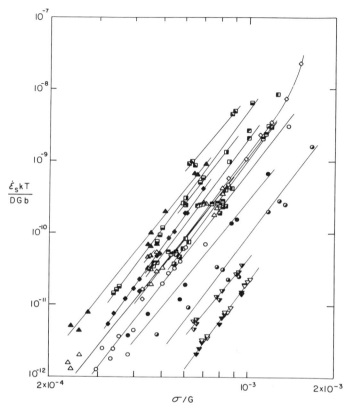

Fig. 8. Steady state creep rates of binary solid-solution alloys correlated by Eq. (9). ▣,
Ni–12% Fe; ▪, Ni–21% Fe; ◪, Ni–26% Fe; ◪, Ni–31% Fe; ▯, Ni–41% Fe; ▯, Ni–61% Fe;
◣, Ni–80% Fe; ▲, Ni–10% Cr; ▲, Ni–20% Cr; ▲, Ni–24% Cr; △, Ni–30% Cr; ○,
Ni–13% Cu; ◑, Ni–46% Cu; ●, Ni–79% Cu; ▼, Cu–10% Zn; ▼, Cu–20% N; ▽,
Cu–31% Zn; ◇, Fe–5.8% Si; ◆, AgMg.

dislocation climb mechanism as it does for the fcc metals. This effect may be
responsible for the slightly lower creep rates of the alloys, as it is generally
observed in most metals that stacking fault width increases with alloying.
Alloying of such metals should thus be accompanied by a decrease in creep
rate and a small increase in n.

Dislocation climb-controlled creep is frequently observed in bcc alloys
and in hcp alloys, although few data for hcp alloys have yet been generated.
It seems probable that the creep behavior of bcc and hcp alloys is not unlike
that of the pure metals and fcc alloys whenever the climb mechanism is
rate-controlling.

D. *Viscous Glide Creep in Solid-Solution Alloys*

The climb mechanism is rate-controlling in solid-solution alloys whenever dislocation glide is rapid. As shown in the previous section, it is characterized by values of n which range from 4.2 to 7. But some solid solution alloys display values of n during creep which are much lower, $3 \leq n \leq 3.6$. This type of behavior is obtained whenever the creep rate is controlled by viscous glide of dislocations, as first demonstrated by Weertman (1957b).

A number of different dislocation–solute atom mechanisms may be responsible for producing viscous glide creep in solid-solution alloys. Any interaction mechanism that can reduce the rate of gliding dislocations to a value which is so low that glide becomes slower than dislocation-climb recovery can produce such creep behavior. In principle several mechanisms can do this. When solute atoms are bound to dislocations through the Cottrell (1953), Suzuki (1955), or other analogous interaction mechanisms (Fleischer, 1961, 1963), glide can proceed no faster than the rate of solute atom diffusion. When disorder is introduced into a crystal by glide of a dislocation, the steady state glide velocity is limited by the rate at which chemical diffusion can reinstate order behind the gliding dislocation. In such alloys glide is diffusion-controlled, and is consequently much slower than glide in pure metals. In some metals, at least, this can lead to glide-controlled creep.

A straightforward derivation for the rate of glide-controlled creep, which does not invoke fixed dislocation sources or glide of dislocation arrays, has been made by Bird *et al.* (1969), for the simple case where all dislocations within a crystal drag a dilute Cottrell atmosphere. According to their analysis the steady state strain rate is given by

$$\dot{\varepsilon}_s = (1/6\alpha^2)(DGb/kT)(\sigma/G)^3 \qquad (13)$$

where α is a constant equal to about 0.6. Equation (13) can be rewritten as

$$\dot{\varepsilon}_s kT/DGb = 0.5(\sigma/G)^3 \qquad (14)$$

which is analogous to Eq. (9) with $A = 0.5$ and $n = 3$.

This result is similar to that originally derived by Weertman. The only difference is that the preexponential term $A = 0.5$ is independent of solute concentration and independent of the details of the Cottrell interaction mechanism. This latter difference is significant, for it suggests that the rate of glide-controlled creep should be given approximately by Eq. (14). The creep rates of solid solution alloys that show a $n \simeq 3$ power stress dependence has been correlated with Eq. (14), and presented in Fig. 9.

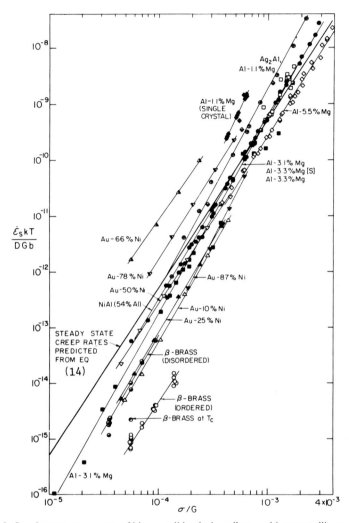

Fig. 9. Steady state creep rates of binary solid-solution alloys and intermetallic compounds which creep by the viscous glide mechanism [Eq. (14)]. $\triangle$, Au–10% Ni; $\blacktriangle$, Au–25% Ni; $\triangledown$, Au–50% Ni; $\blacktriangle$, Au–66% Ni; $\blacktriangledown$, Au–78% Ni; $\blacktriangledown$, Au–87% Ni; $\diamondsuit$, Al–5.5% Mg; $\square$, Al–3.3% Mg[S]; $\blacksquare$, Al–3.3% Mg; $\blacksquare$, Al–3.1% Mg; $\blacklozenge$, Al–1.1% Mg[S]; $\diamondsuit$, Al–1.1% Mg; $\oslash$, βCuZn (disordered)[S]; $\bigcirc$, β'-CuZn (disordered)[S]; $\oslash$, β-CuZn (at T_c)[S]; $\bullet$, NiAl (54% Al); $\oplus$, Ag$_2$Al[S].

The most striking aspect of Fig. 9 is the close agreement found between the experimentally measured steady state creep rates and those predicted from Eq. (14). The other interesting aspect of Fig. 9 is that n is often greater than the theoretically suggested value of 3. As alloying content increases, n veers from values characteristic of climb-controlled creep toward the limiting value of 3 that is predicted from simple models of viscous glide creep. This indicates that the mechanism of viscous glide creep is sensitive to solute atom concentration. Presently formulated models of viscous glide creep are not yet sophisticated enough to predict such behavior. It is perhaps best to think of them as representative of the limiting case of viscous glide creep.

Up to the present it has not been possible to predict which solid solution alloys should creep by the viscous glide mechanism and which should creep by dislocation climb. In attempting such predictions it should be kept in mind that two criteria must be simultaneously satisfied before viscous glide creep can replace dislocation climb as the rate-controlling mechanism during high-temperature creep:

1. Solute atoms must be bound to gliding dislocations through some solute atom–dislocation interaction or ordering mechanism.

2. The resulting rate of dislocation glide must be slow enough to be rate-controlling during steady state creep. This criterion is met only when the steady state creep rate, $\dot{\varepsilon}_s$, as given by Eq. (12), is greater than the steady state creep rate given by Eq. (14).

In the past it has not been possible to accurately formulate the second criterion. With the correlations suggested here, it may be possible to formulate this criterion with sufficient accuracy to make predictions on the creep mechanisms of solid solution alloys feasible.

The presentation in Section IV has been restricted to pure metals and solid-solution alloys where some progress has been made in understanding the creep phenomenon. There is also the commercially important area of high-temperature creep of dispersion-strengthened alloys. The creep deformation in these alloys is often a complex phenomenon that is not well understood. The creep behavior is so sensitive to the type and nature of the dispersion that different processing histories often lead to creep by entirely different mechanisms even in the same alloy. There is no completely satisfactory theory for creep in dispersion-hardened alloys. Often the experimental activation energy of creep for such alloys is several times greater (Meyers and Sherby, 1961–1962) than that for self-diffusion and the stress dependence of the creep rate has unusually high values, e.g. $n = 40$ (Wilcox and Clauer, 1966). The status of the existing theories and the rather difficult to correlate experimental results have been discussed by Bird *et al.* (1969) and by Sherby and Burke (1967) and will not be further dealt with here.

V. Microstructural Aspects in Creep

Correlations between steady state creep at high temperatures, grain size, and substructure will be discussed in this section.

A. *Effect of Grain Size*

The results of early investigations on effects of grain size on high-temperature creep have been reviewed by Conrad (1961) and Garafalo (1965). The early reported data seemed to give rather inconclusive results. Whereas some data suggested that the creep rates decreased with increasing grain size, others suggested that the opposite trend was valid. Garafalo (1965) suggested that the reported trends implied the existence of a minimum steady state creep rate at some intermediate "optimum" grain size.

The data documented in an earlier chapter, excluding, of course, that on Nabarro–Herring creep, refute the concept that the steady state creep rate for either the dislocation climb or the viscous drag mechanism depends appreciably upon grain size. The steady state creep rates for single crystals of Al (Fig. 6) and Al–Mg alloys (Fig. 9) do not differ by more than a factor of 2 from those of polycrystalline test specimens when tested over a wide range of σ/G.

Below a critical grain size the steady state creep rate increases with decreasing grain size. The major evidence reported by Barrett et al. (1967) for randomly oriented Cu is shown in Fig. 10. Although Garafalo et al. (1964) suggested that an optimum grain size for creep resistance is obtained in austenitic stainless steels, their data were also plotted in Fig. 10. Within the normal scatter band these data also suggest that as the grain size is decreased, the steady state creep rate remains constant until the grain size is reduced below a critical value, whereupon the creep rate increases with additional reduction in grain size. The fact that the critical grain size is about the same for Cu and austenitic stainless steel is undoubtedly a coincidence.

Although the rate of Nabarro–Herring creep increases with decreasing grain size it is easily demonstrated that it cannot be responsible for the above-mentioned grain size effects. Its contribution to the total creep rate is negligible for the values of σ/G at which the above discussed grain size is obtained.

Current evidence thus suggests that the effect of grain size *per se* on the secondary creep rate is small except for very low stresses and very fine grain sizes. Increased creep rates with decreasing grain size appear primarily

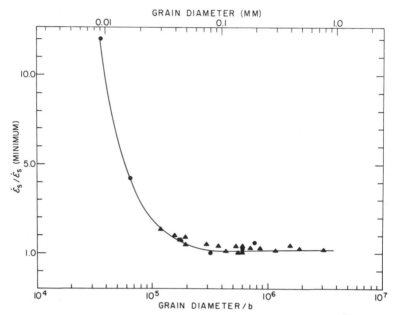

Fig. 10. Effect of grain size on steady state creep rate at constant temperature and stress. ●, Austenitic stainless steel $(T = 977 \text{ K}, \quad \sigma/G = 1.14 \times 10^{-3})$; ▲, Cu $(T = 769 \text{ K}, \sigma/G = 6.0 \times 10^{-4})$.

attributable to grain boundary sliding. Auxiliary effects arising from texture and impurities might account for some of the trends and inconsistencies previously reported on grain size effects.

B. Creep Induced Substructures

The decreasing creep rate over the primary stage for constant conditions of stress and temperature must necessarily be a reflection of the substructural changes that take place: the constant creep rate over the secondary stage of creep demands, equally, that all significant substructural features must have reached a steady state. It is not yet clear, however, just what substructural details might be significant in this respect. Immediately upon loading dislocation entanglements and cells are produced which bear some resemblance to those observed in strain-hardened metals at low temperatures. During the primary stage of creep these cells and entanglements disperse either through recrystallization or what is more pertinent here through dynamic recovery processes such as cross-slip and climb. The dispersal of the original entanglements and cells, although probably facilitated by cross-slip, is largely due to the climb of edge dislocations. Cross-slip is known to be dependent on the stacking fault energy which might, for this reason as well as others, play some role in creep.

Three substructural features are accessible for study by electron transmission microscopy, etch-pitting techniques, optical microscopy, and X-ray analyses. They might be classified as follows: (1) Formation of subgrain during creep and the size of subgrain during the steady state. (2) The disposition and density of dislocations in the interior of subgrains. (3) The nature of the subboundary and the misorientation across them.

1. FORMATION AND SIZE OF SUBGRAINS

Overwhelming evidence now makes it possible to conclude that well-defined subgrains are formed during the creep of metals and solid-solution alloys wherever the dislocation climb mechanism is rate-controlling. Bird *et al.* (1969) in their review pointed out that subgrains had been observed in single as well as polycrystals. It appears that whenever climb is rate-controlling, subgrain formation always accompanies creep in single phase metals and alloys.

A number of isolated observations suggest that subgrains do not form when the rate-controlling creep mechanism is different from that of dislocation climb. Subgrains do not form in hcp Mg when creep takes place at temperatures in excess of $0.75\,T_m$ (McG. Tegart, 1961), where an unidentified mechanism replaces climb as the rate-controlling process. Harper *et al.* (1958) have reported that subgrains do not form in Al when creep takes place at temperatures very close to the melting point under extremely low stresses. Creep under these conditions is diffusion-controlled, but n equals 1 instead of 4.4, suggesting that another crystallographic creep mechanism is operative. The experimental work of Vandervoort and Barmore (1968) suggests that subgrains do not form in W–25% Re alloy where viscous glide creep is the rate-controlling process.

During steady state creep the subgrain structure remains in the steady state although there is no doubt that individual subgrains are in a constant state of change throughout the duration of creep. Bird *et al.* (1969) suggested in their review that the mean subgrain size, δ, is independent of temperature, strain, and usually of grain size. The available data reveal that subgrain size varies strongly with stress and decreases about linearly with the reciprocal of σ/G as shown in Fig. 11. It is significant (Dorn and Mote, 1963) that the same subgrain size is developed in a given metal undergoing steady state creep, regardless of its previous work-hardened or precrept condition.

The effect of stacking fault width on the steady state creep rate for the climb mechanism suggests that the subgrain size might also be influenced by

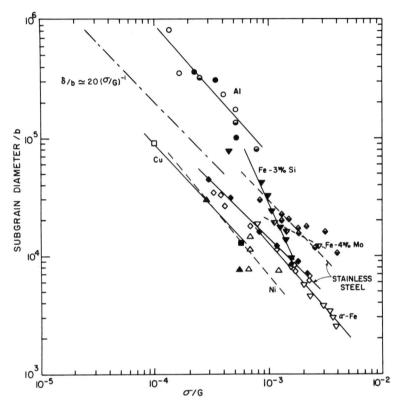

Fig. 11. Stress dependence of subgrain size in steady state creep. Metal and method (opt—optical, X-ray, ep—etch-pitting, em—electron microscopy): ○, Al, opt; ◔, Al-0.05 wt % Fe, opt; ●, Al-5 wt % Zn, X-ray; ▽, α-Fe, ep; ▼, Fe-3 wt % Si, ep; ▼, Fe-4 wt % Mo, ep; △, Ni, em; ▲, Ni, ep; □, Cu, ep; ■, Cu, em; ◇, austenitic stainless steel (8 wt % Ni), —; ◈, austenitic stainless steel (11 wt % Ni), em; ◆, austenitic stainless steel (14 wt % Ni), —.

this parameter. However, the evidence presented in Fig. 11 does not confirm this concept. Data by D. H. Warrington (University of Sheffield, private communication, 1969) on austenitic stainless steel show, for example, that subgrains of identical size are obtained in alloys in which Gb/Γ doubles from a value of somewhere near 350 to about 700. The subgrain size of Al in Fig. 11 appears to be significantly greater than those of other metals, but it is not clear whether this observation is reliable. Measurements of subgrain size in Al were taken from early substructural investigations using polarized light optical microscopy. Later investigations (Wong *et al.*, 1967) suggest that optically observed subgrains do not correspond well to the real subgrain structure when the subgrain size is small. Discounting the Al data in Fig. 11, the mean subgrain size for all of the polycrystalline metals investigated is given by

$$\delta/b = 20(\sigma/G)^{-1} \tag{15}$$

to within a factor of 4. More detailed investigations will be required to establish the accuracy and validity of this equation for other metals and alloys and for wider ranges of σ/G.

The detailed studies of Clauer et al. (1970) show clearly that climb recovery modifies the substructure that is produced during creep deformation, but is not directly responsible for originating the cellular pattern in the first place. This suggests that slip processes are primarily responsible for determining the arrangement of dislocations during plastic deformation, and that climb recovery only modifies this fundamental pattern during the process of formation of subgrains.

2. DISLOCATION DENSITY WITHIN SUBGRAINS

As previously discussed, the arrangement of dislocation within subgrains after creep is similar in appearance to the arrangements that are observed in cells after low temperature deformation. An even greater similarity is found between the high- and low-temperature dislocation substructures when quantitative counts of dislocation densities are made. It has been well established that in many metals at low temperatures the density of dislocations, ρ, varies with the square of the flow stress according to the relation

$$\sigma/2 = \alpha Gb\sqrt{\rho} \tag{16}$$

where the shear modulus of elasticity is that which applies at the test temperature, and α is a constant equal to about 0.5 (Mitchell, 1964). Figure 12 (Bird et al., 1969) shows that this relation also exists for a wide variety of metals during high-temperature creep. The figure contains most known measurements of dislocation densities that have been made within subgrains after steady state creep. Discounting the etch pit data, which are known to be unreliable in Fe at higher values of ρ, it appears that the average value of $(\rho)^{1/2}b$ at the steady state for metals and alloys that creep by the dislocation climb mechanism is only slightly lower at the same values of σ/G than those measured following low-temperature deformation. The mean value of $\alpha \simeq 0.6$ applies to steady state creep by climb in the reported cases. It is significant that the W–Re alloy, which creeps by the viscous glide mechanism, also obeys the same relation.

The similarities which are found between dislocation substructures after high- and low-temperature deformation within subgrain or cell interiors suggests that dislocation substructures achieve a mechanical equilibrium as a result of dislocation interactions whenever plastic deformation occurs.

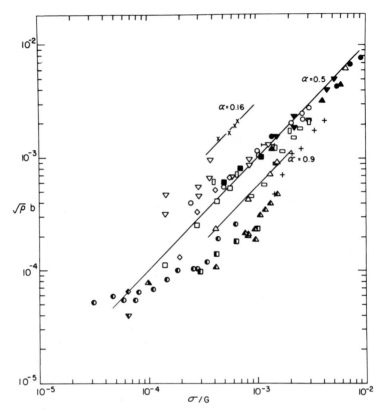

Fig. 12. Stress dependence of dislocation density, ρ. High-temperature creep, ρ measured by electron microscopy: $\diamond$, W; $\square$, W–25 wt % Re (viscous drag); $\bigcirc$, Fe–35 wt % Ni–20 wt % Cr; $\triangledown$, Fe; $\triangle$, Fe–3 wt % Si; $\square$, Fe–4 wt % Mo; $\square$, Fe–0.75 wt % Mn–0.01 wt % N; $+$, Fe–35 wt % Ni–20 wt % Cr and $\times$, Ni–2.66 Vol % ThO$_2$, dispersions. ρ measured from etch-pitting: $\blacktriangle$, Fe–3 wt % Si (same specimens as $\triangle$ above); $\blacksquare$, Fe–3.1 wt % Si; $\mathbf{\bullet}$, Fe–3.1 wt % Si (textured); $\blacktriangle$, Cu; $\blacktriangledown$, Mo $\varepsilon = 0.42$; $\diamondsuit$, Mo $\varepsilon = 0.22$. Low-temperature deformation, ρ measured by electron microscopy: $\blacktriangle$, Cu; $\blacksquare$, Cu; $\blacktriangledown$, Cu–2.2 wt % Al; $\bullet$, Ag.

This has long been recognized in the strain hardening theories at low temperatures, but has not always been included in theories of dislocation climb-controlled creep.

The observations which are presently available show that two of the assumptions made in early climb theories about the dislocation substructure during creep cannot be justified easily. Piled-up arrays of edge dislocations from sources on adjacent slip planes are never observed. Observations also show no experimental justification for the prevalent theoretical assumption that dislocations are generated at fixed sources. Instead, they suggest that there are many possible sources for dislocations, both during creep and low-temperature deformation. Dislocations in the interior of subgrains undoubtedly act as sources of additional dislocation segments as they cross-slip and climb (Jonas *et al.*, 1968). Subgrains and grain boundaries also could serve as sources and sinks for dislocations.

3. SUBGRAIN BOUNDARIES

Subgrains in polycrystalline metals are distinguishable from each other by well-formed tilt boundaries, regular hexagonal dislocation networks, and irregular networks which have a mixed edge-screw character. In subsequent investigations on the creep of single crystals of Mo, however, Clauer *et al.* (1970) have demonstrated that tilt boundaries predominate.

It appears that the early concept of subgrain formation through polygonization of edge dislocations under stress is not wholly tenable. Some subgrain boundaries develop through the climb recovery of dislocation tangles which form upon initial straining of a metal. Others develop during the course of creep by coalescence of other subgrain boundaries. Yet the process ultimately responsible for subgrain formation is still unexplained.

Another unresolved issue is the role that subgrain boundaries play in the theories of climb-controlled creep. Early investigations suggested that the angle across the subgrain boundaries continued to increase over the secondary stage of creep (McLean, 1952–1953). More recent studies (Wong *et al.*, 1967; Hardwick and McG. Tegart, 1961–1962), covering examples that involve extensive deformation, suggest that the subgrain misorientation eventually reaches some constant limiting value. This observation is consistent with the concept that the significant substructural details must remain substantially constant over the secondary stage of creep.

The need for a more detailed study on subgrain misorientation during secondary stage creep (including the possible effects of stresses, stacking fault energies, and purity on the degree of misorientation) should be emphasized here. If, for example, the misorientation of the subgrains increases over the secondary stage of creep, it becomes apparent that such misorientation is not a significant substructural variable since it does not affect the creep rate. But if the misorientation remains constant over much of the secondary stage it may be a significant variable and then should be taken into consideration.

C. *Reflections on Substructure*

Although the effect of stress alone has been emphasized in this discussion, it appears likely that crystal structure, which dictates what type of dislocations are present, the shear modulus of elasticity, which in part determines dislocation interactions, and the stacking fault energy might also be significant in establishing the substructural features during steady-state creep. The emphasis given here to the stress merely reflects the fact that some rather tangible, although incomplete evidence exists on the effect of this variable.

The review of substructure suggested the difficulty of experimental justification for piled-up arrays of edge dislocations in creep and for the assumption that dislocations are generated at fixed sources. In a recent paper Weertman (1972) has suggested several reasons as to why pileups are not usually observed. According to his viewpoint, at high temperatures, any straight dislocation is inherently unstable and will turn into a more irregular form producing entanglements. Alternatively, in the presence of an oscillating internal stress field, such as might arise from the presence of subgrain boundaries, a large pileup of dislocations can break up into smaller subgroups of dislocations. Additionally, at the low stress levels encountered in high-temperature creep, there may not be too many dislocations in any one pileup and these may be further relaxed during unloading of a creep specimen. Weertman also discussed the possibility that during high-temperature creep, the number of active dislocation sources may become as small as possible. The smallest number of sources would be that existing in well annealed material. According to him, the source density may decrease to this value at any stress level and may become independent of stress during creep.

A complete theory for steady-state creep should be able not only to account for the observed steady-state creep rate in terms of the variables of σ, G, D, T, Gb/Γ, etc., but should equally well be able to predict the pertinent steady state substructures. Obviously, such substructures form an integral part of a realistic model for steady state creep. On the other hand, it appears as if this task is so formidable in view of the versatility of dislocation reactions that it will not be possible over a considerable period of time to predict the steady state substructure from first principles. Over this time interval the less pretentious objective of predicting the effects of σ, G, D, T, Gb/Γ on $\dot{\varepsilon}_s$, based on experimentally established substructural knowledge, may be fruitful.

It is interesting now to speculate on which of the three above-mentioned substructural features might be most important in steady state creep. Dorn and Mote (1963) have demonstrated that when, after establishing steady state at a high stress, the stress is abruptly decreased, the substructure at first remains close to that developed by the higher initial stress. Under these conditions the initial creep rate should nearly correspond to that for a structure which has a high density of dislocations in the subgrain and a fine subgrain size. During the period over which the dislocation density decreases and the subgrain size increases to that appropriate to the lower stress, the creep rate increases to its new steady value. If creep were due exclusively to the dislocations in the body of the subgrain, the creep rate should have decreased and not increased during the transition period. On this basis it appears that the creep rate may not be so much dependent on the dislocations in the body of the subgrains. The observed trend suggests that creep rates might increase as a result of increases in the subgrain size. According to the data presented in Fig. 11, however, the subgrain diameter in fcc metals does not appear to increase with an increase in stacking fault energy to the same degree as does the steady state creep rate. Other factors such as unique dislocation reactions, purity, etc. might also influence the subgrain size, and in this way account for the observed spread in $\dot{\varepsilon}_s kT/DGb$ at constant values of σ/G for a series of metals. Earlier theories on steady state creep did not consider this possibility that subgrain size and the nature of the subgrain boundary may be significant in affecting the creep rate. The recent models discussed by Weertman (1968, 1972) on subgrain creep indicate significant progress of this aspect.

Another topic related to the microstructural aspects of creep deformation is grain boundary sliding. As stated earlier, it is not important for the discussion of creep rates for specimens that do not have too fine grain size and that deform by climb-controlled or viscous glide-controlled creep. It is important, of course, for all grain sizes in the tertiary stage of creep because it promotes cavitation in grain boundaries which then leads to rupture. This cavitation phenomenon will not be discussed here. In very fine-grained specimens, at low stress levels, however, grain boundary sliding leads to increasing creep rate. In this range the creep rate is often controlled by the rate of grain boundary sliding and accompanying diffusional accommodation. This will be reviewed briefly in the next section.

VI. Creep in Very Fine-Grained Structure

The high-temperature deformation of very fine-grained material is the subject of considerable research at present, because it includes the technologically important area of superplasticity. A large number of fine-grained metals and alloys often show extremely large neck-free elongation when deformed in superplastic mode at elevated temperatures and at low strain rates. Most of the recent investigations suggest that there are two prerequisites for the manifestation of superplastic behavior: (1) a temperature above approximately 0.4 of the melting point of matrix in absolute degrees; and (2) a fine and stable grain size that does not change significantly during the elevated temperature deformation.

The first criterion essentially refers to the fact that superplasticity is controlled by diffusional processes. The details of these processes will be discussed later. The second criterion is often satisfied by the pinning of the grain boundaries by fine particles or it can be obtained in a two-phase mixture of equiaxed grains. In the latter case both recrystallization and grain growth tend to be restricted by the presence of the duplex phase structure.

A. Microstructural Characteristics

The microstructure of superplastic alloys after deformation typically reveals that (1) there are essentially no large changes in shape or size of grains; (2) there are large relative rotations of adjacent grains; and (3) voids are not commonly seen at the grain boundaries when they are deformed in the superplastic range of temperatures and strain rates (Mukherjee, 1975).

There appear to be some significant differences between superplasticity and usual high-temperature creep behavior. The latter is usually characterized by the presence of dislocation cell structure (Bird *et al.*, 1969) and a transient or primary creep stage (Amin *et al.*, 1970). Both are absent in the typical superplasticity. On the other hand, the superplastic strain rate depends strongly on the grain size whereas the typical creep deformation rate is not (Bird *et al.*, 1969) greatly affected by changes in grain size.

B. Effect of Temperature, Stress, and Grain Size

The steady-state deformation rate of fine-grained metals is a thermally activated process and can often be described by Eq. (9), where D is the appropriate diffusivity and A should now include the grain size effect. The advantage of retaining the formalism of Eq. (9) is that it can also be used for superplasticity as well as for diffusional creep through lattice or through grain boundary, as will be discussed in the next section.

In the case of superplasticity, the stress dependence of strain rate typically (Ball and Hutchison, 1969) has a value of $n = 2$. Hayden *et al.* (1967) have observed that the $n = 2$ power dependence often extends over four orders of magnitude of strain rates. This implies that it is very unlikely that the superplastic $n = 2$ is a transition region between high ($n \simeq 5$) and low ($n \simeq 1$) values of n. Woodford (1969) has analyzed the various experimental data in the literature that supports $n = 1$ and he suggests that not all of the data can satisfactorily claim $n = 1$ dependence. Some of these results may cover a transition region in which $n = 2$.

Every experimental datum on superplasticity reveals that the strain rate increases with decrease in grain size. The analyses presented in the literature include inverse second power or inverse third power dependence of strain rate on grain size. The majority of more recent work, however, supports (Mukherjee, 1972) an inverse second power dependence.

The thermally activated nature of the superplastic deformation manifests itself in the temperature dependence of the strain rate. The measured activation energy values of deformation quoted in the literature generally fall into two categories: activation energies that are similar to those for grain boundary diffusion, and those that are comparable to the energy for lattice diffusion. These have been reviewed by Mukherjee (1971, 1972) and Johnson (1970).

C. Rate-Controlling Deformation Mechanisms

Although a large volume of experimental data on superplasticity now exists, there is still a considerable amount of disagreement regarding the precise details of the rate-controlling mechanism in such a deformation process. The various theories (for a review, see Johnson, 1970; Mukherjee, 1971, 1972; Packer and Sherby, 1967; Hayden et al., 1967) are based on (1) stress induced transport of vacancy through lattice, (2) stress induced vacancy transport through grain boundary, (3) dislocation movement in grain interior, (4) combined process of dislocation climb and slip, and (5) grain boundary sliding. Individual experimental investigation can support each of these mechanisms but there is difficulty in using these models in a completely general fashion for all materials.

The major difficulty in mechanisms (1) and (2) is that they predict a linear relationship between strain rate and stress which is rarely observed in superplasticity. These models also cannot explain why the grain shape remains equiaxed in unidirectional tensile superplastic deformation. Mechanism (3) would elongate the grains, unless it is accompanied by a recrystallization process, or grain boundary migration. These processes are not characteristic of superplastically deformed microstructure. Mechanism (4) has been put forward as a controlling mechanism for models where the activation energy is that for lattice diffusion, which is certainly not always observed. The same mechanism has been advanced in models which predict either an inverse first power or third power dependence of the strain rate. As discussed already, the vast majority of experimental data seem to reveal an inverse second power dependence instead. With respect to the grain boundary sliding models, sliding cannot occur continuously on all boundaries without some accompanying deformation within the grains themselves. Gifkins (1967) has pointed out that diffusion along grain boundaries could be an appropriate accommodation mechanism capable of transferring material across triple points in fine-grained materials. Recently Raj and Ashby (1971) have reviewed in detail the various aspects of grain boundary sliding and diffusional creep. Their analysis shows that if diffusion is the only mechanism of accommodating the incompatibilities caused by grain boundary sliding, diffusional flow and sliding are not independent. They are interrelated and the resulting deformation is described by the combined contribution of Nabarro (1948)–Herring (1950) and Coble (1963) creep. Although such a mechanism with its one power stress dependence could certainly control the deformation rate in fine-grained polycrystalline materials, it probably lies in a region separate from that of superplasticity which has typically a second power stress dependence of the strain rate.

In another paper, Ashby and Verrall (1973) suggested a mechanism for producing the large strains commonly encountered in superplasticity. It differs fundamentally from a combination of Nabarro–Herring and Coble creep in a topological sense; grains switch their neighbors and do not elongate significantly. A constitutive equation is derived from the model that resembles the Nabarro–Herring–Coble equation but predicts strain rates which are approximately one order of magnitude faster. The flow behavior of superplastic alloys has been explained as the superposition of this mechanism and ordinary power law dislocation creep. This model is a significant advance from earlier models and *in situ* dynamic observations (Naziri *et al.*, 1972) in transmission electron microscopy do support the grain-switching events. However, the superposition principle that is involved in order to explain superplastic deformation may not be entirely correct. Experimental investigations by Misro and Mukherjee (1972) and by Hayden *et al.* (1967) show that the second power stress dependence extends over quite a large range of strain rates (four orders of magnitude). Hence it is very likely that superplasticity is not a transition region between two different mechanisms as has been envisaged by Ashby and Verral, but is a mechanism by itself. A studied review of available data suggests that a realistic model for superplastic deformation rate should show a second power dependence on stress, an inverse second power dependence on grain size and an activation energy of the process probably close to that for grain boundary diffusion.

The author has proposed a model (Mukherjee, 1971) that is based on the concept that grain boundary sliding is the predominant mode of deformation and that the sliding is controlled by some dislocation motion within the grains and their eventual climb into the opposite grain boundary. A somewhat modified version of the above model will be presented here. No grain boundary surface is absolutely planar. Hence during the process of grain boundary sliding in superplasticity, the particularly large ledges or protrusions on the grain surface would cause most of the obstruction to the sliding process. The local shear stress concentration thus set up at the obstructing ledge can be large enough for dislocations to be generated at that site. The dislocations thus generated can travel along the most favorable slip plane in the blocked grain and pile up at the opposite grain boundary. Eventually the back stress of the pileup group would prevent the ledge from acting as a dislocation source. As pointed out by Friedel (1964), the leading dislocation in the pileup which is blocked at the grain boundary can climb into and along that grain boundary surface if the temperature is high enough. Hence, the sliding rate will be governed by the rate of climb of these dislocations to the appropriate annihilation sites along the grain boundaries. The basic postulates of the model find support from the work of Lin and McLean (1968). They suggest from the available evidence that during grain boundary sliding the dislocations primarily enter the boundary region by slip. McLean (1970) has shown from geometrical reasoning that grain boundary sliding and diffusion can combine to give nearly unlimited deformation. He suggested that such deformation can result from dislocations moving through the boundary and then through the grain and glide or, in the case of superplasticity, climb through the grain boundary again and so on across the specimen. The sequence of events has the general effect of moving one pair of grains between two others which become separated longitudinally, thus lengthening the specimen. The sequence can repeat at many places until the specimen is only a few grains wide, producing the large extensions that typify superplasticity. The readjustment that occurs after four grains contact each other during the process makes the grains that elongated longitudinally previously, equiaxed again and hence there should not be any marked departure from equiaxed shape of grains.

According to the model of Mukherjee (1971), during superplastic deformation, the compatibility between adjacent grains is achieved by diffusion-controlled climb of dislocation along the grain boundaries. This allows for the repeated accommodation that is necessary for the operation of grain boundary sliding as a unit process. As long as this is happening, slip need not take place on five independent slip systems (as is normally required in order to maintain compatibility during deformation of polycrystals). At any instant, dislocation pileups may move on only one set of slip planes in a given grain. In the very fine precipitate-free grains, often encountered in superplasticity, such pileups that completely traverse the grain may relax upon unloading or during thinning of the specimen for electron microscopy. This may have been one of the reasons for the lack of evidence of dislocation activity in deformed specimens in literature on superplasticity. However, recent investigations have in many cases demonstrated unambiguously the presence of slip activity in grain interior in superplastic deformation (Melton and Edington, 1973; Misro and Mukherjee, 1972).

The absence of grain elongation is accounted for in the same way as that proposed by McLean (170), namely a combination of grain boundary sliding, some shear deformation of grain interior, and glide climb motion of dislocations along the grain boundaries, leading to switching events among adjacent neighbors at various places. The model predicts that during typically superplastic deformation, dislocation cell structure will not be formed. This is because the subgrain or cell size in creep depends (see Fig. 11) approximately inversely on the stress, and at the very low stress levels encountered in superplasticity the cell size is nearly the same as the grain size. There would be no primary stage in the strain–time curve because there will be no adjustments of the density of dislocation entanglements in terms of dispersals of dislocation cells etc. that typifies the transient or primary stage of usual high temperature creep. Hence, there would be no increase in work hardening and no slowing down of the instantaneous strain rates in the initial stage. Since the only barriers to dislocation motion are the grain boundaries, the density of such barriers per unit volume will remain constant and hence one would anticipate a steady state deformation rate right from the beginning in superplasticity. Typical superplastic deformation is no longer expected to occur at higher stresses when dislocation tangles and/or cell structure begin to be observed in the deformed specimen. When this happens, dislocation pileups are prevented by such tangles etc. from going completely across the grain to the annihilation sites at the opposite grain boundary. The typically large deformation in superplasticity is then replaced by more modest deformation of usual high-temperature creep. These trends in the theoretical model are in agreement with most of the experimental observation in superplasticity. The model predicts a superplastic strain rate $\dot{\varepsilon}_s$, given by the dimensionless equation

$$\dot{\varepsilon}_s kT/D_b Gb = A'(\sigma/G)^2 (b/d)^2 \qquad (17)$$

where D_b is the grain boundary diffusivity, G the shear modulus, b the Burgers vector, σ the stress, d the grain size, and A' is a constant that depends on the extent of the unit of sliding. The value of A' was given earlier as being between 2 and 25 (Mukherjee, 1971). A recent recalculation of the strain rate, by the author, that incorporates the neighbor switching event during superplasticity, increases the value of the unit of sliding and raises the value of this constant to between 75 and 150, with an average value of 100. It is substantially similar to the values given by Ball and Hutchison (1969). In the following section, the contribution of superplasticity to the total creep rate will be compared with that due to other high-temperature creep mechanisms.

VII. Correlations of Creep Mechanisms

In the preceding sections of this report various identifiable diffusion-controlled creep mechanisms were isolated for individual examination relative to the various factors that affect their steady state creep rates. In the present section a comparison will be made of these mechanisms in order to reveal the ranges of conditions over which each mechanism predominates and how the various mechanisms are related to one another. This will be done in such a way as to provide a unified semiquantitative picture of the creep resistance of metals and alloys when diffusion-controlled creep mechanisms are operative. Although major emphasis will be given to those creep mechanisms that have now been firmly established as operable, additional possible mechanisms will also be described.

The steady state creep rates, as given by $\dot{\varepsilon}_s kT/DGb$ versus σ/G, for a number of mechanisms which are volume-diffusion controlled are shown in Fig.13. With the exception of the low stress data for Al, shown by datum

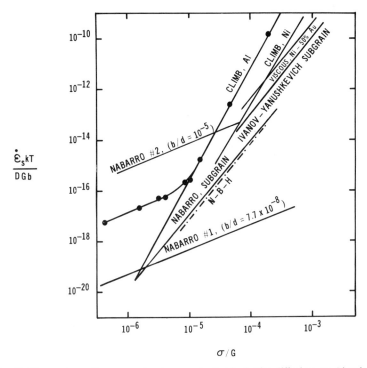

Fig. 13. Comparison of creep mechanisms controlled by lattice diffusion. ●, Al polycrystal data (Harper, Shepard, and Dorn, 1958).

points, the various curves are representative of types of mechanisms and are discussed in more specific terms of alloys solely for purposes of identification.

For metals and alloys in which dislocations glide rapidly, the mechanism is climb-controlled as typified by the Ni curve. When solid-solution alloying results in a viscous glide, the viscous creep mechanism controls as typified by the Ni–50% Au curve. Consequently, the creep rate for viscous glide is invariably below that for the climb-controlled mechanism when the differences in the two operative diffusivities are taken into consideration. As previously described, however, climb always occurs in alloys where the creep rate is viscous glide-controlled. The value of $\dot{\varepsilon}_s kT/DGb$ for the climb mechanism, however, falls below that for the viscous glide mechanism for $\sigma/G \leq 10^{-4}$. Since the climb-controlled and the viscous glide-controlled mechanisms are mutually exclusive, only the slower one can control the creep rate. Consequently it is anticipated that alloys which exhibit viscous glide mechanisms at high values of σ/G will revert to climb-controlled mechanisms when σ/G is decreased somewhat below about 10^{-4}. A convincing proof of such transition is given by the experimental data of Linga Murty et al. (1972). They showed that in Al–3% Mg alloy at stresses above $\sigma/G = 1.2 \times 10^{-4}$, high-temperature creep was controlled by viscous glide. The rate-controlling mechanism switched to that for dislocation climb below that stress level. At yet lower stress an as yet unidentified mechanism was operative, that was similar to Harper–Dorn (1957) creep for Al. This will be discussed later.

Typical curves illustrating the correlations with Nabarro–Herring creep for two different grain sizes are also shown in Fig. 13. Since it provides an independent mechanism, the total creep rates that are operative during the climb, or viscous glide, must include the contribution from Nabarro–Herring creep. As seen from the curves of Fig. 13, this contribution is small for high values of σ/G. At very low values of σ/G, however, depending on grain size, the contributions of creep controlled by the climb or viscous glide mechanisms are small and Nabarro–Herring creep predominates. As shown by the broken curve, the transition from predominantly climb to predominantly Nabarro–Herring creep occurs over less than one order of magnitude change in σ/G.

Recently Weertman (1972) has reviewed the various high-temperature creeps produced by dislocation motion. He concludes that a third power stress dependence of the creep rate is the most "natural" law for steady state creep based on dislocation climb. He specifically considered the more recent developments in the theory of subgrain creep. As mentioned in Section V, the role of subgrains has not received the attention it deserves in the development of creep theories and as such Weertman's recent contribution fills a much neglected gap.

Friedel (1964) and Lifshitz (1963) originally suggested that subgrains as well as grains might undertake Nabarro–Herring creep. The requisites for the operation of such subgrain creep are (a) the dislocations making up the boundary must be good sources or sinks for point defects and (b) the dislocation spacing must be at least one order of magnitude smaller than the subgrain size. The conditions under which this might be possible were studied in detail by Weertman (1968, 1972). The Nabarro subgrain creep rate is given by

$$\dot{\varepsilon}_s kT/DGb = 4(b/\delta)^2(\sigma/G) \tag{18}$$

If one assumes that the subgrain size, δ, follows the stress dependence depicted by Eq. (15), then the stress dependence obeys a third power law. This relationship is shown by the solid curve marked Nabarro, subgrain in Fig. 13. As shown it cannot effectively increase the creep rates obtained for the predominantly climb-controlled mechanism at high values of σ/G. The subgrains will be so large at lower values of σ/G that it appears to be unimportant in that case as well.

Nabarro (1967) has proposed another theory for high-temperature creep based on the climb of dislocations from Bardeen–Herring sources. Unlike the Weertman dislocation-climb model, it does not require glide of dislocations to provide the strain; in N–B–H creep the strain arises from mass transport of atoms to and from climbing Bardeen–Herring sources by a vacancy diffusion mechanism. More recently Weertman (1968, 1972) discussed the theoretical basis for this mechanism. The spacing between the dislocations making up a subgrain boundary should be comparable to the subgrain size. The creep rate for this mechanism is given by

$$\dot{\varepsilon}_s kT/DGb = [1/12 \ln (8G/\pi\sigma)](\sigma/G)^3 \tag{19}$$

The expected trend for this mechanism, as shown by the dotted and dashed line labeled N–B–H in Fig. 13, falls below the line "Viscous" for viscous creep of Ni–50% Au alloy. A recent reformulation of N–B–H creep mechanism by Weertman (1972) gives the creep rate for this mechanism as

$$\dot{\varepsilon}_s kT/DGb = (0.1\pi\beta^2\Omega/b^3)(\sigma/G)^3 \tag{20}$$

where $\beta = 2\alpha$ of Eq. (16) and is approximately equal to 1, and the atomic volume $\Omega \simeq 0.7b^3$ for cubic and hcp metals. This leads to the following steady state creep rate:

$$\dot{\varepsilon}_s kT/DGb \simeq 0.2(\sigma/G)^3 \tag{21}$$

and gives a creep rate for the N–B–H mechanism one order of magnitude higher than that given by Eq. (19). It seems to the author that the special requirements for the operation of the N–B–H, i.e. the regularity and size of the Frank network, are not easily satisfied so that glide plus climb-controlled creep or viscous glide creep should prevail, since glide of dislocations disqualifies the N–B–H model in most cases of interest.

Recently several authors (Weertman, 1972; Ivanov and Yanushkevich, 1964; Blum, 1971) have discussed somewhat different dislocation climb models of creep of subgrains. In these models the subgrain boundary acts as the site where dislocations annihilate each other. In the Ivanov and Yanush-kevich (1964) and Weertman (1972) models, dislocations of opposite character approach the subboundary in opposite directions on parallel slip planes. The dislocations may have been created at sources within the subgrains or at other subgrain boundaries. Dislocation annihilation occurs when dislocations of opposite character meet by climbing along the subboundary. The creep rate for the Ivanov–Yanushkevich mechanism is given by

$$\dot{\varepsilon}_s kT/DGb \simeq 0.15(\sigma/G)^3 \qquad (22)$$

The relationship is shown in Fig. 13. The remarkable result is that subgrain size actually does not appear in the equation. A third power creep equation is retained. Again it appears that it cannot effectively increase the creep rates obtained for the predominantly climb-controlled (Weertman, 1957a) mechanism at high values of σ/G. At lower values of σ/G, it could conceivably be the controlling mechanism.

Blum (1971) has modified the Ivanov–Yanushkevich analysis in an effort to obtain a stress dependence of the creep rate greater than three. He made the assumption that the subgrain walls have a finite width a. The creep rate according to Blum's analysis is given by

$$\dot{\varepsilon}_s kT/DGb \simeq 0.15(a/b)(\sigma/G)^4 \qquad (23)$$

Blum obtained this fourth power creep law by assuming that the width of the subgrain boundary does not depend on stress. If, however, a reasonable stress dependence for the subboundary width a is assumed, which is more likely, then Eq. (23) becomes almost indistinguishable from the Ivanov–Yanushkevich model, given by Eq. (22), yielding a third power creep law.

Weertman (1972) in his recent review has considered the creep rate that may occur for subgrain creep when groups of piled-up dislocations (having opposite signs in neighboring groups) enter a subgrain boundary of zero width. His analysis shows that such a model predicts much higher creep rates than are experimentally observed. This suggests that subgrains may not be very effective barriers to dislocations. There does not appear to be enough definitive experimental data to properly check the validity of these various subgrain creep models. The need for critical experimental work in this aspect should be emphasized.

Although there is considerable literature now on volume-diffusion controlled high-temperature creep, there are some experimental data that cannot be correlated satisfactorily with presently known mechanisms. Data on the creep of polycrystalline Al by Harper *et al.* (1958) are illustrative of this point. Their data are shown by the datum points on the curve for Al in Fig. 13. Above $\sigma/G \simeq 10^{-5}$, the data agree well with that for climb-controlled creep where $n = 4.4$. Below $\sigma/G \simeq 10^{-5}$, however, the creep data give the stress exponent $n = 1$ suggestive of Nabarro creep. The Nabarro creep curve for the grain size that was employed, namely $b/d = 7.7 \times 10^{-8}$, is shown in Fig. 13 as "Nabarro No. 1." Since it falls about three orders of magnitude below that obtained experimentally, the observed creep could not have occurred by the Nabarro mechanism. This was also confirmed by the fact that the strains were about the same within the grains as across the grain boundaries. Furthermore, a single crystal test gave almost the same creep rate as did the polycrystalline aggregate. Obviously, some as yet unidentified volume-diffusion mechanism is operative. This mechanism is not restricted to the case of Al. Recently, Mohammed *et al.* (1973) have shown the operation of Harper–Dorn creep in pure Pb and Sn over a range of normalized stress σ/G, roughly identical to the range over which this type of creep is found in Al. Some progress has recently been made by Barrett *et al.* (1972) and by Linga Murty *et al.* (1972) in understanding the Harper–Dorn creep based on climb-controlled generation of dislocation from a fixed density of sources, and on a modified version of jogged screw dislocations, respectively.

Coble (1963) has shown that stress-directed diffusion of vacancies can take place by grain boundary diffusion as well as by Nabarro volume diffusion. The creep rate by this mechanism increases with the reciprocal of the cube of the grain diameter according to

$$\dot{\varepsilon}_s kT/D_b Gb \simeq 48(b/d)^3(\sigma/G) \tag{24}$$

where the thickness of the grain boundary is taken as about $2b$ and D_b is the grain boundary diffusivity. This grain boundary diffusion mechanism should predominate over the volume diffusion Nabarro–Herring mechanism when

$$48(b/d)D_b/D > 7 \tag{25}$$

If vacancy generation and absorption is highly efficient the total creep rate is given by

$$\dot{\varepsilon}_s kT/DGb = [7(b/d)^2 + 48(D_b/D)(b/d)^3]\sigma/G \tag{26}$$

which is the sum of that arising from Nabarro–Herring and Coble creep.

The recent investigations of Misro (1971), Misro and Mukherjee (1972), and Vaidya et al. (1973) have confirmed the operation of Coble creep in a Zn–22% Al eutectoid alloy at stresses below 10^{-5} G. The reanalysis of the Sn–5% Bi creep data of Alden (1967) by Mukherjee (1972) suggests that the deformation of a comparatively coarse-grained (1.6×10^{-4} in.) specimen is probably controlled by the combined Coble–Nabarro–Herring creep at stress levels below 10^{-3} G.

In an earlier section we have discussed superplasticity as one of the rate-controlling mechanisms that is important in fine-grained metals and alloys. The correlation between superplasticity and Coble creep (as an illustration) might be based on the assumption of the nominal validity of the model for superplasticity proposed in the earlier section. The creep rate due to stress-directed diffusion of vacancies along the grain boundary, i.e. Coble creep, is given by Eq. (24). The predicted curves for correlation are shown in Fig. 14. To permit ready comparison, Eqs. (9) or (12), referring to the creep curves for the dislocation climb mechanism, can be rewritten as

$$\varepsilon_s kT/D_b Gb = \text{const} \ (\sigma/G)^n (D/D_b) \tag{27}$$

Assuming (Friedel, 1964) that the activation energy for grain boundary diffusion, H_b, is equal to one-half of that for lattice diffusion, H_d, $D/D_b = e^{-H_b/RT}$. The three curves for climb of Al ($n = 4.4$), as shown in Fig. 14, correspond to three temperatures: $900°$ K, $D/D_b = 8.2 \times 10^{-5}$, $H_b/RT = 9.4$; $700°$ K, $D/D_b = 5.5 \times 10^{-6}$, $H_b/RT = 2.1$; and $500°$ K, $D/D_b = 4.13 \times 10^{-8}$, $H_b/RT = 17$. The creep due to superplasticity is additive to that due to climb. Hence, about one-half order of magnitude below the point of intersection of the curves for climb and superplasticity, climb-controlled creep no longer significantly affects the total creep rate. Similarly, at one-half order of magnitude of stress level above the point of intersection, creep due to superplasticity is negligible. The resultant creep rate between these limits is shown by the fine dashed lines in Fig. 14 which range from creep due to climb for a particular temperature (or H_b/RT) into that for superplasticity for a particular value of b/d. It is seen that for a given grain size the transition from superplasticity to creep by climb occurs at increasing values of σ/G as the temperature for creep by climb decreases. The figure also explains in a straightforward manner the experimentally observed (Alden, 1967) fact that in superplasticity (a) the flow stress at constant strain rate decreases with decreasing grain size, and (b) at constant stress the strain rate increases with decreasing grain size. For superplastic experiments conducted at a constant temperature but with two different grain sizes, the transition from superplastic to climb-controlled creep will occur at a higher stress level for the finer grain size specimen. This is in accord with experimental observation (Alden, 1967).

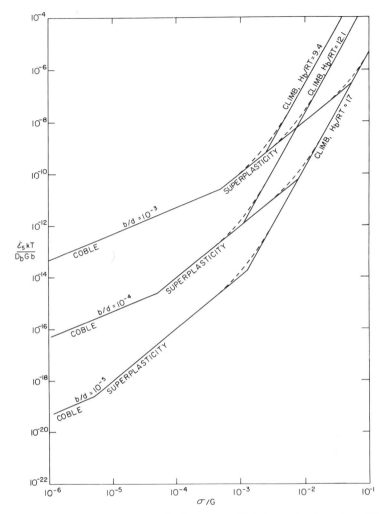

Fig. 14. Comparison of creep mechanisms controlled by grain boundary diffusion (climb-controlled creep is also shown for three different temperatures).

A firm experimental confirmation (Misro and Mukherjee, 1972) of the type of correlation depicted by Fig. 14 is shown in Fig. 15. The experimental results are shown by the datum points and the solid curves in regions A, B, and C represent the Coble creep (Eq. 24), superplasticity (Eq. 17), and climb-controlled creep (Eq. 27), respectively. The microstructural observation was also in accord with the operation of these individual mechanisms. There was no evidence of normal primary creep in regions A and B whereas

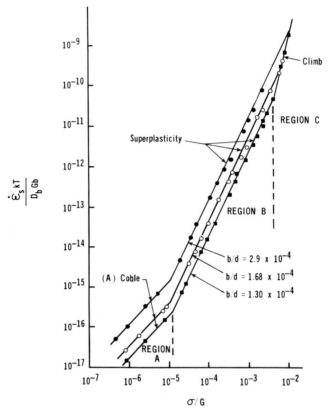

Fig. 15. Correlation between experimental data and operative creep mechanisms in Al–Zn eutectoid alloy ($T = 423$ K). ●, Very fine-grained ($d = 0.97\ \mu$); ○, normal coarse-grained ($d = 1.66\ \mu$); ■, coarse-grained ($d = 2.16\ \mu$); ——, theoretical predictions.

region C was characterized by the usual decelerating primary. Dislocation tangles and cell structure were not observed in regions A and B but were clearly present in steady state deformation in region C. Careful substructural investigation using transmission electron microscopy demonstrated that in region B, the only evidence of dislocation activity was in the form of pileups emanating from one side of the grain boundary and extending across the grain to the boundary on the other side. As expected, there was no effect of grain size in the climb-controlled region C, where results from all the three different grain sizes fell on the same curve.

Because of the difficulties in obtaining correct values of grain boundary diffusion energy and the approximation involved in estimating this value from the energy of lattice diffusion, the above correlation is only expected to reveal the trends. However, the agreement that is shown in Figs. 14 and 15 is encouraging. Most experimental observations on superplasticity agree with the suggested trend of decreasing values of n from that appropriate for dislocation climb ($n \simeq 5$ to 7) to that of $n \simeq 2$ which applies in the super-plastic range. It appears very likely that at least some of the controversy in the literature has occurred due to the attempts that are often made to interpret microstructural and mechanical data, collected from a stress–temperature–grain size–strain rate regime that falls outside the operation of superplasticity. Some of this disagreement includes published data on varia-tion of strain rate sensitivity m with temperature (Nicholson, 1972) and with strain rate (Johnson, 1970) and variation of activation energy (Naziri et al., 1973) with stress and temperature. It is evident that if the strain rate sensiti-vity parameter m (which equals $1/n$) is measured over the entire range of strain rates investigated (Fig. 15), its value would have changed from $m = 0.28$ in region C to $m = 0.5$ in region B and finally to $m = 1$ in region A, and would not appear to be a constant. But as shown in Fig. 15, it is indeed a constant with a value of $m = 1/n = 1/2 = 0.5$ in region B, where superplasti-city is the rate-controlling mechanism. Similarly if the activation energy is measured in region A or B and is compared to that obtain in region C, they will differ approximately by a factor of 2 as noted by Naziri et al. (1973). It should be mentioned that various workers in the past have shown a sigmoi-dal relationship in log–log plot of the stress versus the strain rate, suggesting that at low stresses m decreases again from a value of $m = 0.5$ to a value less than that and often equals 0.2. The present experimental work as well as other and more recent investigations (Misro, 1971; Vaidya et al., 1973), where a good deal of care was exercised in the measurement of stresses (particularly at low stress level) employing double-shear specimens, do not reveal this sigmoidal relationship. Furthermore, this supposed decrease in m at low stress level has been explained in the literature due to the presence of a back stress (Backhofen et al., 1967). It is difficult to see the origin of such back stress, in the microstructure of superplastic material, that supposedly makes itself significant only in the low strain rate region. Another suggestion (Ashby and Verrall, 1973; Hazzledine and Newbury, 1973), e.g. a threshold stress for grain boundary sliding at low strain rates, which is probably controlled by interface reaction, has a sounder justification and deserves serious consideration.

In conclusion, the above discussion is intended to emphasize the fact that although finer grain size, higher temperature, and lower strain rates all favor superplasticity, the actual manifestation of the typically large superplastic deformation will depend on the *combined* nature of the grain size, the temperature, and the stress (or equivalently the strain rate).

VIII. A Universal Law for Transient Creep

A. Empirical Relations

In this section a universal law for high-temperature diffusion-controlled transient creep will be introduced. The presentation will make use of the materials already discussed in Sections III, IV, and V. Most of the early investigations (Dorn, 1954) on high-temperature creep showed that

$$\varepsilon = \varepsilon[t \exp{(-Q_c/RT)}] \text{ for } \sigma = \text{const} \qquad (28)$$

over both the transient and steady state stages. Here, ε is the total strain, which is the sum of the initial strain ε_0 upon stressing at $t = 0$ and the creep strain up to time t. The apparent activation energy, Q_c, for creep was shown to be insensitive to stress, strain, grain size, and temperature. The activation energy agreed very well with that for diffusion. In subsequent years, extensive experimental evidence confirmed the validity of Eq. (28), but not much progress was achieved to advance further the understanding of the transient stage of creep. It is only over the last few years that several important observations have been made that have encouraged a more thorough analysis of transient creep.

For many materials the creep strain is found to follow Andrade's (1910) empirical equation, namely

$$\varepsilon - \varepsilon_0 = \dot{\varepsilon}_s t + Bt^{1/3}, \qquad (29)$$

where $\dot{\varepsilon}_s$ is the steady state creep rate and B is a constant. The formulation often appears to be in fair agreement with high-temperature creep data. One objection to the universal application of Andrade's equation concerns its prediction of infinite initial creep rates. Although it is admittedly difficult to measure the initial creep rates accurately, many carefully conducted experiments suggest that physically they must be finite.

An alternative formulation that has been employed is

$$\varepsilon - \varepsilon_0 = \dot{\varepsilon}_s t + \varepsilon_T[1 - \exp{(-rt)}], \qquad (30)$$

where ε_T is the limiting transient strain and r is the ratio of the transient creep rate to the transient creep strain. This relation was first suggested on a purely empirical basis by McVetty (1934). It has been applied subsequently by Garofalo (1965), Conway and Mullikin (1966), and Evans and Wilshire (1968). This empirical formulation was given a theoretical justification by Amin et al. (1970). Using a unimolecular rate kinetics, Amin et al. (1970)

presented a unified analysis for high-temperature creep that incorporates the effect of both temperature and stress on the shape of transient creep curves. Their approach will be briefly reviewed here.

B. Substructural Changes during Transient Creep

The review by Bird *et al.* (1969) of high-temperature creep reveals that only those metals and alloys that undergo initial straining upon stressing exhibit the usual normal transient stage of creep during which the creep rate progressively decreases to that for the steady state. Those alloys which do not undertake significant initial straining at $t = 0$ upon application of stress do not exhibit the usual normal transient stage. Instead, they display either a brief normal transient or a brief inverted transient, or no transient behavior at all.

Therefore, normal transient creep rates appear to result from changes in the substructure from that produced immediately upon stressing to that pertaining to the steady state. The substructures which are formed upon initial straining at creep temperatures closely resemble those which are developed during strain hardening over stage III deformation at lower temperatures (Gupta and Strutt, 1967; Livingston, 1962; Hammad and Nix, 1966). Most of the dislocations are arranged in rough cellular patterns, the walls of which are composed of dislocation entanglements. Some cells contain a few randomly meandering dislocations in their interior.

Etch-pitting studies by Gupta and Strutt (1967) and Clauer *et al.* (1970) and electron microscopy studies by Hammad and Nix (1966), Clauer *et al.* (1970), and Garofalo *et al.* (1963) have shown that dislocations rearrange themselves during the early part of transient creep as a result of the extra degree of freedom resulting from dislocation climb which becomes facile at high temperatures. Consequently, the entanglements disperse and adjacent cells coalesce to produce more sharply delineated subgrains which have a volume of about eight times that of the original cells. Over the secondary stage of creep an invariant substructure persists which depends only on the magnitude of the applied stress. It consists of subgrains separated from one another by low-angle boundaries within which there exist a few isolated and randomly meandering dislocations, as discussed in Section V.

The substructural changes that take place during normal transient creep are extremely complex. This is clearly revealed by the careful investigations of Clauer *et al.* (1970). Changes in the different substructural details over the transient stage are obviously interrelated. The dispersal of entanglements, the buildup of subboundaries, the changes in density of dislocations, and the alterations in misorientations across subboundaries, etc. are not mutually independent. Among these various substructural details, only the changes in dislocation density have as yet been placed on a quantitative basis.

C. Analysis

The substructural changes taking place during transient creep are so complex as to preclude any detailed mechanistic analysis of the process at this time. Although the dislocation density changes over the transient stage of creep, it alone cannot account for the much greater percentage decrease in the creep rate. The major substructural change, as pointed out by Gupta and Strutt (1967), centers about the dispersal of entanglements but the relation between such dispersal and the transient creep rate is not yet understood. Undoubtedly all substructural changes during high-temperature transient creep are interrelated and dependent upon the dislocation climb mechanism. Amin *et al.* therefore suggested that transient as well as steady state creep is controlled by the rate of climb of dislocations. This is consistent with the validity of Eq. (28). Furthermore they also assumed that the reactions taking place obey the laws of unimolecular kinetics. Their assumption will be verified *a posteriori*.

Assuming that dislocation climb kinetics control transient creep, the rate constant can be given in terms of the temperature and stress as $K\dot{\varepsilon}_s$, where $\dot{\varepsilon}_s$ is the steady state creep rate as controlled by climb and K is a constant. As shown by Webster *et al.* (1969), the unimolecular can be formulated as

$$(d/dt)(\dot{\varepsilon} - \dot{\varepsilon}_s) = -K\dot{\varepsilon}_s(\dot{\varepsilon} - \dot{\varepsilon}_s) \tag{31}$$

Integrating Eq. (31) twice one obtains

$$\varepsilon = \varepsilon_0 + \dot{\varepsilon}_s t + [(\dot{\varepsilon}_i - \dot{\varepsilon}_s)/(K\dot{\varepsilon}_s)][1 - \exp(-K\dot{\varepsilon}_s t)] \tag{32}$$

where $\dot{\varepsilon}_i$ is the initial creep rate at $t = 0$, ε_0 is the initial strain upon stressing, and

$$(\dot{\varepsilon}_i - \dot{\varepsilon}_s)/K\dot{\varepsilon}_s = \varepsilon_T \tag{33}$$

is the transient strain. Since the transient and steady state stages have the same energy of activation and also the same kinetics of reaction as implied by Eq. (28), it follows necessarily that $\dot{\varepsilon}_1 = \beta\dot{\varepsilon}_s$, where $\beta = \text{const} > 1$. Therefore, the total transient strain

$$\varepsilon_T = (\dot{\varepsilon}_i - \dot{\varepsilon}_s)/K\dot{\varepsilon}_s = (\beta - 1)/K \tag{34}$$

where β, K are constants independent of stress and temperature. A significant implication of Eq. (34) is that although $\dot{\varepsilon}_i$ and ε_s are both functions of stress, ε_T is a constant and like β and K is independent of stress and temperature.

According to the above analysis, the following must apply: (1) the initial strain upon stressing, ε_0, is athermal and depends on the original state of the metal or alloy and the value of σ/G where σ is the stress and G is the shear modulus; (2) the ratio of the initial to secondary creep rates, β, is a constant independent of stress, temperature, and initial strain; (3) the rate of dispersal of the entanglements depends on the same function of stress and temperature as does the secondary creep rate; (4) for a given metal or alloy the transient strain, ε_T, is constant independent of stress or temperature; and (5) for a given metal or alloy there exists a universal high-temperature transient and steady state creep curve of the form

$$\varepsilon - \varepsilon_0 = f(\dot{\varepsilon}_s t) = \phi[(\sigma/G)^n (DGb/kT)t] \tag{35}$$

where the function f is derived from the general equation

$$\dot{\varepsilon}_s kT/DGb = \text{const} \times (\sigma/G)^n$$

shown to be valid from a general survey by Mukherjee et al. (1969) on secondary creep, and discussed in Section IV. Here, D is the diffusivity, b the Burgers vector, n the stress dependence of secondary creep rate, k the Boltzmann's constant, and T is absolute temperature. The relation depicted in Eq. (35) is an extension of Eq. (28), which now incorporates the effects of stress as well as temperature on the shape of the transient creep curve.

D. Correlations with Experimental Data

Amin et al. (1970) made a careful survey of all the experimental results on high-temperature transient creep where the pertinent data on diffusivity and shear modulus were also available.

The validity of Eq. (32) is shown by the creep curve depicted in Fig. 16 for

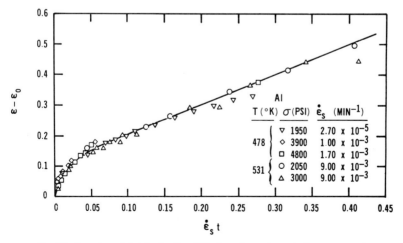

Fig. 16. Experimental data for Eq. (32), time-dependent strain, $\varepsilon - \varepsilon_0$ vs $\dot{\varepsilon}_s t$.

experimental data (Sherby *et al.*, 1956) for two different temperatures and five different stress levels. The curve proves that the total transient strain ε_T is indeed a constant for the material and is independent of stress and temperature. Other metals and alloys investigated by Amin *et al.* showed much the same trend, i.e. that, regardless of temperature and stress, the data for each material fall well onto a single curve which agrees quite well with the solid curve depicting Eq. (32). The temperature and stress dependence of the secondary creep rates agreed quite well with the equation for steady state creep rate controlled by the dislocation climb mechanism [Eq. (12)].

A typical example of the universal creep curve for the case of polycrystalline Ni is shown by the datum points of Fig. 17. Regardless of stress and

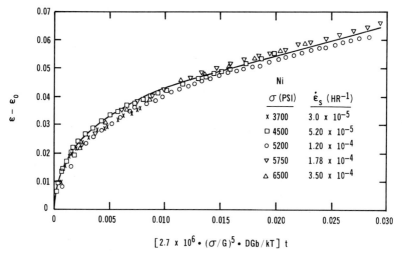

Fig. 17. The universal creep curve for nickel ($T = 973$ K).

temperature all data fall well onto a single curve. The solid line refers to the theoretical expression given by Eq. (35). An even more convincing proof of the validity of the universal creep curve is shown in Fig. 18 for Nb where the experimental data cover ranges of temperature as well as stress and the secondary creep rates vary by two orders of magnitude. The remaining cases that were examined by Amin *et al.* were about equally consistent and illustrated the good agreement between the theory and the experimental results. They also showed that the initial creep rates, $\dot{\varepsilon}_i$, for a series of stresses and temperatures are greater than the secondary creep rates, $\dot{\varepsilon}_s$, by a constant factor β. Both $\dot{\varepsilon}_i$ and $\dot{\varepsilon}_s$ are dependent on the substructural details that develop and hence are functions of stress. Their difference, however, is a constant regardless of stress and temperature and is not a function of state.

They calculated values for the constants K, β, and ε_T from their analysis of available experimental data. Whereas the transient strain ε_T exhibits rather pronounced variations from one case to another, the rate constant K seems to be somewhat less variable. Undoubtedly both depend on significant substructural details. Up to the present, however, no consistent trends in variations of K or β have been uncovered relative to effect of grain size, stacking fault energies, or any other pertinent structural details.

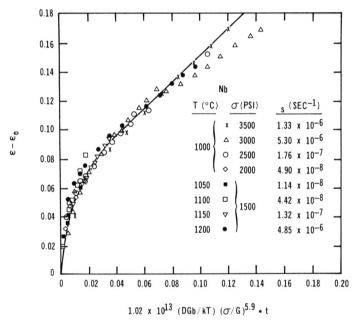

Fig. 18. The universal creep curve for niobium.

Evans and Wilshire (1968) noted that Eq. (32) becomes inaccurate for creep of polycrystalline iron at low stresses. Similar discrepancies, which are observed in the creep of austenitic stainless steel, have been shown by Webster *et al.* (1969) to arise from the increased contribution of grain boundary sliding to the total creep rate at the low applied stresses. When grain boundary sliding prevails, $K\dot{\varepsilon}_s$ is no longer the pertinent rate constant, since it refers to a creep mechanism resulting from climb alone, and under that circumstance ε_T is not expected to be a constant independent of stress. Once grain boundary sliding is better understood quantitatively, this factor might also be taken into consideration so as to account completely for high-temperature transient creep at low stresses in polycrystals as well.

IX. Reflections on Some Specific Aspects of Creep

A. Role of Jogged Screw Models

The presentation of mechanical and microstructural data, outlined in Sections III, IV, and V, emphasized the role played by climb-controlled dislocation models in the analysis of creep data for pure metal and alloys when $n \simeq 5$. Suggestions have been made from time to time that other dislocation–creep models may equally well explain such data on creep. Among them, the dislocation–jog models have the largest number of proponents. Jogs on a screw dislocation are always edge type and if the screw dislocation is forced to move, it is easily shown that such jogs cannot maintain their positions in the dislocation by glide in their own slip planes. That is possible only by climb of jogs. Jogs on edge dislocation, however, are not limited to move by climb motion. This is because the orientation of the slip planes of the jogs is such that they can move along with the rest of the dislocation line by slip. Hence the jog models are necessarily restricted to motion of the jogged screw dislocations only. Many authors have analyzed thermally-activated motion of jogged screw dislocations as a possible rate-controlling mechanism in high-temperature plastic deformation. These theories have been reviewed by Mukherjee *et al.* (1969) and by Lagneborg (1972a). Some of the earlier models of jogged screw dislocations (Hirsch and Warrington, 1961; Barrett and Nix, 1965) need extensive improvements. The problem becomes extremely difficult (Mukherjee *et al.*, 1969) because of moving sources, ill-defined boundary conditions, and competition between pipe and volume diffusion, types of jogs, jog–jog interactions, effects of dissociation, the stress dependence of interjog distance, etc. However, in recent years there has been renewed interest in jogged screw models. Weertman (1972) has discussed a modified jogged screw model for transient and steady state creep. Modéer and Lagneborg (1971), in their electron microscopic observations in stainless steel, have reemphasized that climb-controlled motion of jogs often controls the continued motion of dislocations in the microstructure.

B. Enhancement of Creep Rates at High Stress Levels

The mechanical data presented in Section III suggest that the power law depiction of the steady state creep rate as a function of stress is valid over the entire stress range. However, as pointed out by Sherby and Young (1972), Sherby and Burke (1967), and by Weertman (1972), at stress levels higher than about $5 \times 10^{-4}G$ to $10^{-3}G$, power law creep breaks down and the creep rates are higher than that predicted from an extrapolation from lower stress (power law) regions. In Weertman's (1968) climb-controlled creep model, such a breakdown is expected to occur when the expression $\sigma\Omega/KT$ in the equation for climb of edge dislocation is of the order of or greater than 1. (Here Ω is the atomic volume.) This leads to a value of stress at which the breakdown should occur of $\sigma \simeq 3 \times 10^{-2}G$, which is about an order of magnitude higher than observed experimentally. Weertman (1972) has suggested that the experimental observation could be reconciled to the prediction of the dislocation creep model if a dislocation pileup or some sort of stress concentration factor is involved to raise the creep rate to the required value above that predicted by the power law relation.

A somewhat different explanation has been put forward by Sherby and Young (1972). According to their interpretation the power law breakdown may be associated with two contributing factors: (a) increased contribution from dislocation pipe diffusion and (b) excess vacancy generation. At higher stress levels, the increased creep rate is also associated (Robinson and Sherby, 1969) with an activation energy of creep that is about three-fourths of that for lattice self-diffusion. Robinson et al. have suggested that under these conditions diffusion via dislocation core makes a significant contribution to the total diffusivity and an effective diffusion coefficient can be defined as:

$$D_{\text{eff}} = D_v f_v + D_c f_c \tag{36}$$

Here D_v and D_c are the diffusion coefficient through volume and through dislocation core, respectively, and f_v and f_c are the fractions of atom sites associated with each type of diffusion. The value of f_v is essentially unity. The value of f_c is determined by the dislocation density, ρ, since $f_c = a_c \rho$, where a_c is the cross-sectional area of the dislocation core in which fast diffusion is taking place. The dislocation density, ρ, can be estimated from Eq. (16). The parameter a_c is equal to the ratio of the number of atoms contributing to dislocation short circuit diffusion at a given dislocation site (assumed equal to 10, see Shewmon, 1963) to the number of atoms per square centimeter. A plot of $\dot{\varepsilon}/D_{\text{eff}}$ vs σ/E for aluminum is shown in Fig. 19, due to Sherby and Young (1972). The correlation obtained is certainly encouraging and would suggest that the phenomenological relation given by Eq. (36) has physical justification although no creep theory has yet been developed which incorporates an effective diffusion coefficient.

It is interesting to note, as first pointed out by Sherby and Young (1972), that the deviation from the power law relation begins at $\dot{\varepsilon}/D_{\text{eff}} = 10^9 \text{ cm}^{-2}$. Annealing studies of the electrical resistivity by Armstrong et al. (1973) suggest that significant excess concentrations of vacancies are detected after quenching from warm working only when $\dot{\varepsilon}/D_v$ is larger than 10^9 cm^{-2}. If

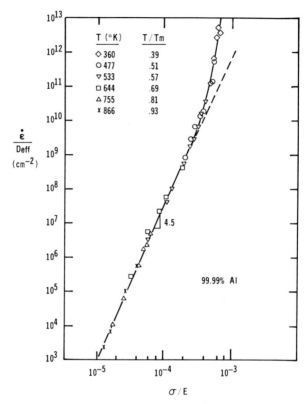

Fig. 19. Enhancement of the steady state creep rate at higher stress level and correlation using effective diffusivity (after Sherby and Young, 1972).

excess vacancy generation dominates dislocation pipe diffusion, one might expect the activation energy of creep to equal that for vacancy migration. The relative role played by the excess vacancies has to take into account the lifetime of excess vacancies, which would depend on the nature of the vacancy source, vacancy sinks, and the details of the diffusive path. As suggested by Sherby and Young (1972), further creep and diffusion studies in the range $\dot{\varepsilon}/D_v > 10^9$ should help to resolve the relative importance of the role played by excess vacancies and by dislocation pipe diffusion.

C. Recovery–Creep Models and Internal Stress

As early as 1926, Bailey (1926) proposed that creep resulted from recovery of the strain-hardened state. Orowan (1946–1947), twenty years later, applied this concept in his discussions on the origin and nature of creep.

Although the recovery concept of creep seems to have been widely held over the intervening years, no serious experimental confirmation of this assumption was attempted until stimulated by the investigations of Mitra and McLean (1966). Interest in this phenomenological concept of creep stems from the fact that it is not subject to the limitations of any specific creep or recovery model, and so might be widely applicable.

Orowan expressed the significant relationship in terms of the flow stress, σ, as

$$d\sigma = (\partial\sigma/\partial t)_\varepsilon \, dt + (\partial\sigma/\partial\varepsilon)_t \, d\varepsilon \qquad (37)$$

which assumes the existence of a mechanical equation of state. For creep under a constant stress, therefore,

$$\dot\varepsilon = (\partial\varepsilon/\partial t)_\sigma = -(\partial\sigma/\partial t)_\varepsilon/(\partial\sigma/\partial\varepsilon)_t = r/h \qquad (38)$$

where r is the rate of recovery and h refers to the rate of strain hardening. If independent experimental determinations of $\dot\varepsilon$, r, and h verify Eq. (38), the validity of the recovery model will have been established.

In the technique employed by Mitra and McLean (1966), they decreased the stress by $\Delta\sigma$ after steady state creep was established. After a time Δt, the specimen seemed to give a steady state creep rate for the reduced stress, from which $\Delta\sigma/\Delta t$ was determined for the limit of $\Delta\sigma \to 0$. This was done for a series of stresses, from which $(\partial\sigma/\partial t)_\varepsilon$ was evaluated. The rate of strain hardening was established by room temperature tensile testing of the structure that had been produced by steady state creep. This was converted to the slope $d\sigma/d\varepsilon$ at the test temperature by introducing the effect of temperature on the modulus of elasticity. Whereas h decreased somewhat as the stress for steady-state creep was increased, r increased rapidly with the creep stress. The relationship given by Eq. (38) for $\dot\varepsilon$ was crudely obeyed. The recovery process following a reduction in stress has previously been investigated by Bayce et al. (1960). Their work suggested that the recovery curve following such stress reduction is distinctly more complicated than that which is commonly assumed.

Although the earlier creep models did not describe the details of the deformation process, the more recent formulations, specifically those given by Mitra and McLean (1966), Lagneborg (1969, 1972b), and Nix and co-workers (Gasca-Neri et al., 1970; Barrett et al., 1970; Gasca-Neri and Nix, 1974), treat the strengthening and recovery mechanisms in some detail. In the models of Mitra and McLean and also Lagneborg, the creep process consists of consecutive events of recovery and strain hardening of a three-dimensional dislocation network. The thermally-activated breaking of the attractive junctions release dislocations which then move, providing strain increments and also an increase in the internal stress, i.e. strain hardening, since the expanding dislocation increases its length and therefore the dislocation density. Simultaneously, recovery of the dislocation network by a gradual growth of the larger meshes takes place. Eventually a situation is reached where the rate of recovery, r, exactly matches the rate of strain hardening, h. In this steady state situation, the creep rate and the dislocation density are shown to remain constant. Mitra and McLean (1966) treated the parameters h and r as constants which is an oversimplification. The recovery rate, r, is expected to increase during the primary stage, since the decreasing mesh sizes for the increasing dislocation density gradually provide a larger driving force for recovery. This concept has been incorporated by Lagneborg (1969, 1972b) in the form of a time-dependent variation of dislocation density in the primary stage.

In the later development of recovery–creep models, a distinction is often made between internal stress and effective stress, in the context of the hardening and the recovery process. For example, Ahlquist et al. (1970) suggested a phenomenological theory of steady state creep which explicitly includes the development of an average internal back stress. The theory is based on the premise that dislocation glide and recovery are separate kinetic processes, driven by different components of the applied stress, the effective stress, and internal stress, respectively. It is argued that the stress and temperature dependence for steady state creep is, in general, a complex function of the separate stress and temperature dependences of dislocation glide and recovery. According to these models, as the strain hardening proceeds, the internal stress increases and the effective stress which drives the glide process diminishes accordingly. If the crystal strain hardens to the point at which the internal stress is equal to the applied stress, the effective stress becomes zero and plastic deformation ceases. Hence the internal stress must always remain smaller than the applied stress in order to have continued plastic deformation. Consequently, the internal stress is established by the balance between strain hardening and recovery. The central point is that the internal stress reaches that value which allows the glide process to occur at a rate sufficient to exactly compensate, through strain hardening, the decrement in the internal stress by recovery events.

It may be recalled that the data presented in Section III suggested that in the case of climb-controlled creep, a constant activation energy and a constant stress dependence of the creep rate, n, are obtained for most pure metals and alloys at high temperatures. Lagneborg (1972b) interpreted these results as consequences of applied stress $\simeq$ internal stress, i.e. the effective stress is zero. Similarly, Ahlquist et al. (1970) suggested that in glide models (e.g., viscous creep) the contributions from the back stress are often negligible, in which case the effective stress equals the applied stress. This raises the question as to how to decide whether glide or recovery is the rate-controlling

step in the overall creep process. Lagneborg (1972b) and Ahlquist *et al.*
(1970) seem to suggest that it is a matter of degree, depending upon the
relative magnitudes of the effective stress and internal stress, i.e. by determining where one must attribute most of the driving force—to glide or to the
recovery process.

During the last few years a variety of experimental techniques have been
proposed for measuring the mean internal back stress. These techniques
have been summarized by Lagneborg (1972a) and by Ahlquist and Nix
(1971). They may be divided into two categories: (a) the level of mean
internal stress is measured by incrementally reducing the applied stress
during a stress relaxation until further stress relaxation produces the condition $d\sigma/dt = 0$, and (b) the applied stress is reduced instantaneously and
directly (by trial and error) to that level at which the plastic strain rate is
zero. The second technique developed by Nix and co-workers (Ahlquist and
Nix, 1971; Solomon and Nix, 1970) can be applied to tensile or creep deformation in one of two ways. In the first, called stress transient dip test, the
applied stress is reduced to the level at which $(d\sigma/dt)_{\varepsilon} = 0$. An alternate
approach, called the strain transient dip test, involves reducing the applied
stress to the level at which $(d\varepsilon/dt)_{\sigma} = 0$. The basis of the technique in the
category (b) is that the plastic strain rate should be zero when the applied
stress for a given structural state is reduced in such a way that it is exactly
equal to the internal stress. Hence the effective stress, i.e. the driving force for
glide, is made zero. All of the above techniques are based on the assumption
that the structural parameters which determine the level of internal stress are
constant during the stress reduction. The techniques in the category (a)
make the further assumption that the structural parameters are also constant during stress relaxation. Solomon *et al.* (1970) have suggested that
during high-temperature deformation the latter assumption is incorrect.

Some of the recent investigations tend to cast some doubts (Davies and
Wilshire, 1971) on the general applicability of the stress reduction technique.
From their recent experiments Lloyd and McElroy (1974) have presented
experimental evidence which indicates that the measurement of recovery
creep parameters and internal stress values by strain transient dip tests may
be often of questionable significance. They suggested that the agreement of
measured recovery and work hardening rates with the Bailey–Orowan equation may be fortuitous and that it is a consequence of the anelastic contribution to strain transients. Their experimental recovery rates have been shown
to be a more complex function of anelastic strain and creep rate, and consequently also of applied stress. Similarly, internal stress as measured by the
dip tests is suggested to be a complicated function of anelastic strain and
strain rate and hence applied stress. Lloyd and McElroy (1974) conclude
that internal stress is unlikely to correspond to any real material parameter,

though misleading correlations with creep behavior may arise, as with re-covery and work hardening rates, and the concept may need revision.

Some of the other experimental results tend to support the viewpoint of Lloyd et al. In recent creep experiments on internally oxidized Cu–Al single crystals, Lloyd et al. (1973) showed that creep can occur at stresses below the average internal stress. Their observation supported the view that the mobile dislocation density is the most important parameter in creep and the con-cept of internal stress was not significant. Their measurement of the radii of curved dislocations in crept $Cu-Al_2O_3$ single crystals indicates that these bowing dislocations experience a stress equal to the applied stress and not the effective stress (i.e. applied stress minus the internal stress). It should be added that on theoretical grounds Weertman (1972) has criticized the use of an internal stress in the case of glide-controlled viscous creep. Similarly Bird et al. (1969) have suggested that it is inappropriate to use the concept of internal stress for all climb-controlled mechanisms. It is the author's opinion that the discussion presented above suggests that although the basic tenets of the recovery–creep model are correct, more definitive experimental work is needed to remove some of the uncertainties in the literature discussed earlier. Also, additional careful experimental investigations that are under-way (W. D. Nix, Stanford University, private communication, 1974), which employ precision strain gages mounted directly onto the specimen having microstrain resolution, an extremely stiff machine with no backlash, microsecond time resolution, and an unloading system that allows the load to be decreased within 10^{-2} sec, will shed more light regarding the accuracy and the significance of the internal stress measurements. These in turn will by very effective inputs to a better formulation of the recovery theory of creep.

X. Summary

Much progress has been made over the past twenty years in rationalizing the creep behavior of metals and alloys above about one-half of their melting temperature. Many significant details, however, still require clarification. The following generalizations are now fairly well established.

1. The apparent activation energies for high-temperature creep of metals and dilute solid solutions are independent of the creep strain and insensitive to the applied stress; they are in close agreement with the activation energies for self-diffusion. The agreement becomes even better when the temperature dependence of modulus is taken into account and the true enthalpy of activation of creep is estimated.

2. The changes in the creep rate with time under constant conditions of the independent variables of stress and temperature are reflections of substructural changes that accompany creep; such substructural changes are insensitive to the temperature and depend almost exclusively on the time and the stress divided by the modulus of elasticity.

3. Under steady-state creep the significant substructural details remain constant. More creep resistant substructures are produced by higher stresses. Thus the steady state creep rate depends not only on the stress but also on the steady state substructure produced in the test metal by that stress.

4. Attempts to present a unified picture of high-temperature creep have been made difficult by the effects of other factors such as modulus, grain size, subgrain size, and stacking fault energies. Perhaps the most significant of these is the modulus effect which can be incorporated into the analysis on a quasi-theoretical basis. In this way a more universal comparison of the secondary creep rates of different metals can be realized. Nevertheless, even when this is done, the secondary creep rates still differ from metal to metal by several orders of magnitude, due to the effects of other factors.

5. Although early investigations suggested that grain size effects may be large, recent evidence suggests that in the absence of grain boundary sliding (noticeable especially in very fine-grain sized specimens), the effects are small.

6. In very fine-grained specimens, Nabarro, Coble, and superplastic creep mechanisms often make the dominant contribution to overall creep rate. The relative role played by these mechanisms is determined by the temperature, stress, grain size, and the values for the diffusivity via lattice and via grain boundary.

7. The steady state creep rates from all diffusion-controlled mechanisms can be correlated by the expression $\dot{\varepsilon}_s kT/DGb = A(\sigma/G)^n$. The various diffusion-controlled mechanisms differ from one another only with regard to (a) the diffusivity D that is involved, (b) the value of A and the structural factors that are incorporated in it, and (c) the value of n. This is summarized in Table III. Not included in this table are the mechanisms of Nabarro creep in subgrains, Nabarro–Bardeen–Herring climb creep of Frank–Read sources, and the creep models of Ivanov–Yanushkevich and of Blum due to annihilation of dislocations of opposite character by climb along subgrain boundary walls. These mechanisms have not yet been confirmed experimentally.

8. In the case of diffusion-controlled climb models, a universal law for high-temperature creep has been presented that incorporates the effect of both temperature and stress on the shape of transient creep curves.

Despite the advances that have been made toward an understanding of

TABLE III

CREEP MECHANISMS

Mechanism	Diffusivity D	A	n
Nabarro	Volume by vacancy exchange	$7(b/d)^2$	1
Coble	Grain boundary by vacancy exchange	$50(b/d)^3$	1
Weertman climb (fcc)	Volume by vacancy exchange	2.5×10^6	4.2–5.5, increasing with Gb/Γ
Weertman climb (hcp)	Volume by vacancy exchange	$\simeq 2.5 \times 10^6$, variable	4.5 but variable $4.0 \le n \le 7$
Weertman climb (hcp)	Volume by vacancy exchange	$\simeq 2.5 \times 10^6$	$3.0 \le n \le 5.5$
Weertman viscous glide in solid-solution alloy	Chemical interdiffusivity	3	$3.0 \le n \le 3.5$
Superplastic creep	Grain boundary by vacancy exchange	$\simeq 100(b/d)^2$	2
Harper–Dorn in Al single and polycrystals, also in Pb and Sn	Volume by vacancy exchange	1.35×10^{-11}	1

high-temperature creep, many important issues remain unresolved. For the usual ranges of σ/G they concern: (a) the origin of the variance of n that is observed in pure metals undertaking the climb mechanism, (b) the reason for high values of n in dispersion-strengthened alloys, and (c) the reason why viscous glide mechanisms are rate-controlling in some solid solution alloys but not in others.

In the low σ/G level, there is need for clarification of several major issues. These are the nature and origin of Harper–Dorn creep, the improvement in the details of superplastic creep mechanism, and the existence, if any, of a threshold stress at the very low stress levels. As mentioned in the text, recent investigations have already shown the transition from climb mechanism to superplasticity and also the transition from superplasticity to Coble creep at the theoretically predicted values of σ/G.

Most of the recent development in creep theories places the emphasis on substructural changes attending creep. More work still needs to be done concerning the disposition and densities of dislocations, the subgrain sizes, and the misorientation of subgrains. The recently proposed creep theories based on subgrains are yet to be confirmed by careful experimental data. Data on the substructure developed during creep of alloys undertaking the viscous glide mechanism is rather scarce. Such information should do much to clarify the reasons for differences in transient creep for these two types of behavior. In the high stress level where the power law formulation breaks down, definitive investigations are necessary in order to ascertain the relative roles played by pileups acting as an aid to stress concentration, by increased diffusivity via the dislocation core, and the role of excess vacancies. With reference to the recovery–creep models, more careful experimental work is now in progress, and should shed light on the significance and extent of the role played by internal stress and the effective stress in the context of the pertinent hardening and recovery mechanisms.

In conclusion, high-temperature creep is dependent on a host of inter-related and complex phenomena. It is highly structure-sensitive and the creep rates of materials can often be changed up to 10^6 times or more by introduction of microstructural and substructural modifications. Because of this highly structure-sensitive nature of creep processes, electron microscopy coupled with additional tools of examination will be essential and invaluable aids in unraveling this complex subject. A more thorough investigation is needed on the substructures that are developed, with specific emphasis on their significance to creep mechanism. These observations will be essential input to the formulation of more realistic theories of high-temperature creep that will eventually be developed. Ultimately, such new theories must "predict" the substructure as well as the mechanical behavior.

ACKNOWLEDGMENTS

The author would like to express his gratitude to the late Professor John Emil Dorn, who introduced him to the theory of rate processes in crystal plasticity in general and to creep mechanisms in particular. He would also like to express his thanks to his former students, especially to Mr. James E. Bird, who was coauthor of earlier reviews, and to Dr. S. Misro for their contribution to the experimental aspect of the program. Thanks are also due to Professors J. Weertman and O. D. Sherby for their kind permission to use material from their papers prior to publication, and to Professor W. D. Nix for a stimulating discussion on the issues presented in Section IX. This investigation was supported in part by NSF Grant No. GH-42736.

References

Ahlquist, C. N., and Nix, W. D. (1971). *Acta Met.* **19**, 373.
Ahlquist, C. N., Gasca-Neri, R., and Nix, W. D. (1970). *Acta Met.* **18**, 663.
Alden, T. H. (1967). *Acta Met.* **15**, 469.
Amin, K. E., Mukherjee, A. K., and Dorn, J. E. (1970). *J. Mech. Phys. Solids* **18**, 413–426.
Andrade, E. N. (1910). *Proc. Roy. Soc., Ser. A* **84**, 1.
Armstrong, P. E., Green, Q. V., Sherby, O. D., and Zukas, E. G. (1973). *Acta Met.* **21**, 1319.
Ashby, M. F. and Verrall, R. A. (1973). *Acta Met.* **21**, 149.
Backofen, W. A., Azzarto, F. J., Murty, G. S., and Zehr, S. W. (1967). "Ductility," p. 279. Chapman & Hall, London.

Bailey, R. W. (1926). *J. Inst. Metals* **35**, 27.

Ball, A., and Hutchison, M. M. (1969). *Metal Sci. J.* **3**, 1.

Barrett, C. R., and Nix, W. D. (1965). *Acta Met.* **13**, 1247.

Barrett, C. R., and Sherby, O. D. (1964). *Trans. AIME* **230**, 1322.

Barrett, C. R., Lytton, J. L., and Sherby, O. D. (1967). *Trans. AIME* **239**, 170.

Barrett, C. R., Ahlquist, C. N., and Nix, W. D. (1970). *Metal Sci. J.* **4**, 41.

Barrett, C. R., Muehleisen, E. C., and Nix, W. D. (1972). *Mater. Sci. Eng.* **10**, 33.

Bayce, A. E., Ludeman, W. D., Shepard, L. A., and Dorn, J. E. (1960). *Trans. Amer. Soc. Metals* **52**, 451.

Bird, J. E., Mukherjee, A. K., and Dorn, J. E. (1969). "Quantitative Relation Between Properties and Microstructure," pp. 255–342. Israel Univ. Press, Jerusalem.

Blum, W. (1971). *Phys. Status Solidi B* **45**, 561.

Clauer, A. H., Wilcox, B. A., and Hirth, J. P. (1970). *Acta Met.* **18**, 381.

Coble, R. L. (1963). *J. Appl. Phys.* **34**, 1679.

Conrad, H. (1961). *In* "Mechanical Behavior of Materials at Elevated Temperatures" (J. E. Dorn, ed.), p. 218. McGraw-Hill, New York.

Conway, J. B., and Mullikin, M. J. (1966). *Trans. AIME* **236**, 1496.

Cottrell, A. H. (1953). "Dislocations and Plastic Flow in Crystals," p. 136. Oxford Univ. Press (Clarendon), London and New York.

Davies, P. W., and Wilshire, B. (1971). *Scr. Met.* **5**, 475.

Davies, R. G. (1963). *Trans. AIME* **227**, 665.

Dorn, J. E. (1954). *J. Mech. Phys. Solids* **3**, 85.

Dorn, J. E. (1962). "Progress in Understanding High-Temperature Creep," Horace W. Gillett Memorial lecture. Amer. Soc. Test. Metals, Philadelphia, Pennsylvania.

Dorn, J. E., and Mote, J. (1963). "High Temperature Structures and Materials," pp. 95–168. Pergamon, Oxford.

Evans, W. J., and Wilshire, B. (1968). *Trans. AIME* **242**, 1303.

Fleischer, R. L. (1961). *Acta Met.* **9**, 996.

Fleischer, R. L. (1963). *Acta Met.* **11**, 203.

Flinn, J. E., and Gilbert, E. R. (1968). *Trans. Metal Soc. AIME, Pap.* **A68-40**.

Friedel, J. (1964). "Dislocations." Pergamon, Oxford.

Garafalo, F. (1965). "Fundamentals of Creep and Creep Rupture in Metals." Macmillan, New York.

Garafalo, I., Richmond, C., Domis, W. F., and von Gemmingen, F. (1963). *In* "Joint International Conference on Creep," pp. 1–31. Inst. Mech. Eng., London.

Garafalo, F., Domis, W., and von Gemmingen, F. (1964). *Trans. AIME* **230**, 1460.

Gasca-Neri, R., and Nix, W. D. (1974). *Acta Met.* **22**, 257.

Gasca-Neri, R., Ahlquist, C. N., and Nix, W. D. (1970). *Acta Met.* **18**, 655.

Gibbs, G. B. (1966). *Phil. Mag.* [8] **13**, 589.

Gifkins, R. C. (1967). *J. Inst. Metals* **95**, 373.

Gilbert, E. R., and Munson, D. E. (1965). *Trans. AIME* **233**, 429.

Glen, J. (1958). *J. Iron Steel Inst., London* **189**, 333.

Gupta, V. P., and Strutt, P. R. (1967). *Can. J. Phys.* **45**, 1213.

Hammad, F. H., and Nix, W. D. (1966). *ASM (Amer. Soc. Metals) Trans. Quart.* **59**, 94.

Hardwick, D., and McG. Tegart, W. J. (1961–1962). *J. Inst. Metals* **90**, 17.

Harper, J. G., and Dorn, J. E. (1957). *Acta Met.* **5**, 654.

Harper, J. G., Shepard, L. A., and Dorn, J. E. (1958). *Acta Met.* **6**, 509.

Hayden, H. W., Gibson, R. C., Merrick, H. F., and Brophy, J. H. (1967). *ASM (Amer. Soc. Metals) Trans. Quart.* **60**, 3.

Hazlett, T., and Hansen, R. D. (1954). *Trans. Amer. Soc. Metals* **47**, 508.

Hazzledine, P. M., and Newbury, D. E. (1973). *In* "Proceedings of the 3rd International Conference on the Strength of Metals and Alloys," p. 202. Institute of Metals, London.

Herring, C. (1950). *J. Appl. Phys.* **21**, 437.

Hirsch, P. B., and Warrington, D. H. (1961). *Phil. Mag.* [8] **6**, 735.

Ivanov, L. I., and Yanushkevich, V. A. (1964). *Phys. Metals Metallogr. (USSR)* **17**, 102.

Johnson, R. H. (1970). *Met. Rev.* **15**, (**n146**), 115.

Jonas, J. J., McQueen, H. J., and Demianczuk, D. W. (1968). "Deformation Under Hot Working Conditions," Spec. Rep. No. 108, p. 97. Iron and Steel Institute, London.

Lagneborg, R. (1969). *Metal. Sci. J.* **3**, 161.

Lagneborg, R. (1972a). *Int. Met. Rev.* **17**, 130–146.

Lagneborg, R. (1972b). "A Modified Recovery-Creep Model and its Evaluation," Paper No. MS270, pp. 127–133. Institute of Metals, London.

Lawley, A., Coll, J. A., and Cahn, R. W. (1960). *Trans. AIME* **218**, 166.

Lifshitz, I. M. (1963). *Sov. Phys.—JETP* **17**, 909.

Lin, T. L., and McLean, D. (1968). *Metal Sci. J.* **2**, 108.

Linga Murty, K., Mohamed, F. A., and Dorn, J. E. (1972). *Acta Met.* **20**, 1009.

Livingston, J. D. (1962). *Acta Met.* **10**, 229.

Lloyd, G. J., and McElroy, R. J. (1974). *Acta Met.* **22**, 339.

Lloyd, G. J., McElroy, R. J., and Martin, J. W. (1973). *In* "Proceedings of the 3rd International Conference on the Strength of Metals and Alloys," p. 185. Institute of Metals, London.

McG. Tegart, W. J. (1961). *Acta Met.* **9**, 614.

McLean, D. (1952–1953). *J. Inst. Metals* **81**, 293.

McLean, D. (1966). *Rep. Progr. Phys.* **29**, Part 1, 1.

McLean, D. (1970). *Metal Sci. J.* **4**, 144.

McLean, D., and Hale, H. F. (1961). "Structural Processes in Creep," p. 19. Iron and Steel Institute, London.

McVetty, P. G. (1934). *Mech. Eng.* **56**, 149.

Melton, K. N., and Edington, J. W. (1973). *Metal Sci. J.* **7**, 172.

Meyers, C. L., and Sherby, O. D. (1961–1962). *J. Inst. Metals* **90**, 380.

Misro, S. C. (1971). Ph.D. Dissertation, University of California, Davis.

Misro, S. C., and Mukherjee, A. K. (1972). *In* "Proceedings of John Dorn Memorial Symposium." Amer. Soc. Metals, Metals Park, Ohio (to be published).

Mitchell, T. E. (1964). *Progr. Appl. Mater. Res.* **6**, 117.

Mitra, S. K., and McLean, D. (1966). *Proc. Roy. Soc. Ser. A* **295**, 288.

Modéer, B., and Lagneborg, R. (1971). *Jernkontorets Ann.* **155**, 363.

Mohammed, F. A., Murty, K. L., and Morris, J. W., Jr. (1973). *Met. Trans.* **4**, 935.

Mukherjee, A. K. (1971). *Mat. Sci. Eng.* **8**, 83.

Mukherjee, A. K. (1972). "Mechanical Behavior of Materials," Vol. III, pp. 52–63. Soc. Mater. Sci., Japan.

Mukherjee, A. K. (1975). *Proc. Bolton-Landing Conf. Grain Boundaries, 4th, 1974* pp.93–106.

Mukherjee, A. K., Bird, J. E., and Dorn, J. E. (1969). *ASM (Amer. Soc. Metals) Trans. Quart.* **62**, 155–179.

Nabarro, F. R. N. (1948). *Rep. Conf. Strength Solids, 1947*, pp. 75–90.

Nabarro, F. R. N. (1967). *Phil. Mag.* [8] **16**, 231.

Naziri, H., Pearce, R., Henderson Brown, M., and Hale, K. F. (1973). *J. Microsc.* **97**, 229.

Nicholson, R. B. (1972). *In* "Electron Microscopy and Structure of Materials," (G. Thomas, R. M. Fulrath, and R. M. Fisher, eds.), p. 689. Univ. of California Press, Berkeley.

Orowan, E. (1946–1947). *J. West. Scot. Iron Steel Inst.* **54**, 45.

Packer, C. M., and Sherby, O. D. (1967). *ASM (Amer. Soc. Metals) Trans. Quart.* **60**, 21.

Raj, R., and Ashby, M. F. (1971). *Met. Trans.* **2**, 1113.

Raymond, L., and Dorn, J. E. (1964). *Trans. AIME* **230**, 560.

Robinson, S. L., and Sherby, O. D. (1969). *Acta Met.* **17**, 109.

Sellars, C. M., and Quarrell, A. G. (1961–1962). *J. Inst. Metals* **90**, 329.

Sherby, O. D. (1962). *Acta Met.* **10**, 135.

Sherby, O. D., and Burke, P. M. (1967). *Progr. Mater. Sci.* **13**, 325–390.

Sherby, O. D., and Young, C. M. (1972). *In* " Proceedings of John Dorn Memorial Symposium."
Amer. Soc. Metals, Metals Park, Ohio (to be published).

Sherby, O. D., Trozera, T. A., and Dorn, J. E. (1956). *Proc. ASTM* **56**, 784.

Shewmon, P. G. (1963). "Diffusion in Solids," p. 176. McGraw-Hill, New York.

Solomon, A. A., and Nix, W. D. (1970). *Acta Met.* **18**, 863.

Solomon, A. A., Ahlquist, C. N., and Nix, W. D. (1970). *Scr. Met.* **4**, 231.

Suzuki, H. (1955). *Sci. Rep. Res. Inst., Tohoku Univ., Ser. A* **7**, 194.

Vaidya, M. L., Linga Murty, K., and Dorn, J. E. (1973). *Acta Met.* **21**, 1615.

Vandervoort, R. R., and Barmore, W. L. (1968). *Plansee Proc., Pap. Plansee Semin., 6th, 1968,*
p. 108. (F. Benesovsky, ed.), June, 24–28, Austria.

Webster, G. A., and Piearcey, B. J. (1967). *Metal Sci. J.* **1**, 97.

Webster, G. A., Cox, A. P. D., and Dorn, J. E. (1969). *Metal Sci. J.* **3**, 221.

Weertman, J. (1957a). *J. Appl. Phys.* **28**, 362.

Weertman, J. (1957b). *J. Appl. Phys.* **28**, 1185.

Weertman, J. (1968). *ASM (Amer. Soc. Metals) Trans. Quart.* **61**, 681–694.

Weertman, J. (1972). *In* " Proceedings of John E. Dorn Memorial Symposium." Amer. Soc.
Metals, Metals Park, Ohio (to be published).

Wilcox, B. A., and Clauer, A. H. (1966). *Trans. AIME* **236**, 570.

Wong, W. A., McQueen, H. J., and Jonas, J. J. (1967). *J. Inst. Metals* **95**, 129.

Woodford, D. A. (1969). *Mater. Sci. Eng.* **4**, 146.

Review Topics in Superplasticity

THOMAS H. ALDEN

Department of Metallurgy
University of British Columbia
Vancouver, British Columbia, Canada

I. Introduction

A. History and Scope

The time devoted to the study of superplastic alloys and their uses has been brief, about ten years in Europe and North America, yet in that time progress has been made at a rate which is perhaps unique in the history of physical metallurgy. In 1962, when the first English review appeared (Underwood, 1962), knowledge of superplasticity was primitive. Certain alloys, mainly eutectics, were known to exhibit extreme elongation when extended at high temperature. The reason for this effect, the mechanism and its relation to microstructure, and the potential applications were all unknown or unexplored. Yet in 1973, all these things are fairly well known and, in addition, the various mechanical properties of a great variety of superplastic alloys have been studied.

The highlights of this era include demonstration of: (1) the high strain rate sensitivity of the flow stress as the essential and unique characteristic of superplasticity (Backofen et al., 1964), (2) the strong variation of properties with grain size and strain rate (Avery and Backofen, 1965; Alden, 1967), (3) the importance of grain boundary sliding in the deformation mechanism (Alden, 1967, 1969a; Holt, 1968), (4) the vacuum forming of sheet into dies (Fields, 1965), (5) superplastic stainless steels (Hayden et al., 1967), and (6) the theory of diffusion-accommodated grain boundary sliding (Ashby and Verrall, 1973).

It is not the intent in this article to review these many achievements, but instead to discuss some subjects previously neglected, to present results from several unpublished studies, and to attempt a new synthesis of published data which may further clarify the phenomonology and mechanism of superplasticity. Among the topics discussed are techniques of grain refinement, grain growth, alloy-specific properties, creep behavior, and the effect of temperature on the strain rate sensitivity. An attempt is made also to evaluate our understanding of these phenomena.

B. Rate Sensitivity and Elongation

Avery and Backofen (1965) established a correlation between an increasing strain rate sensitivity of the flow stress $m = d \log \sigma / d \log \dot{\varepsilon}$ and increasing elongation in the Pb–Sn eutectic (Fig. 1). This relation is explained theoretically in terms of a "strain rate hardening" of material deforming in necked regions of the sample (Hart, 1967a), and is now widely known and understood. However, Avery and Stuart (1967) showed also that the *initial geometry* of the sample strongly affects its elongation, particularly at large m.

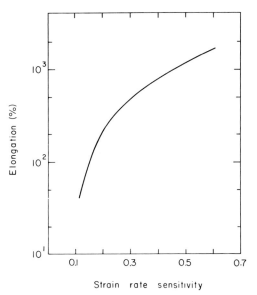

Fig. 1. Experimental elongation to rupture versus strain rate sensitivity for Sn–Pb eutectic. Both strain rate and grain size were varied (Avery and Backofen, 1965).

If the sample area is A_0, but regions of smaller area αA_0 exist because of error or dimensional tolerance, then the "failure strain," defined by a neck size βA, depends on the value of α (Fig. 2). The equation for these curves

$$\% \text{ el} = 100\{[(1 - \beta^{1/m})^m/(1 - \alpha^{1/m})] - 1\}$$

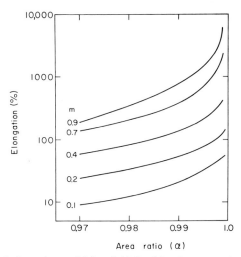

Fig. 2. Theoretical elongation to "failure" (defined by the area ratio $\beta = 0.9$) for various initial area ratios α and values of m (Avery and Stuart, 1967).

derived by Avery and Stuart (1967) does not seem to have been tested experimentally.

A problem as yet unsolved, but certainly related to the above, is to determine the effect of *structural inhomogeneity* on the elongation. The analyses assume uniform microstructure and properties, but regions of slightly different flow stress or rate sensitivity certainly exist and might well be regions of potential failure. This could be a significant metallurgical problem in commercial superplastic forming.

C. Mechanical and Microstructural Characteristics of Superplasticity

For completeness, a summary of the mechanical and microstructural features of superplasticity is presented. For more detail, the reader is referred to several reviews (Johnson, 1970; Davies *et al.*, 1970; Nicholson, 1972; Baudelet, 1971) and to the literature.

1. Strain rate sensitivity: The flow stress σ is a strong function of the strain rate $\dot{\varepsilon}$. In the power law relation

$$\sigma = k\dot{\varepsilon}^m$$

m has the value roughly of 0.3–0.8. The strain dependence of σ is relatively weak (Backofen *et al.*, 1964; Holt and Backofen, 1966; Alden, 1967; Cline and Alden, 1967).

2. Temperature dependence: The strain rate at constant stress increases with increasing temperature but not strongly. Measured activation energies are definitely less than for bulk diffusion and perhaps appropriate for grain boundary diffusion (Cline and Alden, 1967; Alden, 1968; Cook, 1968; Donaldson, 1971).

3. Creep: The creep rate is constant for constant stress; no strain rate transients are observed. The rate is independent of the immediate prior stress and strain history (Surges, 1969).

4. Strain hardening: Samples subject to superplastic strains do not experience an increase in low temperature yield stress relative to annealed, unstrained material (Alden, 1968).

5. Dislocation microstructure: Few dislocations or cells are observed in deformed samples by electron microscopy (Hayden *et al.*, 1967; Lee, 1969).

6. Grain shape: The grain elongation is very much less than the sample elongation. The number of grains in the cross section decreases with strain (Alden, 1967; Alden and Schadler, 1968; Donaldson, 1971; Watts and Stowell, 1971).

7. Texture: No texture is created by large extension; existing texture is destroyed (Lee, 1971; Cutler and Edington, 1971).

8. Grain boundary sliding: Evidence exists showing extensive grain boundary sliding and rotation during deformation, including semiquantitative studies using grids or marker lines and *in situ* observation (Alden, 1967; Holt, 1968; Lee, 1969; Nicholson, 1972).

D. Deformation Mechanism

A comparison of this list of the characteristics of superplastic flow (Section I,C) with a similar list which could be prepared for high temperature deformation of coarse-grained metals and alloys (Garofalo, 1966) shows *dissimilarity, point by point*. To note a few examples, coarse-grained metals are relatively rate insensitive, show creep transients, and develop cell structures during deformation. Indeed, the phenomenology is so diverse, one must conclude that the atomic mechanisms of deformation are also diverse. This rules out the popular early notion (Hayden *et al.*, 1967; Hayden and Brophy, 1968; Ball and Hutchison, 1969) that modifications of theories developed for recovery creep (i.e., dislocation creep) would be useful for superplasticity (Alden, 1969b).

Of the remaining theories, grain boundary sliding (Alden, 1969a; Lee, 1970) probably with diffusional accommodation (Ashby and Verrall, 1973) is almost certainly the dominant mechanism (although slip accommodation cannot be ruled out as a parallel or alternate process). Diffusional creep (Nabarro, 1948; Coble, 1963) is attractive for several reasons, mainly for the prediction of mechanical behavior (Avery and Backofen, 1965; Zehr and Backofen, 1968), but it fails to account for the microstructural characteristics of the deformation (Section VII,A).

E. Stress–Strain Rate Behavior

The stress level required for the deformation of superplastic alloys depends primarily on the *strain rate* and only weakly on the strain. It is now conventional to describe this behavior by means of a log σ–log $\dot{\varepsilon}$ curve (Holt and Backofen, 1966; Alden, 1967; Zehr and Backofen, 1968).

The nature of these curves is usually stated to be "S-shaped" or "three-stage" and indeed most superplastic alloys show this characteristic (Fig. 3a). The stress rises steadily with strain rate, but first with a moderate slope (stage 1), through a region of high slope and a maximum (stage 2) before the slope decreases rather abruptly to stage 3. Stage 2 (at least) is definitely not linear. It is a preoccupation of the theory to interpret this three-stage behavior. Various mechanisms have been discussed which, when combined

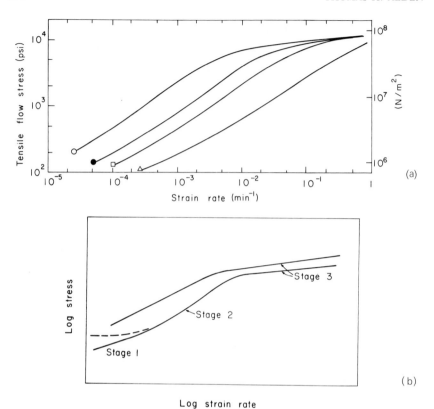

Fig. 3. (a) Logarithmic stress–strain rate curves for Sn–5Bi of several grain sizes. The curves are gently S-shaped (Alden, 1967). $\bar{L}$: $\triangle$, 1.2 μ; $\square$, 2.3 μ; $\bullet$, 3.5 μ; $\bigcirc$, 5.5 μ. (b) Schematic log σ–log $\dot{\varepsilon}$ curves observed in superplastic alloys. Three-stage behavior typified by Sn alloys, Cd–Pb alloys, and Al–Cu eutectic; two-stage behavior by Zn–Al eutectoid and Fe–Ni–Cr alloys. The separation is not always clear because stage 1 may be difficult to identify.

dependently and/or independently can explain the shape of the curve (Zehr and Backofen, 1968; Hart, 1967b; Alden, 1969a; Ashby and Verrall, 1973).

There are, however, notable exceptions to the S-shaped log σ–log $\dot{\varepsilon}$ curve. In Fe–Ni–Cr (Hayden *et al.*, 1967), Pb–5Cd† (Alden, 1968), and Cd–Bi eutectic alloys (Turton, 1971), the data are well represented by two *straight lines*, one with low slope at high strain rates and a second of higher slope at low strain rates (Fig. 3b). Zn–22Al seems to be an intermediate case. Below 200° C, there is little increase in slope as the strain rate is increased from low rates. At 250° C, the increase is variously reported to be strong (Backofen *et al.*, 1964), moderate (Alden and Schadler, 1968), or weak (Nicholson, 1972).

---

† Pb–5Cd contains 95 wt % Pb, 5 wt % Cd.

As will be shown (Section VII,G), the understanding of these straight lines of moderate slope ($0.3 \lesssim m \lesssim 0.6$) in terms of deformation mechanism is particularly difficult.

F. Superplastic Materials

At the time of the review paper by Underwood (1962), the known superplastic alloys were the Zn–Al eutectoid and various eutectics. To some extent, the notion has persisted that *only* eutectic and eutectoid alloys can exhibit superplastic properties. However Alden (1967) showed superplasticity in a dilute precipitation alloy, Sn–5Bi, and concurrently in a *solid solution alloy*, Sn–1Bi (Alden, 1966). By this time, it was fairly clear that *small grain size* rather than composition *per se* was the controlling factor (Avery and Backofen, 1965), and this conclusion was further reinforced by the identification of rate-sensitive deformation in fine-grained alloys across virtually the entire Pb–Sn alloy system (Cline and Alden, 1967). At constant grain size L (but necessarily different thermal-mechanical histories), similar $\log \sigma$–$\log \dot{\varepsilon}$ curves were measured for several alloys ranging from pure Sn, through the eutectic to Pb–19Sn.

These results had the further important effect of clarifying the separate role of *intercrystalline* and *interphase* boundaries (Alden and Schadler, 1968). Interphase boundaries *migrate slowly* and are responsible for maintaining a *stable, fine grain size*. Here (in part) lies the superiority of eutectic alloys. However, the *mechanical effects* of the two kinds of boundaries are similar, as the above results indicate.

For a discussion of specific superplastic alloys, the reader is referred to the review by Johnson (1970).

G. Applications

The utility of superplastic alloys is associated with special methods of deformation processing and, in the case of high melting range alloys, with the exceptional room temperature mechanical properties of the ultrafine-grained material (Gibson *et al.*, 1968). *Sheet forming* by vacuum or pressure is *slow*, but complex parts can be made often in a single operation with a relatively inexpensive female die (Johnson, 1969). The net result is a cost saving for certain components which are manufactured in fairly small numbers, say up to 50,000. Currently, Zn–Al and Ti alloys are being used in this way. Less well known is the excellent response of superplastic alloys to *closed die forging* (Stewart, 1973). Again, complex parts may be formed, slowly but in a single operation. The finished part has exceptional dimensional accuracy, surface smoothness, and nondirectional properties. The forging loads are comparatively small.

II. Production of Fine Grain Size

It is well-known that superplasticity is enhanced by a homogeneous fine grain size. There are several methods which have been used to obtain fine-grained microstructures, but all involve *conversion* to fine grains by hot working or phase transformation and *retention* of fine grains by control of the phase composition and distribution.

A. Eutectic Alloys

The most common superplastic alloys are eutectics. The reason is that eutectics have a *fine structure* in the *cast state*, without the necessity of special procedures like quenching from the melt. The structure is "fine" in the sense that the phases are intimately mixed and the dispersed phase is small in scale (in two dimensions at least) (Fig. 4a). In the cast condition, however, this structure is *not* fine-grained and, as a result, *not superplastic*. For example, in the cast Pb–Sn eutectic, the tin matrix contains very few grain boundaries (Cline and Alden, 1967). Confusion exists about this obvious reason for the lack of superplasticity in cast alloys. It is most likely related to the problem of *etching* grain boundaries in the matrix phase. Etchants which reveal the two-phase structure often do not reveal boundaries in the individual phases. In the Pb–Sn eutectic, Sn grains are visible in polarized light, but are not easily etched.

Near eutectic compositions will contain relatively large grains of pro-eutectic phase. These structures are inhomogeneous and are inferior to eutectic structures.

The fine structure of the cast eutectic can be converted to a *fine-grained* structure by *hot working*, usually extrusion or rolling. The recrystallized structure is geometrically similar to a *pure metal structure*, except that some of the grains are of the original matrix phase and some of the dispersed phase (Fig. 4b,c). (Often, there is elongation of the phases in the working direction.) In contrast to pure metals, however, the structure is *grain size stable*; the grains resist coarsening during annealing or superplastic deformation. The origin of the stability is the relative immobility of the many interphase boundaries in the structure. In order for migration of these boundaries to occur, diffusion is required over a distance roughly equal to the grain size.

The distinction between the deformation behavior of cast and recrystallized microstructures explains the unique appearance of cast tensile specimens of Al–Cu eutectic deformed at high temperature (Holt and Back-ofen, 1966). Initially, the samples have low rate sensitivity and the deformation is confined to a neck. As a result, the small volume of material in the neck recrystallizes, becomes superplastic, and subsequently exhibits extreme elongation before rupture.

Some of the superplastic eutectics are mixtures of *terminal phases*, while

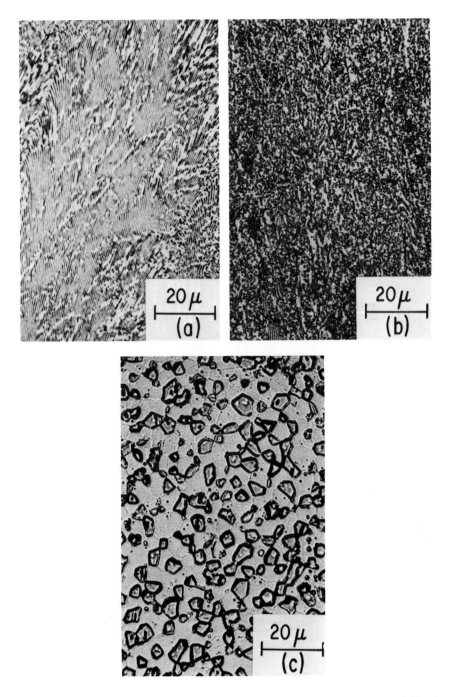

Fig. 4. Microstructure of Mg–Al eutectic in condition (a) as cast, (b) extruded at 300° C, and (c) extruded and annealed 12 hr at 400° C (Lee, 1969).

others contain *intermetallic compounds.* Some examples of the former are Sn–Pb, Sn–Bi, Pb–Cd, Bi–Cd, Ag–Cu, and Zn–Al; and of the latter Al–33Cu, Cu–9.5Al–4Fe, Mg–33Al, and Co–10Al. The superplastic behavior of these two classes is not dramatically different, but the alloys containing compounds may be brittle at room temperature and so not attractive for many applications.

B. Eutectoid Alloys

Despite the usual assertion, *eutectoid* superplastic alloys are rare. Only one important example exists, namely the Zn–Al eutectoid. When cooled at moderate rates from above the invariant temperature, 275° C, the micro-

Fig. 5. Fine-grained microstructure of Zn–Al eutectoid alloy, solution treated at 375° C, quenched and transformed at 0° C. The roughened phase is zinc. The aluminum grain boundaries are not etched (Alden and Schadler, 1968).

structure of this alloy is similar to cast eutectic structures (Alden and Schadler, 1968). In this condition, it is not fine-grained nor superplastic but, again, this condition can be achieved by hot working. Alternatively, even smaller grains can be produced by water quenching from around 300° C. Decomposition of the α' phase (FCC) proceeds at room temperature and the decomposition product is a very fine-grained mix of zinc and aluminum-rich grains (Fig.5) (Alden and Schadler, 1968). This extraordinary reaction has been studied by dilatometry and hardness measurements and by optical and X-ray metallography, but is not completely understood (Garwood and Hopkins, 1952–1953). It has been suggested that the decomposition is spinodal (Nuttall and Nicholson, 1968), but the metallographic study shows it to be a discontinuous reaction, proceeding by the growth of nuclei formed at prior α' grain boundaries. Considering its simplicity as a means for producing fine grain size, the search for a similar reaction in other alloys might be worthwhile.

C. Precipitation Alloys

The final major class of superplastic alloys exhibit precipitation. Many are relatively dilute. Ideally, these alloys are single phase at the initial temperature of hot working and undergo precipitation either immediately following

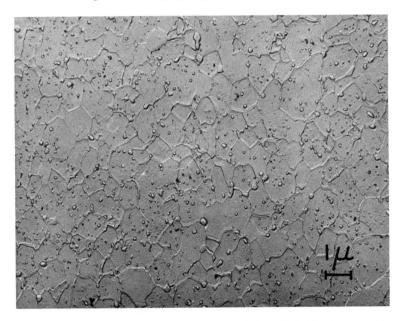

Fig. 6. Fine-grained microstructure of Zn–1Al extruded at 60° C. Aluminum particles are dispersed in the zinc matrix (Turner, 1971).

the working or during working at progressively lower temperatures. One may think that recrystallization and precipitation occur "simultaneously" and thus the growth of the new grains is inhibited. Variations on this procedure involving also cold work and annealing are possible, of course. Materials of this type are Sn–5Bi, Pb–5Cd, Ni–Fe–Cr alloys, dilute Zn–Al alloys, and probably Ti alloys (Fig. 6).

Dilute alloys with an insoluble second phase may be superplastic after hot working if they are initially chill cast. An example is Cd–8Pb (Donaldson, 1971). However, the microstructure of this alloy is definitely inferior in homogeneity to the precipitation alloys. In the cast state the lead phase is fairly massive and becomes elongated as "stringers" of particles in the working direction. As a result, the cadmium grains are also elongated (Fig. 7).

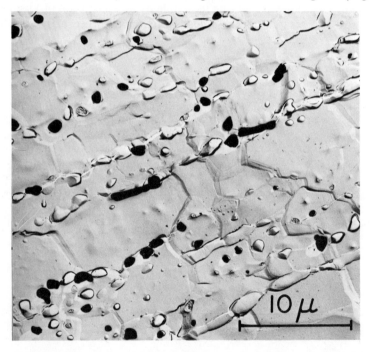

Fig. 7. Microstructure of extruded Cd–8Pb alloy showing stringers of Pb particles and somewhat elongated Cd grains (Donaldson, 1971).

D. Multiple Phase Change

Steel which is repeatedly heated and cooled through the $\alpha \leftrightarrows \gamma$ phase transformation becomes fine-grained (Schadler, 1968). This procedure is perhaps neglected as a method of grain refinement; however, it is time-consuming. Note that these transformations are not occurring under load, a procedure which, if repeated, produces large strains directly by a possibly separate mechanism termed "transformation plasticity" (Johnson, 1970).

III. Microstructural Changes during Deformation

A. Grain Growth

Because superplastic deformation involves considerable holding time of ultrafine-grained structures at high temperature, it is anticipated that grain growth will occur. Indeed, there are few cases in which such growth, when sought, has not been observed. The significant fact, however, is that the growth usually occurs at an *enhanced rate during deformation*. Comparisons of the grain size between deformed samples and samples statically annealed for the deformation time show that the former have larger grain size, e.g. 4 μ vs. 1.6 μ in a Sn–5Bi sample strained 1000% in about 3 hr (Alden, 1967).

Good evidence for the enhancement of grain size by strain was found in the variation of microstructure along the length of a single, deformed tapered sample. (The material is Al–Cu eutectic prepared from eutectic powder. The phases are Al and $CuAl_2$.) The grain size increases progressively toward the smaller end of the sample, which suffers a larger strain (Fig.8) (Watts and Stowell, 1971).

Isolating the effect of *strain rate* is more difficult because equal strain requires different time by as much as 10^5 in a usual investigation of superplasticity. One approach is to define a "growth enhancement parameter," $\Delta L/L_a$, where ΔL is the *difference* between the grain size of a deformed sample and a sample annealed at the deformation temperature for an equal time, and L_a is the annealed grain size. In Sn–1Bi, it is found that $\Delta L/L_a$ increases with strain, but with varying slope depending on the strain rate (Fig.9) (Clark and Alden, 1973). The result is summarized in Fig. 10 which shows that the enhancement of growth achieves a maximum at a strain rate near 10^{-3} min. Concurrent measurement of the mechanical properties shows that the enhancement is a maximum in the *region of maximum strain rate sensitivity*. This result indicates that there is a close relation between the specific mechanism of superplastic deformation and grain growth, and that grain size enhancement is not an important *general* characteristic of plastic deformation.

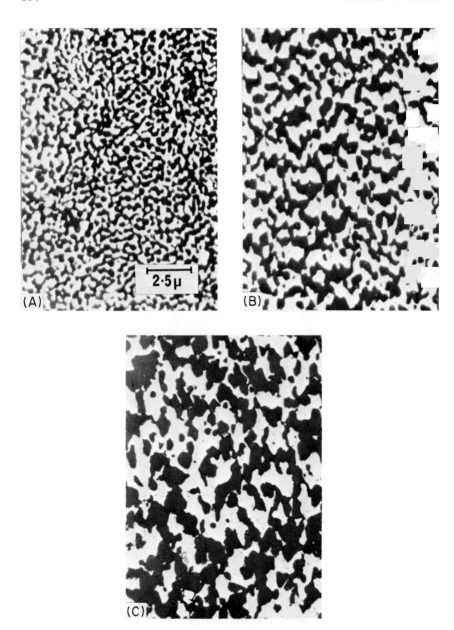

Fig. 8. Grain growth in a tapered sample of Al–Cu eutectic, deformation temperature 520° C. Approximate strains: (a) 0, (b) 1.8, and (c) 3.9 (Watts and Stowell, 1971).

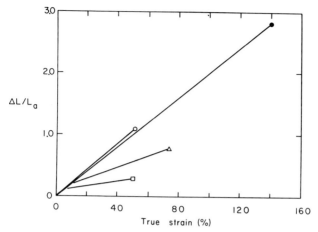

Fig. 9. Strain enhanced grain growth in Sn–1Bi. The enhancement $\Delta L = (L - L_a)$, where L is the grain size after deformation and L_a is the grain size after static annealing for the deformation time. The enhancement varies with strain rate (Clark and Alden, 1973). $\dot{\varepsilon}(min^{-1})$: $\square$, 1.0; $\triangle$, 0.1; $\bullet$, 0.01; $\bigcirc$, 0.002.

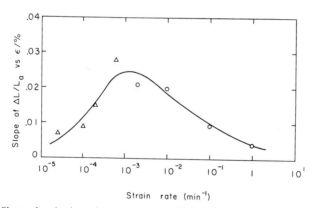

Strain rate (min⁻¹)

Fig. 10. Slope of grain size enhancement curve (cf. Fig. 9) versus strain rate. The maximum grain size enhancement occurs near the strain rate for m_{max}. $\bigcirc$, Tensile tests; $\triangle$, creep tests.

B. Hardening

Grain growth during deformation of superplastic alloys leads to a "strain hardening." This hardening, of course, has nothing to do with the accumulation of dislocations in the microstructure, a process which is specifically absent in superplasticity (Section I,C). Rather, it reflects the inverse relation between grain size and stress at a fixed strain rate (Alden, 1967; Holt and Backofen, 1966).

In order to identify the hardening, it is necessary to deform at constant *strain rate*, not the usual constant *extension rate* (often carelessly called "constant strain rate"). A few such tests have been made, and the hardening is found to be linear with strain (Watts and Stowell, 1971; Suery and Baudelet, 1973). It can be seen that the hardening slope varies with strain rate, but since the *time* is also a principal variable, it is not possible to separate an intrinsic strain rate effect (Fig. 11).

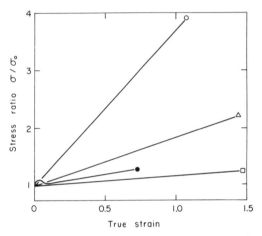

Fig. 11. Increase of flow stress with strain in two superplastic alloys. σ_0 is the initial stress. The increase is ascribed to grain growth (Suery and Baudelet, 1973, Watts and Stowell, 1971). $\bigcirc$, Al–Cu eutectic, $\dot{\varepsilon}(\sec^{-1}) = 6 \times 10^{-5}$; $\triangle$, Al–Cu eutectic, $\dot{\varepsilon} = 6 \times 10^{-4}$; $\square$, Al–Cu eutectic, $\dot{\varepsilon} = 6 \times 10^{-3}$; $\bullet$, 60/40 brass, $\dot{\varepsilon} = 1.7 \times 10^{-4}$.

In constant stress creep tests at low stress, Surges (1969) found a slowly decreasing creep rate. This decrease may also result from grain growth, although the cause was not positively identified.

C. Structural Anisotropy

The microstructure of superplastic alloys which have been extruded or rolled often appears *elongated* along the working direction. It is necessary to distinguish carefully two sources of such "elongation": (1) the grains of each phase may have an aspect ratio *greater than unity*, or (2) the grains may be equiaxed, but *preferentially connected* to other grains of the same phase along the working direction. Because of the previously mentioned difficulty of etching grain boundaries in two-phase structures, it is often difficult to distinguish these effects. In the Al–Cu eutectic alloy (Fig. 12), preferential

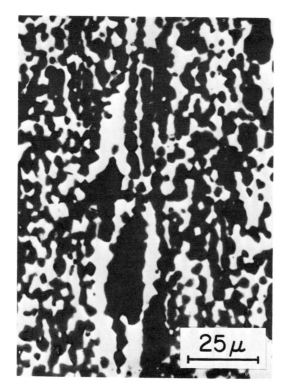

Fig. 12. "Banding" in extruded Al–Cu eutectic. The $CuAl_2$ phase (dark) is fine grained, as shown by the cusps, but is preferentially connected in the working direction (Stowell *et al.*, 1969).

connection seems dominant as shown by the *cusps* in the interphase boundaries (Stowell *et al.*, 1969). In 60/40 brass, significant grain elongation is evident also in the as-extruded structure (Suery and Baudelet, 1973).

Superplastic deformation tends to destroy *both* these sources of directionality; the deformed structures are nearly isotropic. In the Al–Cu alloy, however, the *mechanical effect* of this structure change is negligible because it merely changes to some degree the relative area of intercrystalline and interphase boundary. By contrast, in the brass, changes in the grain aspect ratio lead to changes in flow stress. The flow stress maximum (Fig. 11) corresponds to a maximum in grain elongation (Suery and Baudelet, 1973). As the grains approach an equiaxed shape, the flow stress falls.

It has been emphasized that the deformation structures tend toward isotropy. However, it is significant that grain length/width ratios usually do not achieve unity, but rather values in the range 1.1–1.5 (Watts and Stowell,

1971; Donaldson, 1971; Turner, 1971). These values are, of course, very small in comparison to the length ratios of superplastically deformed tensile samples so that the early conclusion that the grains "remain equiaxed" during superplastic deformation can be understood (Alden, 1967).

IV. Rate-Sensitive Deformation—Material Effect

A. Dilute Alloys

Certain alloys seem to have superior superplastic properties. To establish the validity of this impression, comparison must be done at constant value of the principal parameters, particularly grain size and diffusion coefficients. Fortunately, there are sufficient data on a number of alloys to fulfil (roughly) these requirements. However, lacking adequate diffusion data, particularly for grain boundary diffusion, comparison is done at nearly constant homologous temperature T_H. This procedure should be satisfactory, particularly because the variation of properties with temperature is rather weak.

Studies have been made on many *fairly dilute* alloys in which T_H can be calculated for the base metal. Figure 13 shows the log σ–log $\dot{\varepsilon}$ curves for these alloys with $L \simeq 3 \mu$ and $T_H \simeq 0.55$. Also shown is the variation of strain rate sensitivity determined graphically from these same curves. Looking at the stress–strain rate behavior, the curves for the various alloys, including Pb-, Sn-, Zn-, and Cd-base, lie close together in the central region of maximum slope, i.e. the principal superplastic region. Note also that the strain rate for m_{max} varies by only about one order of magnitude. These observations indicate that the method of comparison is adequate.

The curves in Fig. 13 show that some alloys have excellent superplastic properties and some mediocre. In particular, the strain rate sensitivity curves are diverse. The Sn–5Bi and Zn–0.2Al alloys have m_{max} values nearly twice as large as for the lead alloys; Cd–8Pb is intermediate. Not only is m_{max} larger, but the entire curve for Sn–5Bi lies well above the Pb–5Cd curve over three orders of magnitude of strain rate. A result is that the log σ–log $\dot{\varepsilon}$ curves for these alloys *cross* at intermediate strain rate. At low strain rate the lead alloy supports a higher stress; at high strain rate the tin alloy does the same. At high strain rate, the flow stress ratio σ_{Sn}/σ_{Pb} is almost 4. A partial explanation of this large ratio and the related difference in strain rate sensitivity is based on the difference in the grain size dependence of the stress for slip deformation (Section VII,F).

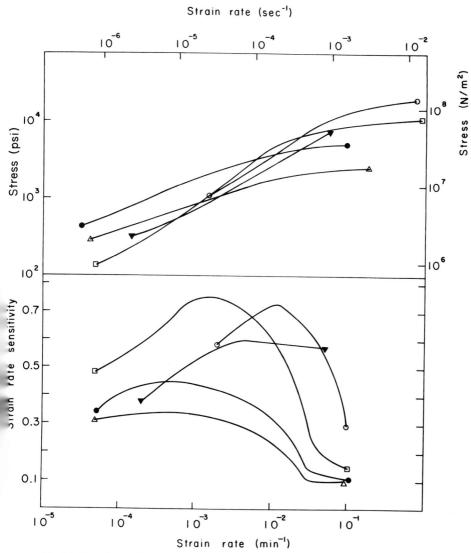

Fig. 13. Experimental logarithmic stress–strain rate curves (upper panel) and strain rate sensitivity curves (lower panel) for various dilute alloys. Parameters other than material are relatively constant. Note substantial differences in rate sensitivity. Upper panel: ○, Zn–0.2Al, $\bar{L} = 3.7$, $T/T_m = 0.54$ (Cook, 1968); ●, Pb–14Sn, $\bar{L} = 2.4$, $T/T_m = 0.54$ (Zehr and Backofen, 1968); △, Pb–5Cd, $\bar{L} = 5.0$, $T/T_m = 0.54$ (Alden, 1968); □, Sn–5Bi; $\bar{L} = 3.5$, $T/T_m = 0.58$ (Alden, 1967); ▼, Cd–8Pb, $\bar{L} = 3.0$, $T/T_m = 0.55$ (Donaldson, 1971). Lower panel: ○, Zn–0.2Al; □, Sn–5Bi; △, Pb–5Cd; ●, Pb–14Sn; ▼, Cd–8Pb.

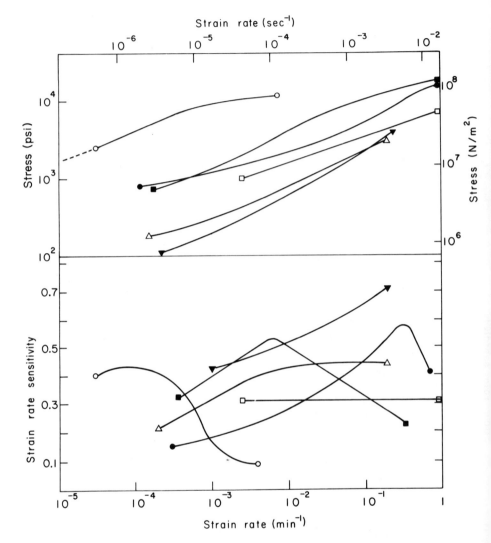

Fig. 14. Similar to Fig. 13 except data are for eutectic alloys. Strain rate at m_{max} varies in addition to the rate sensitivity itself. Upper panel: ■, Ag–28Cu, $L(\mu) = 0.7$, $T(°\text{ C}) = 500$, $T'_H = 0.61$ (Cline and Lee, 1970); △, Pb–18Cd, $L = 3.0$, $T = 85$, $T'_H = 0.60$ (Donaldson, 1971); ○, Al–33Cu, $L = 1.7$, $T = 360$, $T'_H = 0.70$ (Holt and Backofen, 1966); □, Zn–22Al, $L \sim 1.0$, $T = 125$, $T'_H = 0.65$ (Nicholson, 1972); ▼, Sn–38Pb, $L = 2.0$, $T = 44$, $T'_H = 0.60$ (Cline and Alden, 1967); ●, Mg–33Al, $L = 2.2$, $T = 250$, $T'_H = 0.61$ (Lee, 1969). Lower panel: ■, Ag–28Cu; △, Pb–18Cd; ○, Al–33Cu; □, Zn–22Al; ▼, Sn–38Pb; ●, Mg–33Al.

B. Eutectic Alloys

A difficulty in comparing eutectic alloys is to establish a definition for the homologous temperature. It makes no sense to divide the test temperature by the *eutectic temperature*, since in the solid state the individual grains in the structure do not "know" the overall composition of the alloy. Experience with alloys in the Pb–Sn system deformed at 20° C indicated that there is a gradual increase in an *effective homologous* temperature T'_H as the composition varies from Pb-rich to Sn-rich (Pb has the higher melting point). There is nothing special about the eutectic composition (Cline and Alden, 1967). On this basis, the suggested definition for T'_H is

$$T'_H = v_\alpha(T_H)_\alpha + v_\beta(T_H)_\beta$$

where v is the volume fraction.

The differences between the eutectic alloys (and the Zn–Al eutectoid) are not as striking as for the dilute alloys, at least insofar as the strain rate sensitivity curves are concerned (Fig. 14, lower panel). The Pb–18Cd alloy has rather low m, but better than Pb–5Cd. The divergence in position of the log σ–log $\dot\varepsilon$ curves (upper panel) may be largely a horizontal shift, which could be the result of differences in the diffusion coefficients at roughly constant T'_H.

However, the Al–Cu eutectic, even though its properties have been

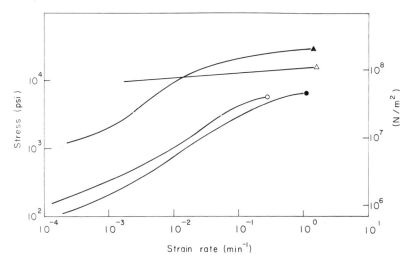

Fig. 15. Logarithmic stress–strain rate curves for dilute and concentrated Zn–Al and Sn–Pb alloys. Aluminum suppresses the superplasticity of zinc, while lead has little effect on the properties of tin. ▲, Zn–0.2Al (Cook, 1968); △, Zn–22Al (Nicholson, 1972); ●, Sn–2Pb (Cline and Alden, 1967); ○, Sn–38Pb (Cline and Alden, 1967).

measured at the highest $T'_H = 0.70$, shows its m_{max} at a surprisingly low strain rate, about four orders of magnitude less than the Mg–Al and Sn–Pb eutectics. Moreover, its strain rate at a constant stress of say 3×10^3 lb/in² is 2 to 3.5 orders slower than the other alloys. This result is difficult to understand theoretically (Section VII,G,2).

The other anomaly is the Zn–Al eutectoid, which has the low and constant $m = 0.31$ at $T'_H = 0.65$. It was shown in Section IV,A that dilute Zn–Al has excellent superplastic properties even at fairly low T_H, and one might have expected similar behavior in the eutectoid. An additional comparison which compounds the difficulty is shown in Fig. 15. The addition of aluminum (43 vol%) to zinc suppresses its superplastic behavior; however, the addition of lead (28 vol %) to tin barely changes the mechanical properties of the tin.

V. Rate-Sensitive Deformation—Temperature Effect

Variation of Strain Rate Sensitivity

The well-known effect of increased temperature is to increase the strain rate for optimum superplasticity. Concurrently, the strain rate at constant stress increases, which indicates that the deformation is thermally activated. Further discussion is presented in Section VI,B.

Less well recognized, but as important as the increase in deformation rate, is an increase in the *slope m* of the log σ–log $\dot{\varepsilon}$ curve with increasing tempera-

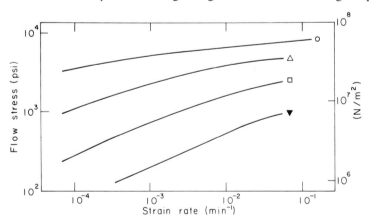

Fig. 16. Apparent strong temperature dependence of m_{max} in Pb–Cd eutectic, $L = 8 \mu$. Actual effect is simply the gradual onset of stage 2 with increasing temperature; true m_{max} cannot be identified without stage 1 data. Data from Donaldson (1971). ○, 0° C; △, 56° C; □, 118° C; ▼, 187° C.

ture. This effect will lead to an enhanced stability of flow and is an additional factor favoring high temperature in superplastic metal forming. In addition, it represents a significant challenge to the theory, which usually predicts $m \simeq 1$ and which has not been shown to be compatible with significantly smaller and variable values of this parameter. However, the variation of m with temperature is not the same in all alloys but, roughly, is weak in one class and strong in another. A demonstration of these effects is the topic of this section.

Because m varies with strain rate, and the strain rate range of study is relatively fixed, it may be difficult to identify an independent temperature effect. An example is the behavior of 8 μ grain size Pb–Cd eutectic (Fig. 16). It appears that m is larger at high temperatures. However, this appearance is largely the result of comparison between *different stages* of deformation (and thus mechanisms of different rate sensitivity). In this case, the decrease in slope with decreasing temperature reflects simply an extension of stage 3 to lower strain rates. In order to make a valid comparison, data are required at still lower strain rates for this alloy, sufficient to reveal the onset of stage 1. These data would permit the determination of m_{max} at each temperature. Fortunately, there are data for many alloys which extend into stage 1. Alternatively, if m is a very weak function of strain rate, values obtained at a constant stress will be adequate. The Zn–Al eutectoid shows this behavior except possibly near the eutectoid temperature (Backofen *et al.*, 1964).

The data shown in Figs. 17 and 18 include those from a variety of alloys of low and high melting range. Some of the alloys are nearly single phase, some are mixtures of two terminal solid solutions, and some are mixtures of a solution and an intermediate phase. The grain size is not constant, but varies between about 1 and 4 μ. In particular alloys, an effort has been made by prior annealing to minimize the variation of grain size during deformation, although it does occur, especially in the dilute alloys. Frequently, metallographic study has established that the growth has been minimal.

The figures show that m_{max} varies from 0.1 to 0.9, although the variation is not that great in any one alloy. All the alloys show a smooth increase of m_{max} with temperature. It is evident from these figures that there are substantial differences among the alloys in the sensitivity of m_{max} to temperature. This sensitivity, which is the slope of the lines in Figs. 17 and 18, is shown in Table I. (No great accuracy is claimed for these numbers, but the differences are sufficiently large that real behavioral differences are evident.) Roughly speaking, the cubic alloys as well as those with Sn-base or Cd-base are temperature insensitive. More sensitive are hexagonal alloys and those containing intermediate phases. Note that the proportion of β phase in the aluminum bronze increases rapidly with temperature and may be responsible for the atypical behavior of this alloy. The temperature sensitivity of m

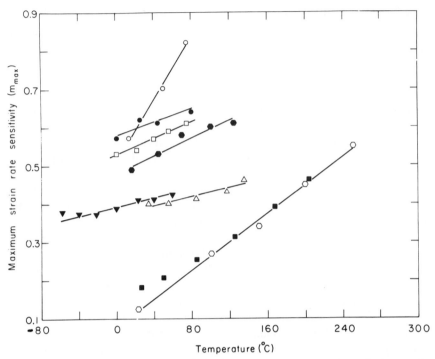

Fig. 17. Increase of m_{max} with temperature for low melting range alloys. The increase is monotonic for all materials, but the coefficient varies. ○, Zn–0.2Al (Cook, 1968); ▼, Zn–1Al (Turner, 1971); ◐, Zn–20Al (Backofen *et al.*, 1964); ■, Zn–22Al (Nicholson, 1972); □, Cd–8Pb, 3 μ (Donaldson, 1971); ○, Cd–8Pb, 8 μ (Donaldson, 1971); △, Pb–17Cd (Donaldson, 1971); ●, Sn–38Pb (Cline and Alden, 1967).

TABLE I

TEMPERATURE VARIATION OF m_{max} (PER $100°$ C)

Alloy	Δm_{max}	Comment
Zn–1Al	0.060	Anomalously small
Pb–17Cd	0.060	
Ag–28Cu	0.075	
Sn–38Pb	0.095	
Cd–5Pb	0.11	
Ti–6Al–4V	0.17	
Al–33Cu	0.18	
Zn–20Al	0.20	
Mg–34Al	0.29	Two points only
Zn–0.2Al	0.40	Anomalously large
Cu–9.5Al–4Fe	0.50	Maximum value; phase structure variable

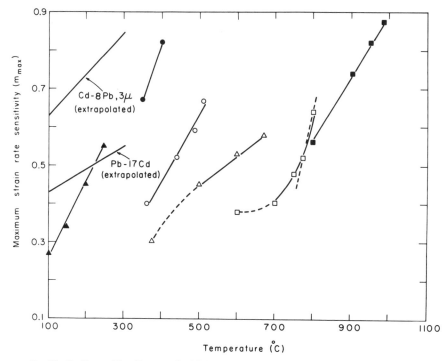

Fig. 18. Similar to Fig. 17 except for high melting range alloys. ▲, Zn–20Al (Backofen *et al.*, 1964); ○, Al–33Cu (Holt and Backofen, 1966); ■, Ti–6Al–4V (Lee and Backofen, 1967); ●, Mg–34Al (Lee, 1969); △, Ag–28Cu (Cline and Lee, 1970); □, Cu–9.5Al–4Fe (Dunlop and Taplin, 1972).

in Zn–0.2Al is probably erroneously large because of grain growth (N. Risebrough, private communication). Zn–1Al is anomalous.

The temperature sensitivity of m for the various alloys might be compared better by reference to a change of homologous temperature rather than temperature itself. This procedure would produce a relative decrease in the sensitivity of low melting alloys in comparison with high melting alloys. The essential distinction between say Ag–28Cu and Al–33Cu or between Pb–17Cd and Zn–20Al would remain, however.

VI. Creep

A. Pb–Sn Eutectic

A comprehensive study of the creep properties of extruded Pb–Sn eutectic, $L = 2\ \mu$, was done by Surges (1969). He measured constant stress creep curves at room temperature at stresses between 100 and 6500 psi. Within this stress range, several distinctive creep curves were identified which could be correlated with the stages of the log σ–log $\dot{\varepsilon}$ curve, shown in Fig. 19. The points in this curve were obtained from the *creep tests* by measurement of the initial slope of the creep curves (i.e. near $t = 0$), except in stage 3 where the secondary creep rate was plotted. Agreement with data from Cline and Alden (1967), who used constant extension rate tests and measured the "steady state" flow stress, is satisfactory.

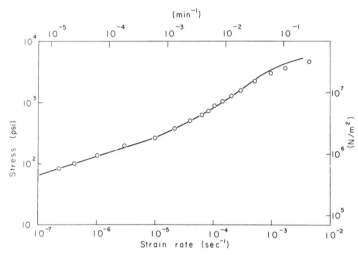

Fig. 19. Creep test determination of log σ–log $\dot{\varepsilon}$ curve for Sn–Pb eutectic. The data fit the curve of "steady state" stress values from Instron tests. Points from Surges (1969); curve from Cline and Alden (1967).

The remarkable feature of creep in the principal superplastic range (stage 2; 500–3000 psi) is the absence of strain rate transients. The creep curves are straight lines which pass through the origin (Fig. 20a). Abrupt changes in stress during creep also fail to initiate transients. On an increase of stress, the strain rate rises abruptly to a higher, constant value; on a decrease of stress, the strain rate falls but again remains constant (Fig. 20b). A periodic excursion to zero stress initiates no contraction ("strain relaxation") and does not affect the subsequent creep rate.

These creep experiments demonstrate most clearly the unique mechanical behavior of superplastic alloys in comparison with coarse-grained alloys. The superplastic alloys follow a *mechanical equation of state* of the form

$$\dot{\varepsilon} = \dot{\varepsilon}(\sigma, T, L)$$

At constant grain size L, the strain rate is independent of strain and, in fact, of the prior thermal mechanical history, given that this history includes only

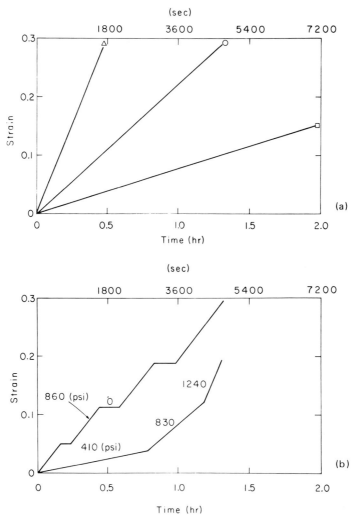

Fig. 20. (a) Creep curves for Sn–Pb eutectic at various stress levels in stage 2, temperature 22° C. The creep rate is constant and there is no loading strain (Surges, 1969). □, 430 psi; ○, 715 psi; △, 1140 psi. (b) Straight line creep curves for Sn–Pb eutectic. The behavior is unaffected by creep history (Surges, 1969).

deformation in stage 2. No other important class of metals and alloys exhibits a deformation law of this simple form.

At higher stress (stage 3), primary creep is evident. The primary strain ε_p was determined by extrapolation of the linear (secondary) portion of the curve. ε_p is nil until stage 3 is definitely established. As the stress is increased

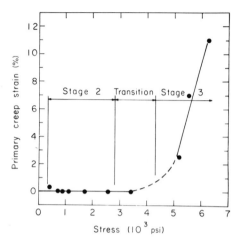

Fig. 21. Primary creep strain in Sn–Pb eutectic at 22° C. There is no primary creep in stage 2, while in stage 3 it increases with increasing stress (Surges, 1969).

in stage 3, the primary strain also increases (Fig. 21). This latter behavior is characteristic also of normal creep (Garofalo, 1966). Strain rate transients are associated with the storage and/or annihilation of obstacle dislocations, i.e. strain hardening and recovery (Friedel, 1966; Alden, 1972). In this

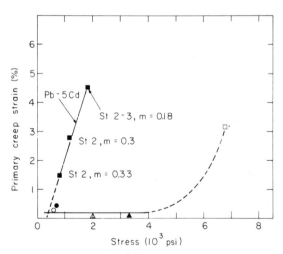

Fig. 22. Primary creep in Pb–Cd eutectic and Pb–5Cd. In tests of the eutectic, the temperature varies. Results similar to Fig. 21 except Pb–5Cd shows some primary creep even though (apparently) in stage 2. Pb–Cd eutectic: ○, 125° C, $m = 0.42$, stage 2; ●, 60° C, $m = 0.36$, stage 2; △, 25° C, $m = 0.30$, stage 2; ▲, 25° C, $m = 0.23$, stage 2–3; □, 0° C, $m = 0.11$, stage 3.

context, it is interesting that measurements of low temperature yield stress following stage 2 deformation indicate no increase relative to the unstrained alloy, whereas stage 3 deformation does produce hardening (Alden, 1968).

Roughly similar results were obtained by Donaldson (1971) in Cd–Pb alloys, although the Pb–5Cd shows a small amount of primary creep, even though apparently in stage 2 ($m = 0.3$) (Fig. 22).

Curious behavior was found at high stress, just below stage 3 (in Pb–Sn, 2800–4500 psi), namely a continuous increase of strain rate with strain (Surges, 1969). Actually, small increases of the order of 10% can be seen even in stage 2 near 40% strain. However, at 3580 psi, the strain rate increased nearly three times at the same strain.

B. Temperature Dependence

The strain rate at constant stress increases with temperature. If one assumes a strain rate equation of the form

$$\dot{\varepsilon} = \dot{\varepsilon}_0(S, \sigma) \exp\left[-Q_s(S, \sigma)/kT\right]$$

where the symbol S represents microstructure, then the usual log $\dot{\varepsilon}$ vs $1/T$ plot at constant stress can be used to determine Q_s, *if the microstructure is also constant*. Evidence already presented indicates that this constancy of structure is a unique feature of superplastic flow (Section I,C). Therefore, it is unnecessary to use the more refined incremental temperature change tests as in the study of activation energy of normal creep (Garofalo, 1966). It is necessary only that the deformation be of a stage 2 nature over the entire temperature range.

Results obtained from constant extension rate tests have been discussed previously (Davies *et al.*, 1970) and with a few exceptions show that Q_s is small relative to the activation energies for self-diffusion. Data of a similar nature for Cd–8Pb were obtained by Donaldson (1971) by means of *creep tests*. The creep rate of a *single sample* was measured at a series of decreasing temperatures, obtained by quenching in fluid baths. At each change of temperature, the sample "immediately" exhibited a new lower strain rate; there were no strain rate transients. Such transients are observed in similar experiments on coarse-grained alloys and indicate rapid temperature induced structure changes. Their absence in the superplastic alloy is further evidence of constancy of structure, or in particular of the failure to develop dislocation networks during the straining. The activation energy for Cd–8Pb, $L = 3\ \mu$, was found to be 13.5 kcal/mol (Fig. 23).

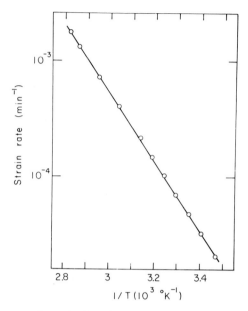

Fig. 23. Creep test determination of activation energy of flow in Cd–8Pb. The points were obtained from a single sample by incremental temperature decreases (Donaldson, 1971).

VII. Discussion

A. Microstructure and the Mechanism of Deformation

Despite the variety of distinctive mechanical and thermal properties of superplastic alloys, it is actually two microstructural observations which most clearly distinguish the deformation mechanism. These observations are: (1) the strong *inverse* dependence of strain rate on grain size, typically $\dot{\varepsilon} \sim 1/L^3$, and (2) the small change in *grain shape* which accompanies large extensions. In particular, there is little grain elongation. The inverse grain size effect indicates a mechanism in which *diffusional shape change* is important. Attempts to reconcile the effect with shape change by slip, whether direct or as an accommodation process for grain boundary sliding, have been *contrived* by comparison. On the other hand, until recently, models invoking diffusional shape change predicted *grain elongation* equivalent to the elongation of the specimen. This dilemma has now been resolved in the model of Ashby and Verrall (1973).

B. *Theory of Ashby and Verrall*

In this theory, change of specimen shape is accomplished not by change of *shape* of the individual grains, but by change of *place*. The place change in turn is achieved by *grain boundary sliding*. This feature is in accord with earlier conclusions about the dominant role of sliding in the deformation mechanism (Section I,C). In order for the grains to remain "stuck together" during this process, they must suffer a transient but complex shape change or *accommodation strain*. The unique feature of this theory is that the accommodation is not by slip or boundary migration, as has usually been supposed, but by *diffusional transport*. In superplastic alloys, the diffusion is mainly through the grain boundaries.

The specific geometrical nature of the model may be understood by reference to Fig. 24. In two dimensions, a set of four hexagonal "grains" is subject to a stress σ. The separation of the centers of the "vertical pair" of grains (2 and 4) in the initial state is $0.87d$ and of the "horizontal pair" (1 and 3) $1.5d$ (Fig. 24a).† In the intermediate state, the grains have suffered their maximum accommodation strain (Fig. 24b) and grains 2 and 4 have *sheared* along the four boundaries of contact with 1 and 3 a distance $u = 0.23d$ (Fig. 24c). This state is symmetrical and the spacing of the grain centers is $1.2d$. Finally, the grains relax to a new equilibrium configuration suffering further sliding and accommodation strain. The extension strain is 1.55. The strain rate equation is

$$\dot{\varepsilon} = (100\Omega/kTd^2)[\sigma - (0.72\Gamma/d)]D_v[1 + (3.3\delta/d)(D_B/D_v)]$$

where Ω is the atomic volume, δ is the width of the diffusion zone along grain boundaries, and Γ is the grain boundary free energy. The first factor in square brackets indicates that the process requires a *threshold stress* as a consequence of transient increases in the *grain boundary area*. As the applied stress nears the threshold stress, m falls to zero. Thereby, the model contains an explanation for stage 1.

Ashby and Verrall conclude that the strain rate is about a factor of seven *faster* than in ordinary diffusional creep (Nabarro, 1948; Coble, 1963). This enhanced rate exists because *less matter* is transported by diffusion over a *shorter distance* per unit strain. Thus, except in unusual circumstances (e.g. bamboo structures) diffusion-accommodated sliding should always be the dominant process of "diffusional flow."

---

† The symbol L, previously used to refer to grain size, is a metallographically determined mean intercept length. It will differ from d by a constant geometrical factor.

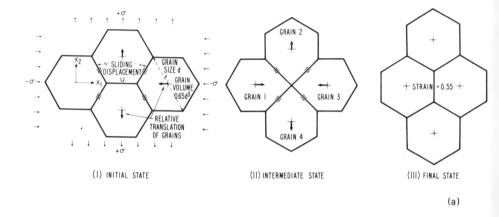

(I) INITIAL STATE   (II) INTERMEDIATE STATE   (III) FINAL STATE

(a)

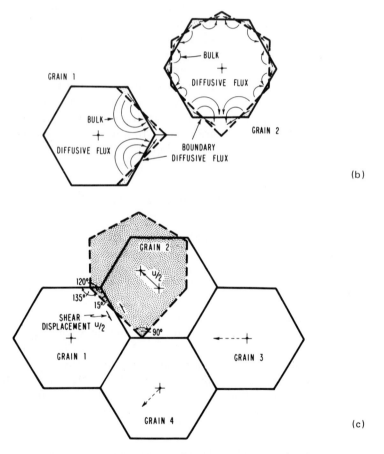

(b)

(c)

Fig. 24. (a) Geometry of deformation by diffusion accommodated grain boundary sliding. (b) Diffusive shape change of the "grains." (c) Sliding displacement (Ashby and Verrall, 1973).

C. Stress–Strain Rate Curves

The author agrees with Ashby and Verrall (1973) that, "The strongest evidence that the present model is a good description of superplasticity is—that the microstructural and topological predictions of the model for large strains match those observed in superplastic alloys." On the other hand, the predictions of mechanical behavior are less obviously correct and must be examined more closely.

The predicted log σ–log $\dot{\varepsilon}$ curve for Pb at 300° K and the experimental curve for Pb–5Cd show poor agreement. The theoretical curve shows a pronounced S-shape and an $m_{max} \simeq 0.95$ (Ashby and Verrall, 1973). Experimentally, Pb–5Cd exhibits a poorly defined stage 1 (if any)[†] and has an $m_{max} \simeq 0.35$ (Alden, 1968; Donaldson, 1971). In other words, theory and experiment are diverse. However, before rejecting the theory on these grounds, one must recognize the vital nature of the processes in stage 1 and/or stage 3 in determining the general shape of the curve and the value of m_{max} in particular. The essential fact is that the Newtonian mechanism occurs as a *transition process* (Alden, 1967; Hart, 1967b); i.e. it is "trapped" between mechanisms of low rate sensitivity. Thereby, m_{max} will be reduced from unity to a lesser value. The *stress range* of the transition defined by the difference in level of the stage 1 and stage 3 curves at some "transition strain rate" is then the *major factor which determines* m_{max}.

For example, if stage 1 reflects the existence of a threshold stress σ_0, Ashby and Verrall calculate $\sigma_0 = 9 \times 10^5$ dyn/cm^2 and $m_{max} \simeq 0.95$ for their particular mechanism, namely the requirement for a reversible increase

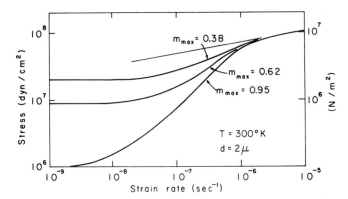

Fig. 25. Variation of theoretical logarithmic stress–strain rate curves and m_{max} with threshold stress σ_0. As σ_0 increases, m_{max} decreases.

[†] The behavior has yet to be examined at very low strain rate.

in grain boundary area; but if some other (unidentified) mechanism operates to give a larger σ_0, the effect on m_{max} can be dramatic. If $\sigma_0 = 9 \times 10^6$ dyn/cm^2, $m_{max} \simeq 0.62$ and if $\sigma_0 = 2 \times 10^7$ dyn/cm^2, $m_{max} \simeq 0.38$ (Fig.25). Otherwise the model should remain unchanged. The conclusion is that the AV model may be correct, both microstructurally and mechanically, if an appropriate mechanism for stage 1 is conceived.

D. Effect of Temperature

The AV theory correctly predicts the increasing strain rate at m_{max} with increasing temperature, but fails to predict the increase in the *magnitude* of m_{max}. The theoretical curves are shown in Fig. 26 for 3 temperatures, 300° K, 400° K and 500° K, for which m_{max} *decreases* from about 0.95 to 0.72. (The m versus $\dot{\varepsilon}$ curves were estimated by graphical measurement). The reason for the decrease is easily seen to be the *decreasing stress range* of the transition between stages 1 and 3. The level of the stage 3 curve decreases with rising temperature, but the surface energy mechanism for stage 1 is temperature independent.

Again one must suppose that the AV mechanism for stage 1 is either incorrect or (less likely) incorrectly described. Assuming the validity of the analysis as described in Section VII,C, the stress level of stage 1 must be *more* temperature dependent than the stress level of stage 3 in order for m_{max} to increase with temperature.

E. Stage 1

Returning to Fig. 25, the upper curve is qualitatively similar to experimental curves for Pb–5Cd (Alden, 1968; Donaldson, 1971) except that in the "theoretical" curve, the decreasing slope in stage 1 is more pronounced. This strong decrease is the result of the *threshold stress* description of stage 1, which is a feature of the AV model, but which has been adopted here *only for ease of calculation*. Alternative models consider stage 1 as a region of reduced but constant rate sensitivity (Hart, 1967b; Avery and Stuart, 1967; Alden 1969a). It is unlikely that the essential conclusion of the previous sections would be changed if this model of stage 1 were adopted. That is, given a Newtonian process for the transition, m_{max} will depend on the stress range of the transition. In addition, if the stress range is limited, as it appears to be for lead, and if the slope of the stage 1 process is not too small, say ≥ 0.25, then the onset of stage 1 should be gradual and not easily identified. This may be the case for Pb–5Cd.

It will be clear to the reader that much of this discussion rests on speculation about stage 1. It is unfortunate that so little is known about this important process, and in particular about its dependence on material and temperature. On the question of whether stage 1 is to be described by a threshold stress or by a region of lowered rate sensitivity, it is the author's judgment that the latter case is more common. There is a general lack of data at very low strain rate. (The occurrence of grain growth for these long testing times is serious.) The perplexing nature of this problem is illustrated by results for the Sn–Pb eutectic. Several workers have shown independently a linear stage 1 with $m \simeq 0.3$ (Cline and Alden, 1967; Zehr and Backofen, 1968; Surges, 1969). Recently, however, Burton (1971) has reported a threshold stress for this alloy.

There is an additional significant theoretical issue related to this question. Constant slope in stage 1 has been related to a requirement for *slip accommodation* to grain boundary sliding in *stage 2* rather than diffusional accommodation (Hart, 1967 ; Alden, 1969). In other words, the details of the stage 2 mechanism may be elucidated by an understanding of stage 1. (Recall, however, that diffusional accommodation is more easily reconciled to the strong inverse effect of grain size on the strain rate.)

F. Effect of Material

Within the same framework, it is possible to understand why, for example, Zn alloys are superior superplastic materials to Pb alloys. The focus here is on the *stress level of stage 3* rather than stage 1, and specifically the issue is the stress required for slip deformation of an ultrafine-grained alloy.

The grain size dependence of the yield stress for slip is usually described by the equation

$$\sigma_y = \sigma_0 + k_y L^{-1/2}$$

For fine grain size, the second term will be dominant and its magnitude will depend on k_y. It has been found that k_y is large for hexagonal metals and small for cubic metals (Armstrong, 1969). This means that the stress level for stage 3 will be high in zinc and a *large stress range* will be available for deformation by the Newtonian (stage 2) process. m_{max} will be large.

The concept is illustrated in Fig. 27. "Theoretical" curves are plotted for lead at $500°$ K (from Fig. 26) and "pseudo-lead" which deforms by slip in stage 3 at greater or lesser stresses. The rate sensitivity falls as slip deformation becomes easier.

Experimental data are available for stage 3 deformation of fine- and coarse-grained lead and tin alloys (Fig. 28). For lead, a decrease of grain size from 500 μ to 5 μ (Pb–5Cd) raises the stress level only slightly. By contrast, 3.5 μ tin (Sn–5Bi) is much harder than 45 μ tin.

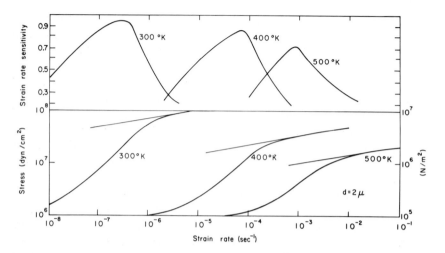

Fig. 26. Theoretical temperature dependent log σ–log $\dot{\varepsilon}$ curves for lead. Contrary to inference from experiment, $m_{\max}$ decreases with temperature.

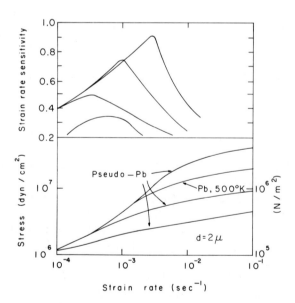

Fig. 27. Effect of stress level in stage 3 on superplastic behavior. The effect is illustrated by calculations for lead and "pseudo-lead" which deforms by slip at greater or lesser stresses.

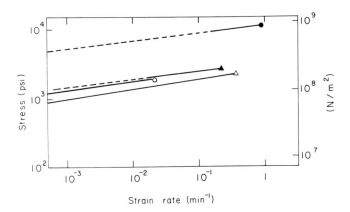

Fig. 28. Experimental grain size effect on the stress level in stage 3. Fine-grained tin is very hard in comparison to lead. This effect increases the stress range for Newtonian deformation in tin and enhances its superplastic properties. ○, Sn, $L = 45\ \mu$, $T = 23°$ C (Cline and Alden, 1967); ●, Sn–5Bi, $L = 3.5$, $T = 23$ (Alden, 1967); △, Pb, $L = 500$, $T = 23$ (Cline and Alden, 1967); ▲, Pb–5Cd, $L = 5.0$, $T = 23$ (Donaldson, 1971).

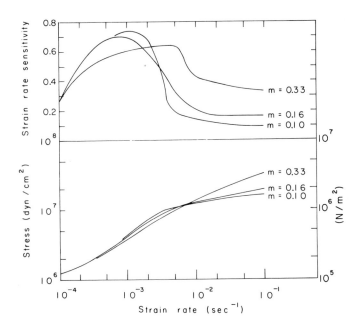

Fig. 29. Dependence of superplastic behavior on the rate sensitivity in stage 3. m_{max} decreases as the stage 3 slope increases.

The rationale for the large k_y in zinc (and some other hexagonal metals) is that hard, nonbasal slip systems are required to a greater extent at small grain size in order to satisfy the compatability requirements. In face-centered cubic metals, these requirements are met by the soft octahedral slip systems. By the same argument, grain size hardening is likely to be important in tetragonal Sn and in *intermediate phases*. Many of these phases have complex crystal structures and may not have five independent, soft slip systems. The excellent properties of the Mg–Al eutectic may be a consequence of this effect. (On the other hand, the Ag–Cu eutectic has better properties than anticipated.)

An additional distinctive feature of zinc is the large rate sensitivity of some nonbasal slip systems (Gilman, 1956). Some slip systems in intermediate phases may have a similar characteristic. However, while this property might increase the slope in stage 3, the effect on stage 2 is apparently a *decrease in maximum slope* (Fig. 29). The "high" slope stage 3 "intrudes" to a greater degree into stage 2 and reduces the slope of the latter.

G. Slip Deformation in Stage 2

All things considered, there is little positive evidence for slip as an important accommodation process in stage 2 and quite a lot of negative evidence. For this reason, the author has chosen to rationalize the various mechanical phenomena in terms of the model of Ashby and Verrall (1973) rather than say, of Hart (1967b). Albeit with considerable speculation, particularly about stage 1, this can be done with some success. However, there are some phenomena, hitherto ignored in this section, which it seems cannot be squeezed into the framework of the AV model.

1. The Zn–Al Eutectoid

The Zn–Al eutectoid alloy does not generally show an S-shaped $\log \sigma$–$\log \dot{\varepsilon}$ curve; the concept of Newtonian deformation in a transition stress range cannot be applicable (Nicholson, 1972). Moreover, the straight lines of "stage 2" vary in slope continuously from about 0.15 at room temperature to 0.45 at 200° C. Yet even at room temperature ($m = 0.1$), a few of the special features of superplastic deformation are evident (Nuttall, 1972). An electron micrograph (Fig. 30) showing grain displacement during deformation, much like that proposed in the AV model, was obtained at 100° C and $m \simeq 0.3$, usually considered a lower limit for superplasticity and far below the predicted $m \simeq 1.0$.

Although further study is required, the impression of a *gradual* onset of superplasticity is strong, including not only the rate sensitivity but the other

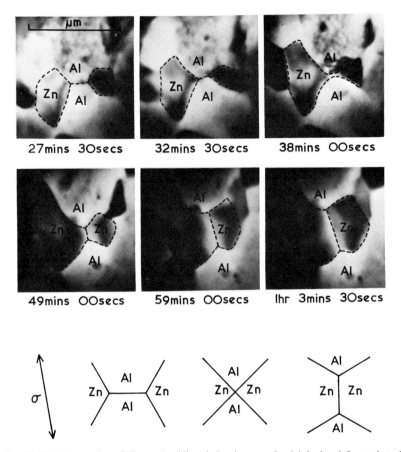

Fig. 30. Relative motion of zinc grains (β) and aluminum grains (α) during deformation of a 4 μ eutectoid foil at 100° C (Naziri *et al.*, 1973).

identifying features of the process (Section I,C). In other words, there may be deformation which is a "little bit superplastic," and this quality may persist over several orders of strain rate (i.e., it shows primary creep but less than the usual amount; there are dislocations stored in the grains but not very many, etc.). It would certainly be easier to understand this effect if slip were generally involved in the deformation but the specific dislocation mechanisms varied. For example, a gradual increase in rate sensitivity with temperature is found in single crystal deformation at high temperature (Weinberg, 1968).

The failure to observe dislocations in the microstructure following superplastic deformation is not definite evidence that dislocations do not move and multiply *during* the deformation. They may be readily absorbed at grain boundaries (by an unknown mechanism) as suggested by the rapid rate of recovery in fine-grained Pb–5Cd (Alden, 1968).

A clue to the behavior of Zn–Al may be that it is a mixture of phases which differ both in *melting point* and *crystal structure*. Near room temperature, the zinc phase may exhibit (relatively) easy diffusion and difficult slip (cf. the dilute Zn–Al alloys at this temperature), whereas aluminum may show the reverse. The "mixed" behavior implied above is then to be associated with a distinctive deformation mechanism in each of the two phases.

The implications of this line of thought are great, since so many superplastic alloys are mixtures of dissimilar phases, whereas the models and analyses of this section treat superplastic alloys as if they were single phase. The diversity of Zn–Al and Sn–Pb (Fig. 15) indicates that the problem is not simple, but Pb and Sn do have more nearly similar melting points.

2. Al–Cu AND Mg–Al EUTECTICS

If superplastic deformation requires only diffusion, the kinetics should be as fast or faster in alloys containing intermetallic compounds as containing only terminal phases. If slip is required, the former will usually require a higher stress, since intermediate phases tend to be difficult to deform by slip. Unfortunately, few data are available. Whether the Al–Cu eutectic deforms less readily than the Mg–Al eutectic (Fig. 14) because the θ phase is harder than the γMg–Al phase, or because Cu atoms diffuse slowly, cannot be decided; but the question is worth asking.

VIII. Summary

The microstructure–property relationships in superplastic alloys are sufficiently well understood for purposes of application, except possibly for the effects of local heterogeneity. The diversity of properties among various alloys and temperature effects can be understood in terms of a varying stress range for a Newtonian transition process (stage 2), although speculation about stage 1 is necessary at this time. Further theoretical study of the average properties of mixtures of dissimilar phases is required.

References

Alden, T. H. (1966). *Trans. AIME* **236**, 1633–1634.
Alden, T. H. (1967). *Acta Met.* **15**, 469–480.
Alden, T. H. (1968). *ASM* (*Amer. Soc. Metals*) *Trans. Quart.* **61**, 559–567.
Alden, T. H. (1969a). *J. Aust. Inst. Metals* **14**, 207–216.

Alden, T. H. (1969b). *Acta Met.* **17**, 1435–1440.
Alden, T. H. (1972). *Phil. Mag.* [8] **25**, 785–811.
Alden, T. H., and Schadler, H. W. (1968). *Trans. AIME* **242**, 825–832.
Armstrong, R. W. (1970). *Proc. Sagamore Army Mater. Res. Conf., 16th, 1969,* pp. 1–28.
Ashby, M. F., and Verrall, R. A. (1973). *Acta Met.* **21**, 149–163.
Avery, D. H., and Backofen, W. A. (1965). *ASM (Amer. Soc. Metals) Trans. Quart.* **58**, 551–562.
Avery, D. H., and Stuart, J. M. (1968). *Proc. Sagamore Army Mater. Res. Conf., 14th, 1967,* pp. 371–390.
Backofen, W. A., Turner, I. R., and Avery, D. H. (1964). *ASM (Amer. Soc. Metals) Trans. Quart.* **57**, 981–989.
Ball, A., and Hutchison, M. M. (1969). *Met. Sci. J.* **3**, 1–6.
Baudelet, B. (1971). *Mem. Sci. Rev. Met.* **68**, 479–487.
Burton, B. (1971). *Scr. Met.* **5**, 669.
Clark, M. A., and Alden, T. H. (1973). *Acta Met.* **21**, 1195–1206.
Cline, H. E., and Alden, T. H. (1967). *Trans. AIME* **239**, 710–714.
Cline, H. E., and Lee, D. (1970). *Acta Met.* **18**, 315–324.
Coble, R. L. (1963). *J. Appl. Phys.* **34**, 1679.
Cook, R. C. (1968). M.A. Sc. Thesis, University of British Columbia.
Cutler, C. P., and Edington, J. W. (1971). *Met. Sci. J.* **5**, 201–205.
Davies, G. J., Edington, J. W., Cutler, C. P., and Padmanabhan, K. A. (1970). *J. Mat. Sci.* **5**, 1091–1101.
Donaldson, K. C. (1971). Ph.D. Thesis, University of British Columbia.
Dunlop, G. L., and Taplin, D. M. R. (1972). *J. Mater. Sci.* **7**, 84–92.
Fields, D. S. (1965). *IBM J. Res. Develop.* **9**, 134.
Friedel, J. (1964). "Dislocations." Addison-Wesley, Reading, Pa.
Garofalo, F. (1966). "Fundamentals of Creep and Creep-Rupture in Metals." Macmillan, New York.
Garwood, R. D., and Hopkins, A. D. (1952–1953). *J. Inst. Metals* **81**, 407.
Gibson, R. C., Hayden, H. W., and Brophy, J. H. (1968). *ASM (Amer. Soc. Metals) Trans. Quart.* **61**, 85–93.
Gilman, J. J. (1956). *Trans. AIME* **206**, 1326–1336.
Hart, E. W. (1967a). *Acta Met.* **15**, 351–355.
Hart, E. W. (1967b). *Acta Met.* **15**, 1545.
Hayden, H. W., and Brophy, J. H. (1968). *ASM Trans. Quart.* **61**, 542.
Hayden, H. W., Gibson, R. C., Merrick, H. F., and Brophy, J. H. (1967). *ASM (Amer. Soc. Metals) Trans. Quart.* **60**, 3–13.
Holt, D. L. (1968). *Trans. AIME* **242**, 25–31.
Holt, D. L., and Backofen, W. A. (1966). *ASM (Amer. Soc. Metals) Trans. Quart.* **59**, 755–767.
Johnson, R. H. (1969). *Des. Eng.* March issue.
Johnson, R. H. (1970). *Met. Rev.* **15**, 115–133.
Lee, D. (1969). *Acta Met.* **17**, 1057.
Lee, D. (1970). *Met. Trans.* **1**, 309–311.
Lee, D. (1971). *J. Inst. Metals* **99**, 66–67.
Lee, D., and Backofen, W. A. (1967). *Trans. AIME* **239**, 1034–1040.
Nabarro, F. R. N. (1948). In "Strength of Solids," p. 75. Physical Society, London.
Naziri, H., Pearce, R., Henderson-Brown, M., and Hale, K. F. (1973). *J. Microsc. (Paris)* **97**, 229–238.
Nicholson, R. B. (1972). In "Electron Microscopy and the Structure of Materials" (G. Thomas, R. M. Fulrath, and R. M. Fisher, eds.), pp. 689–721. Univ. of California Press, Berkeley.
Nuttall, K. (1972). *J. Inst. Metals* **100**, 114–120.
Nuttall, K., and Nicholson, R. B. (1968). *Phil. Mag.* [8] **17**, 1087.

Schadler, H. W. (1968). *Trans. AIME* **242**, 1281.
Stewart, M. J. (1973). *Can. Met. Quart.* **12**, 159–169.
Stowell, M. J., Robertson, J. L., and Watts, B. M. (1969). *Met. Sci. J.* **3**, 41–45.
Suery, M., and Baudelet, B. (1973). *J. Mater. Sci.* **8**, 363–369.
Surges, K. (1969). M.Sc. Thesis, University of British Columbia.
Turner, D. M. (1971). M.Sc. Thesis, University of British Columbia.
Turton, D. (1971). Ph.D. Thesis, University of Manchester.
Underwood, E. E. (1962). *J. Metals* 914–919.
Watts, B. M., and Stowell, M. J. (1971). *J. Mater. Sci.* **6**, 228–237.
Weinberg, F. (1968). *Trans. AIME* **242**, 2111.
Zehr, S. W., and Backofen, W. A. (1968). *ASM* (*Amer. Soc. Metals*) *Trans. Quart.* **61**, 300–313.

Fatigue Deformation of Polymers

P. BEARDMORE and S. RABINOWITZ

Scientific Research Staff
Ford Motor Company
Dearborn, Michigan

I. Introduction

The successful application of engineering materials to structural, load bearing components requires extensive knowledge of (a) the function and stress–strain service environment of the component and (b) the mechanical response of candidate materials to those service demands. One of the most common failure modes experienced in engineering practice occurs in the process of fatigue, namely failure resulting from fluctuating stresses or strains. (*Dynamic* or *cyclic fatigue* is the subject of this treatise and should not be confused with creep or stress–corrosion failure which is often termed *static fatigue* in the literature on glasses and ceramics.) Such engineering failures result from the difficulty of development of safe design procedures to allow for fatigue problems. The results of failure from unanticipated fatigue

267

fractures can be disastrous in terms of both human life and cost. It is indicative of the complexity of the fatigue problem that it has been studied in metals for many decades yet only in recent years has it been possible to realistically design against fatigue. Indeed, this is still not possible in many areas of engineering as evidenced by continuing fatigue failures in practice.

The increasing usage and development of polymers for structural, load bearing applications makes it incumbent upon the design engineer to ascertain that adequate allowance is made in the engineering design for the fatigue properties of the materials. Such data are only sparsely available at present; particularly with regard to the mechanisms involved in the fatigue process and the effects of variables. The development of fatigue data and an understanding of fatigue phenomena in polymers are critical prerequisites to the successful structural application of these materials. Without the generation of such data, such applications will be restricted and will impose a restraint on the usefulness of polymers. In addition, the buildup of knowledge of fatigue phenomena in polymers is necessary to interpret the failure mechanisms when structures do fail.

Failure under a fatigue loading environment may be interpreted in several different ways. In a general sense, it can be specified as being that condition in which the component no longer serves its desired function. Such a definition does not differentiate between separation into two pieces, thermal or mechanical loss of (a certain amount of) strength, or a critical loss in other physical properties. An example of the latter in the case of polymers is the loss of transparency due to the generation of crazes. Thus, in a given technological application, the definition of what constitutes failure depends on which property or properties are critical to that particular application. It is obvious that no singular definition of failure will satisfy all potential applications. For want of a better definition, most studies of fatigue for the purpose of material evaluation rather than for specific applications use some type of strength criterion to define failure. Usually, laboratory data uses separation into two pieces (zero strength) as the criterion for fatigue failure. It is also possible to specify failure as a given percentage of load carrying ability relative to the initial load bearing capability. For the most part, in the work reported here the separation of the specimen into two pieces constitutes failure.

Thermal effects create a particular problem in the cyclic testing of polymers. Adiabatic thermal effects have been shown to be common in fatigue testing of polymers and are always likely to occur unless specific conditions to minimize the effect are maintained. Cyclic loading can result in the dissipation of heat in polymers due to the low damping capacity. In addition, the low heat capacity and conductivity promote the heating effect. With the appropriate combination of load and frequency, it is relatively easy to fail an amorphous polymer in fatigue by melting. Some data on temperature effects have been published (Manson and Hertzberg, 1973), in particular Higuchi and Imai (1970) developed relationships between temperature rise and anelastic strain.

For the purpose of the present article, we are only concerned with true mechanical fatigue failure and thermal effects will not be considered further. In the tests carried out by the present authors, all the data generated were produced such that there were no thermal contributions to the fatigue failures. Other data used here have been selected to minimize the possibility that any thermal effects were present during fatigue. Thus, for present purposes, fatigue is defined as isothermal reversed plastic deformation processes leading to crack initiation followed by incremental crack growth.

The fatigue of polymers may be studied in different ways, as shown schematically in Fig. 1. As indicated by the top arrow, the fatigue life may be

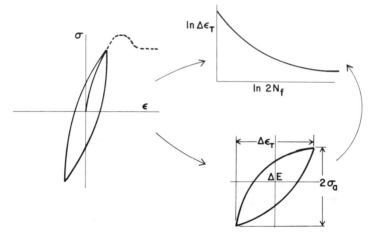

Fig. 1. Schematic of the different approaches to studying fatigue.

evaluated as a function of the constant peak stress or strain, resulting in the conventional S-N curve. This approach takes fatigue fracture as an event rather than a process and is of greatest use for material evaluation and selection for fatigue resistance. Most of the polymer fatigue research in the literature is of this type. This type of S-N representation of fatigue resistance provides no direct information about the mechanism(s) of fatigue failure, i.e. the mechanisms of fatigue crack initiation and crack propagation. However, the micromechanisms of cyclic deformation and fracture can be studied as part of (or in their own right) such S-N studies either in individual polymers or in systems grouped by type. This type of additional information is vital to the materials scientist if polymer microstructures are to be tailored specifically for fatigue applications. With some exceptions, most of the noteworthy efforts with respect to mechanisms of fatigue failure have been directed toward the cyclic growth of fatigue cracks (Manson and Hertzberg, 1973). The limited understanding of the molecular nature of yield and flow in polymers is one of the major reasons for this paucity of research into fatigue mechanisms.

An additional method of studying cyclic deformation and fracture is to consider the fatigue process phenomenologically, and to describe the cyclic behavior in terms of the mechanics of deformation. This is illustrated by the bottom arrow in Fig. 1. This approach seeks to characterize fatigue failure parametrically on a macrolevel, i.e. it characterizes the changes in stress–strain relationships occurring during the fatigue process and gives only peripheral attention to the specifics of deformation in any one material. The phenomenology of cycle dependent deformation in metals has been well characterized in recent years, and great success has been achieved in describing and predicting the fatigue behavior of many metals and alloys. The ultimate advantage of this phenomenological approach is indicated by the arrow on the right in Fig. 1—the use of the mechanics to predict overall fatigue behavior.

An overall understanding of the fatigue behavior of polymers can only be achieved by studying fatigue with all these techniques. Thus, a combination of the S-N and micromechanistic approach together with the phenomenological approach is required. In this chapter, the authors have attempted to present the current state of these separate approaches in the fatigue of polymers. Of necessity, much of the data has had to be drawn from our own studies but the supportive evidence from the literature, where available, is quoted in order to provide as general and complete a picture as possible. In particular, an attempt has been made to develop patterns of behavior and overall characteristics so that some inferences can be drawn regarding the cyclic response of untested polymers.

II. General Fatigue Considerations

The fatigue behavior of rigid polymers can vary widely depending on the testing conditions and the type of polymer being tested. The sensitivity to experimental and material parameters is more pronounced than for (say) metals, and comparison of data should always be made with caution unless these parameters are explicit. In order to develop some "feel" for these factors, the following is intended to provide a summary of the variety of factors necessary to consider in both the experimental conditions and the polymer deformation mechanisms. The necessary references are included for readers sufficiently interested to require further details.

A. Testing Variables

The majority of fatigue studies on polymers have been carried out at room temperature. It should be recognized that room temperature is a relatively high temperature on a homologous temperature basis. Typical glass transition temperatures (T_g) for most common polymers lie in the range 100–250° C. Thus, on a T/T_g scale, room temperature is in the range 0.6–0.8. It is not surprising, therefore, that there is a strong sensitivity to testing conditions—in the equivalent high temperature region for metals (0.6–0.8 T_M), metals show a similar high sensitivity. One example of cognizance of the effective high temperature of testing is the potential creep component which can enter into the fatigue response. The separation of the fatigue and creep components is the subject of wide investigation in the high temperature fatigue of metals. Only a limited effort in similar demarcation has been made to date in polymers (see Section V).

Basically, testing variables can be broken into two major categories, namely (1) the imposed mechanical test conditions, and (2) the testing environment. Discussion of these experimental variables has been given previously (see Dillon, 1950; Andrews, 1969). Consequently the treatment here will be brief and deal only with those factors pertinent to the present work.

In the category of imposed mechanical test conditions lie factors such as controlling parameter (stress or strain) and type of control function (sine wave, saw tooth, square wave, etc.). The choice of control parameter is important since stress control and strain control can produce very different results depending on the strain level. In general, as long as the imposed strain level is nominally elastic (the macroscopic stress–strain curve is essentially linear) then both stress control and strain control are equivalent. However, if the deformation involves significant nonelastic strains then strain control is the most meaningful parameter under which to control fatigue tests to prevent such problems as runaway creep failure. It can be assumed that the tests were always carried out under strain control limits unless specifically stated otherwise. In addition, the control mode was always fully reversed tension–compression. Note that in such a mode, fully reversed strain control limits do not produce a fully reversed stress response in polymers because of the strength differential effect (see Section III).

The rate of imposition of the strain limits, i.e. the strain rate, is also an important part of the control parameter. Many polymers exhibit a high strain rate sensitivity of the deformation behavior and this strain rate (or frequency) can be a major variable particularly in the low cycle fatigue life (high strain) regime. Also, testing at high frequency can produce self-heating of the specimen. In the present tests, particular experimental care was taken to ensure that there were no macroscopic heating effects. Control was maintained such that the temperature rise in the specimen was always less than $5°$ C. In general, all tests were performed at a constant strain rate of $0.01 \sec^{-1}$. Under these conditions, no thermal effects contributed in the fatigue mechanisms and all the data refer to mechanical fatigue failure.

The constant strain rate of $0.01 \sec^{-1}$ was maintained by using a control cycle of the sawtooth form. The use of other types of control function such as a sine wave suffer from the fact that the strain rate varies continuously throughout the cycle; in fact, the relatively large fraction of time spent close to the peak of the sine wave (essentially zero strain rate) promotes any creep component and can change the fatigue life considerably. The sensitivity to different control cycles is something which must ultimately be investigated in depth for polymers since in technological applications the materials are unlikely to undergo well characterized stress–strain reversals. Thus, the use of control cycles to allow rest periods and different sequencing effects will become more important as structural applications increase.

The extreme sensitivity of many polymers to the environment in which they are tested must always be borne in mind in comparison of data. Unless stated otherwise, all the fatigue data reported here are of tests performed in air.

B. Material Variables

The following discussion of material variables and deformation modes is not intended to be complete. It is included so that readers unfamiliar with polymer behavior will have a capsule summary in order to have some feeling for the material variables at play under fatigue testing. For more detailed discussions of polymeric structure and deformation, readers are referred to a recent review (Kambour and Robertson, 1972).

Fatigue failure mechanisms in any material depend upon the occurrence of plastic deformation, albeit on a local scale or a macroscopic scale. Rigid polymers can deform by a variety of processes depending upon the structure of the particular polymer and the testing conditions imposed on the specimen. Some knowledge of the competitive deformation modes is necessary to understand the particular response of a given polymer.

The macroscopic stress–strain behavior of polymers reflects the ability to deform by elastic, anelastic, viscoelastic, and viscoplastic processes. The separation of these processes is not as straightforward as, say, the elastic, anelastic, and plastic components in metals. In particular, that part of the deformation which is equivalent to "plastic" deformation in the fatigue sense cannot be readily assessed. The "memory" effect by which polymers can revert to their original dimensions upon heating a deformed specimen is a profound demonstration of the unique character of these materials.

In terms of microstructure, polymers may be classified as single phase (amorphous polymers), two-phase (semicrystalline polymers, composites), and multicomponent systems (composites). Amorphous, glassy polymers, e.g. polycarbonate (PC), poly(methyl methacrylate) (PMMA), and polystyrene (PS), are single phase materials of nominally homogeneous structure. Recent studies have, however, indicated that the structure may be comprised of localized, ordered regions (Siegmann and Geil, 1970; Yeh, 1972). Since these ordered zones are of the order of 1000 Å in diameter, on a light microscope level of observation the microstructure appears homogeneous. Semicrystalline polymers, e.g. nylon, polypropylene, and polyoxymethylene, consist of two phases—crystalline aggregates in a matrix of amorphous polymer of the same composition. The crystalline regions (spherulites) vary in size and dispersion depending on the thermomechanical history of the material. Likewise, polymer composites consist of two or more phases with an infinite variety of combinations.

Homogeneous flow processes, that is deformation modes occurring throughout the bulk of the material, can occur in different ways depending upon the microstructure of the polymer. In amorphous polymers, segmental motion of the molecular chains is believed to be the basic deformation mode. Precise models of the mechanism are still unconfirmed but the overall kinetics of the yielding process are generally described in terms of the Robertson (1966) theory of yielding. In polymers of heterogeneous microstructure, additional deformation modes come into play (Kambour and Robertson, 1972). In semicrystalline polymers, in addition to deformation within the amorphous matrix, a spectrum of deformation modes is possible within the crystalline spherulites. Deformation modes such as slip, twinning, crystal rotation, unravelling and fibrillation are available to accommodate flow. In any multicomponent system, the specifics of the interfaces between the phases will dictate any additional deformation mode contribution in these regions.

Inhomogeneous flow occurs predominantly in the form of crazes. Crazing is probably the most studied form of deformation in polymers and the complete topic of crazing and fracture has been the subject of a recent, comprehensive review (Rabinowitz and Beardmore, 1972). A craze is a planar defect formed by intense, highly localized plastic deformation. The plane of the craze forms at right angles to the principal tensile stress and effectively stress relieves a volume of surrounding material in a manner similar to a crack. Crazes do not form under compressive stresses. Once formed, crazes grow both in area and in thickness. With respect to the initiation of fracture, crazing is essentially a surface phenomenon. Since the growth of crazes is continuous and time dependent, the part of the craze at or nearest the surface is accordingly at a later stage in its development and is more likely to contain the crack initiation process. The importance of crazes lies in the relationship between crazing and fracture in many amorphous polymers. The so-called brittleness exhibited by these materials is a direct consequence of deformation occurring in the form of crazes rather than as bulk deformation. Thus, in any deformation mode (creep, fatigue, etc.) the development of crazes by the particular testing condition is an important aspect of the resistance of the material in that particular testing mode. For further detailed information on crazing, readers are referred to Rabinowitz and Beardmore (1972) and Kambour (1974).

III. Phenomenological Cyclic Deformation Behavior

The flow properties of a material may be markedly changed by repeated plastic strains. The documentation and description of these transients in flow properties has been well developed for metals and such characterization of the phenomenological cyclic deformation behavior is now used as a standard procedure. The success of this approach in metals in providing material parameters directly relatable to the cyclic response has led to efforts to develop similar cyclic descriptions for polymers. In this section, the current state of this phenomenological description is detailed, with particular emphasis on consistent trends.

A. Transient Deformation Behavior

The deformation resistance of a material may increase (cyclic hardening), decrease (cyclic softening), or remain essentially unchanged (cyclic stability). The specific response of a given material depends upon the initial state of the material and the particular conditions under which testing is performed. Figure 2 is a schematic illustration of cyclic hardening and cyclic softening under control conditions of completely reversed *strain* cycling. In cyclic hardening, the stress (both tension and compression) required to maintain the fixed strain limit increases with increasing number of cycles. Conversely, in cyclic softening the stress required to maintain the fixed strain limit

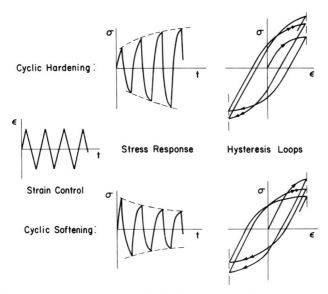

Cyclic Hardening :

Strain Control

Cyclic Softening :

Stress Response   Hysteresis Loops

Fig. 2. Schematic representation of cyclic hardening and cyclic softening under reversed strains.

decreases with increasing number of cycles. In the case of cyclic stability, the stress would remain unchanged throughout the fatigue life. It is immediately obvious from Fig. 2 that the flow properties of a material as determined from a simple monotonic tension or compression test are not representative of the stress–strain relations in fatigue if cyclic hardening or softening occurs. A detailed knowledge of this transient deformation behavior is consequently extremely important to understand (and predict) the response of (say) a structure to the type of random cycling usually experienced in practical applications.

In the schematic representation of Fig. 2, the cyclic response as evidenced by the change in stress response to fixed strain control is depicted. However, it is also possible to study cyclic hardening and softening by controlling a test to fixed *stress* limits and measuring the resulting cyclic strain as a function of stress cycles. Thus, in *stress* control tests, cyclic hardening is manifested as a decrease in strain range; conversely, an increase in strain range indicates cyclic softening. It is important to point out that *stress* control is generally a less desirable mode of testing if significant plastic strain is involved. As will become clear later in this section, unstable testing conditions (runaway creep) can result from *stress* control tests. This problem is alleviated in *strain* control. Also, in practice it is difficult to control *stress* and instead *load* control is used—since *load* control and *stress* control are not equivalent, the problems mentioned above are compounded. (It should be noted that with the sophisticated equipment now available, in particular computer controlled systems, it is feasible to accurately control *stress*.)

At the risk of being repetitious, it should be reemphasized that in all the examples to be cited below, the transient-deformation behavior is induced as a purely mechanical effect—no thermal contribution occurred. Much of the data in this section has been generated by the present authors and very little information has appeared in the literature for comparison purposes. However, there is sufficient published data to confirm the uniformity of our data with other investigators, and thus the overall trends discussed below are believed to be comprehensive. In the remainder of this section on transient deformation behavior, examples of the various types of cyclic response observed in polymers will be discussed, together with some interpretation of

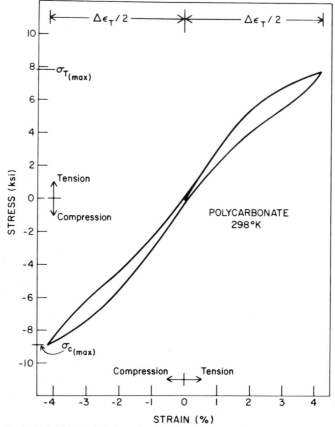

Fig. 3. Typical initial hysteresis loop for ductile polymers—example shown is for PC cycled at 298° K. The various parameters used to describe the test conditions are also indicated.

the behavioral patterns. In order to describe the experimental results, data of the type illustrated schematically in Fig. 2 will be presented. In addition, however, the changes in stress–strain relationships during cycling will also be presented and it is consequently important to define the parameters and indicate some unusual aspects not encountered in the more readily familiar data on metals.

The typical first cycle stress–strain response (hysteresis loop) of a ductile polymer (Rabinowitz and Beardmore, 1974) is shown in Fig. 3. The propeller-type shape of the loop is in marked contrast to the symmetric (about the tip to tip axis) loops obtained in metals. The unusual shape is a direct consequence of the time dependent flow mechanism operative in polymers, and Fig. 3 can be interpreted to mean that the total strain is accommodated primarily as viscoelastic and anelastic strain with relaxation/recovery times comparable to the cycle time. The small amount of strain accommodated as plastic strain—the closure strain of the loop at zero stress—is small.

The parameters describing the cyclic test are indicated in Fig. 3. Note that the peak compressive stress $\sigma_{c(max)}$ is (generally) greater in magnitude than the peak tensile stress $\sigma_{T(max)}$ for equal peak strains. This flow strength differential, called the S-D effect, is characteristic of the uniform nonelastic deformation in ductile polymers. The ratio σ_c/σ_T is the same in the initial cyclic response as it is in monotonic deformation. Again, the occurrence of a significant S-D effect in ductile polymers highlights the inequality of equal strain limit testing versus equal stress limit testing. Unless otherwise noted, all the transient deformation studies described here were carried out under conditions of equal strain limits.

1. Homopolymers

Homopolymers may be subdivided in terms of (molecular) microstructure into semicrystalline and amorphous groups. The different modes of deformation available to the two types are summarized in Section II,B. The transient deformation behavior of homopolymers will be discussed first in terms of crystalline polymer behavior followed by the response of amorphous polymers.

In general, crystalline polymers exhibit a high degree of ductility at room temperature. Typical monotonic stress–strain curves for nylon and polyoxymethylene are shown in Fig. 4. Note the different types of yielding response. In nylon discontinuous yielding (cold drawing) occurs, while in polyoxymethylene yielding occurs smoothly (continuous yielding). (The discontinuous yield pattern of behavior is the most commonly observed pattern for ductile homopolymers.) In Figs. 5 and 6 are shown the experimental strip

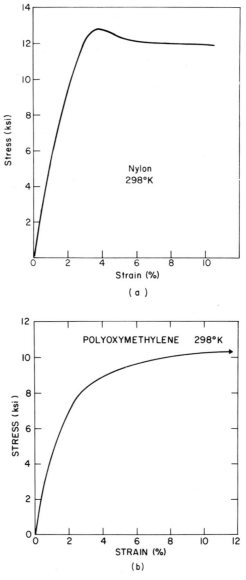

Fig. 4. Tensile stress–strain curves at 298° K for (a) nylon, and (b) polyoxymethylene.

chart readings of strain controlled fatigue tests on polypropylene and nylon, respectively. The lower half of Fig. 5 shows the imposed cyclic variation in strain at constant-strain rate. The upper half of the figure is a continuous trace of the stress response to the imposed strain variation. The envelope of

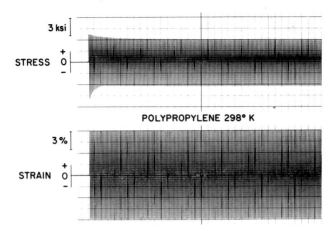

Fig. 5. Reproduction of experimental strip chart recording for polypropylene cycled in strain control at 298° K. (Only the softening region and initial portion of stable stress state is shown.)

the stress trace indicates the cyclic evolution of the stress at peak strain in tension and compression.

Examination of Figs. 5 and 6 reveals the immediate change in stress–strain relations which occur. The stress to maintain the fixed strain decreases in the initial few cycles—termed the transition region—and subsequently attains a new steady state which remains essentially constant for the remainder of the fatigue life prior to the crack propagation stage. The overall response is, obviously, cyclic softening. Confirmatory evidence of cyclic softening in nylon has been published (Tomkins, 1969). The development of a new steady state relation between stress and strain for the majority of the life emphasizes that, just as in metals, the monotonic stress–strain relation is not representative of the mechanical response of crystalline polymers under

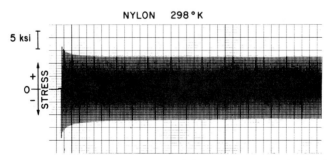

Fig. 6. Reproduction of experimental record of the stress response of nylon in a strain controlled fatigue test at 298° K.

cyclic stresses or strains. As discussed in Section III,B, this new unique relation between cyclic stress and strain can be described and utilized as an important characteristic material property. The general conclusion to be drawn from these data is that (ductile) semicrystalline polymers always cyclically soften.

Typical changes in the hysteresis loop corresponding to the stress decreases of Figs. 5 and 6 are shown in Figs. 7 and 8. The resistance to nonelastic deformation and the cyclic hysteresis undergo marked changes in the transition period. In Figs. 7 and 8 the continuous trace recordings throughout the transition period are reproduced showing the nested hyster-

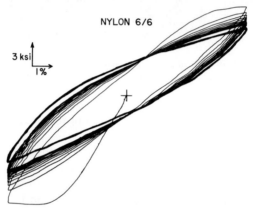

Fig. 7. Typical continuous trace of hysteresis loops for a nylon specimen cycled between fixed strain limits at 298° K showing the immediate continuous softening. The heavy trace corresponds to the cyclic stable steady state (softened) condition.

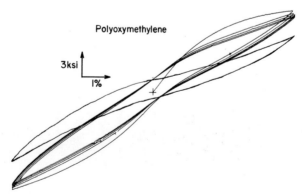

Fig. 8. Typical continuous trace of hysteresis loops for a polyoxymethylene specimen cycled between fixed strain limits at 298° K. A hysteresis loop typical of the cyclic stable steady state is superimposed.

esis loops; the loop characteristic of the stable cyclic state is superimposed to highlight the changes. Note the decrease in the tensile and compression peak stresses and the concomitant changing loop shape.

The amount of softening, as measured by the stress decrease, is a function of the applied peak strain. As the strain increases, the amount of softening increases. At low strains, softening occurs more gradually over a larger number of cycles. At sufficiently low strains no softening occurs, i.e. the stress remains constant at the initial stress level—this generally corresponds to strain levels below the fatigue limit. The way in which the amount of softening as a function of strain level can be characterized, the cyclic stress–strain curve, is described in Section III,B.

Amorphous polymers behave under cyclic loading in a manner similar to semicrystalline polymers provided the overall monotonic properties are similar. Specifically, if an amorphous polymer has a large degree of ductility, the overall cyclic response will be similar to that described above (Rabinowitz and Beardmore, 1974). Thus, the cyclic response of polycarbonate at room temperature and PMMA at temperatures higher than $\sim 40°$ C is cyclic softening. The monotonic stress–strain curve for PC, Fig. 9, demonstrates the large degree of ductility. In Fig. 10, an annotated reproduction of the strip chart recording of a strain controlled fatigue test of PC is shown. Although the overall response of PC is cyclic softening, in a manner analogous to semicrystalline polymers, an additional region can be seen in the transient deformation behavior. Prior to the transition (or softening) region, an incubation period is manifest in which the stress–strain response on the first cycle—the monotonic behavior—is maintained essentially unchanged. After the transition period, the response is exactly comparable to the softening pattern described above.

The incubation period is a variable of the imposed strain amplitude. At high strain amplitudes, softening occurs immediately. In Fig. 11a the nested hysteresis loops tracing the immediate softening are shown—compare this figure with Figs. 7 and 8. The incubation period increases as the imposed peak strain decreases; at the same time, the stress decrease (amount of softening) in the transition decreases. As in semicrystalline polymers, at sufficiently low strains the material remains cyclically stable until a fatigue crack forms. The change in energy dissipation per cycle, amounting to a sixfold increase, is shown in Fig. 11b. Note the differences in hysteresis loop geometry in the cyclically softened, stable state in Figs. 7, 8, and 11. This will be discussed in more detail below relevant to the deformation mechanisms involved in the softening process.

It is fairly clear from the above data that ductile polymers, both semicrystalline and amorphous, undergo significant cyclic softening. However, a large number of amorphous homopolymers cannot be classed as ductile.

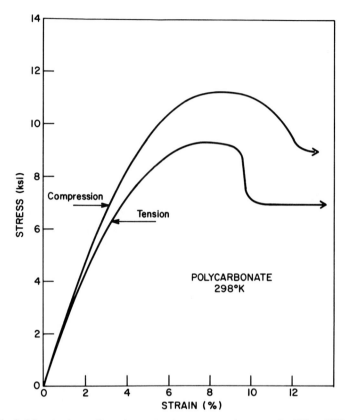

Fig. 9. Monotonic tensile and compressive stress–strain curves for PC at 298° K.

These materials can be classified from completely brittle[†] to semiductile. For the present purposes, we can define completely brittle as polymers exhibiting only elastic behavior to fracture under normal tensile testing conditions. Many epoxies fall into this category. Semiductile polymers are those which demonstrate considerable deviation from elastic behavior but have ductilities less than, say, 10%. Polymethylmethacrylate (PMMA) at room temperature is an example of a semiductile polymer, Fig. 12.

The cyclic response of limited ductibility (semiductile) polymers will be detailed with respect to the behavior of PMMA at room temperature (Rabinowitz and Beardmore, 1974). Most of the tests have been performed on this material. PMMA is intrinsically ductile as evidenced by the high ductility

---

[†] The term brittle is used in a macroscopic sense, i.e. no outward evidence of plastic deformation occurs. It is recognized that plastic deformation on a local scale is always involved in "brittle" fracture.

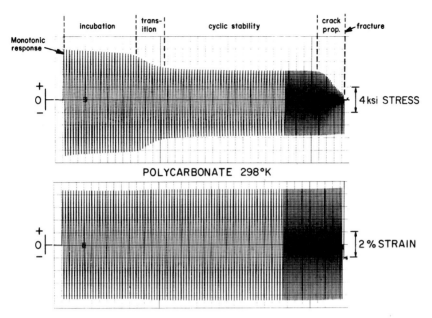

Fig. 10. Reproduction of experimental strip chart recording for PC cycled in strain control at 298° K. (The closer spacing of the cycles towards the end of the test is due to a change in chart speed.)

exhibited in a uniaxial compression test (Beardmore, 1969), in fact, in this sense it is more ductile than PC. However, in uniaxial tensile deformation the intrinsic ductility is never manifest because of the simultaneous generation of crazes. The characteristic craze growth and craze breakdown

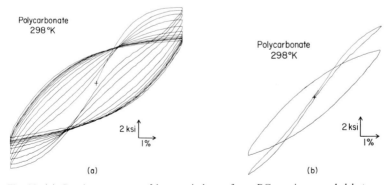

Fig. 11. (a) Continuous trace of hysteresis loops for a PC specimen cycled between high strain limits showing continuous softening to steady state (heavy trace) condition; (b) typical hysteresis loops for PC, showing the initial loop and a loop in the fully cyclically softened condition.

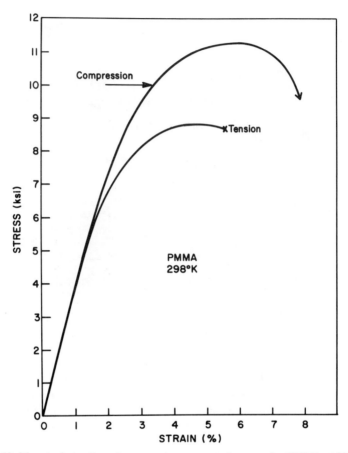

Fig. 12. Monotonic tensile and compressive stress–strain curves for PMMA at 298° K.

processes lead to crack formation and subsequent fracture at low strain, Fig. 12, preempting the continued development of homogenous flow and high ductility.

The typical stress response to constant-strain rate strain controlled cycling of PMMA at room temperature is shown in Fig. 13. A small amount of continuous softening occurs throughout the fatigue life, evidenced by the gradual decrease in the tensile and compressive peak stresses. Relative to the softening exhibited by ductile polymers, the overall softening in PMMA is small. The entire fatigue life is spent in the incubation period such that little softening occurs. Thus, in polymers of limited ductility, a small degree of cyclic softening can be anticipated.

In the case of brittle or semibrittle polymers, the monotonic tensile stress–

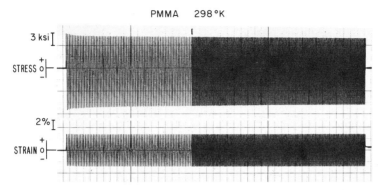

Fig. 13. Reproduction of experimental strip chart recording for PMMA cycled under strain control at 298° K.

strain curve is essentially linear. Thus, in the range of strains of interest in fully reversed cyclic deformation—from zero to ± the tensile fracture strain—the bulk compliance is essentially elastic. Phenomenologically, these polymers are stable in cyclic deformation, i.e. the monotonic stress–strain relation is maintained throughout the fatigue life. The only instability which can occur is at very short lives in the cases where extensive craze growth occurs—this is discussed under anomalous behavior in Section III,A,4. However, this instability can be easily recognized because it is limited to tensile softening only. In general, therefore, polymers exhibiting essentially elastic behavior to fracture will be cyclically stable.

The transitions between the three types of behavior discussed above are not always sharp. A mixed fatigue response could result from testing a polymer under the appropriate conditions of strain rate, frequency, temperature, environment, etc. However, the three general classes of behavior defined above—ductile polymers, limited ductility polymers, and brittle polymers—can be used as a basis to conclude what the cyclic response will be.

2. COMPOSITES

The microstructure of composites is distinct from that of homopolymers in the sense that one or more additional (generally uncompatible) phases are introduced into the material. The presence of these extra component(s) leads to the potential development of deformation modes either in addition to those which occur in homopolymers or the development of specific deformation modes to a greater degree. An example of the first type would be interfacial strain accommodation between the matrix and the dispersed phase; an example of the second is the promotion of crazing to occupy a large volume fraction of the sample by providing craze nucleation sites at matrix/particle interfaces.

The only data on the mechanics of response in fatigue of composite materials have been generated on ABS and glass fiber reinforced nylon (Beardmore and Rabinowitz, 1974). ABS is a typical example of the class of composites in which ductility is promoted by dispersion of rubber particles in a glassy matrix. Glass fiber reinforced nylon is typical of fiber strengthening of a ductile matrix, in which some degree of ductility is maintained in the composite. (Brittle composites, which are elastic up to fracture, behave essentially like brittle homopolymers, i.e. they are cyclically stable.) In Fig. 14 the mono-

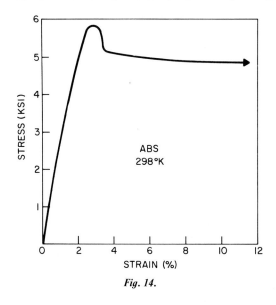

Fig. 14.

tonic stress-strain curve of ABS is shown, and in Fig. 15 the stress response to cycling between fixed strain limits is depicted. Note that immediate cyclic softening occurs as in ductile (heterogeneous microstructure) homopolymers (see Figs. 5 and 6). Thus, although the strain accommodation (ductility) is achieved in a very different mode than in homopolymers, the overall response of cyclic softening is the same. However, there are discrete differences. The cyclically softened ABS never obtains a fully stable relation between the stress and strain. Examination of Fig. 15 shows that the stress decreases slowly throughout the softened "stable" state region. This gradual continuous softening is also accompanied by continuous changes in hysteresis loop shape, Fig. 16. This will be discussed further in Section III,A,3 as evidence of the mechanisms involved.

The tensile behavior of nylon reinforced with 13% glass fibers is shown in

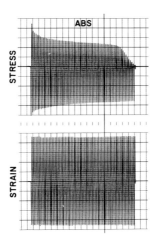

Fig. 15. Reproduction of experimental strip chart recording for ABS cycled under strain control at 298° K.

Fig. 17. The material yields smoothly and exhibits a fracture strain of 14%. In strain controlled cycling, the stress response is identical to that of ductile homopolymers, Fig. 18, namely immediate softening followed by a cyclic steady state for the majority of life prior to crack propagation. The nested hysteresis loop characteristics of the initial softening region are shown in Fig.19. The effect of increasing the volume fracture of the glass fiber to 33% on both the monotonic behavior, Fig. 17, and the stress response is one of degree rather than kind. The hysteresis loops are essentially similar to those shown in Fig. 19. One discernible effect of the increased volume fraction of fiber is the change from a cyclically softened steady state, Fig. 18, to a

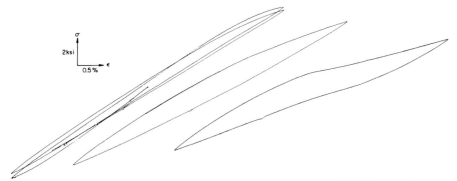

Fig. 16. Typical hysteresis loops for ABS at 298° K showing the initial loop, a loop characteristic of the response after the primary softening stage, and a loop characteristic of the deformation response just prior to the development of a crack (cf. Fig. 15).

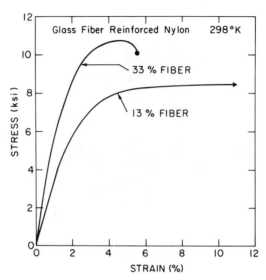

Fig. 17. Tensile stress–strain curves at 298° K for nylon reinforced with 13% and 33% glass fibers.

Fig. 18. Reproduction of experimental record of the stress response of 13% glass fiber reinforced nylon in a strain controlled fatigue test at 298° K.

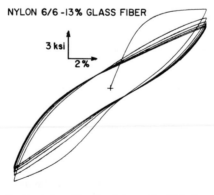

Fig. 19. Typical continuous trace of hysteresis loops for 13% glass fiber reinforced nylon at 298° K—corresponds to initial portion of Fig. 18.

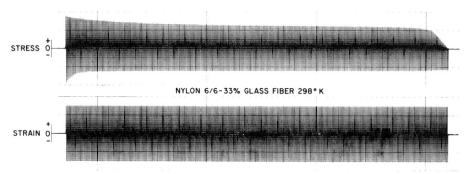

Fig. 20. Reproduction of experimental strip chart recording for 33% glass fiber reinforced nylon cycled under strain control at 298° K.

cyclically softened state showing a gradual continuous stress decrease, Fig. 20, reminiscent of the response shown for ABS, Fig. 16.

The above examples serve to illustrate the generality of the phenomenon of cyclic softening for ductile polymers. The constancy of the softened steady state stress is, however, obviously sensitive to the compliance modes operative in the materials, and the presence of second phases (either in particulate or fiber form) promotes this instability. Further, as evidenced by the volume fraction of fiber effect in nylon, the amount of strain accommodated by the alternative deformation mechanism determines the degree of instability of the cyclic steady state.

3. INTERPRETATION OF BEHAVIORAL PATTERNS

The cyclic stress–strain response, as described above, of ductile polymers is cyclic softening. The transition from the monotonic stress–strain response to a cyclic steady state may be described phenomenologically as a decrease in material resistance to nonelastic strain with reversed deformation. In this section, the various stages of the cyclic response for ductile polymers will be discussed in terms of the possible mechanistic interpretation of the phenomena.

In the transition (softening) region, the stress–strain response changes rapidly; the peak stress decreases and the cyclic hysteresis increases from cycle to cycle (see Fig. 11). It is obvious that the resistance to nonelastic deformation decreases and the cyclic energy dissipation increases continuously through the transition region. Mechanistically, it is important to emphasize that the softening is a true material response and is not a manifestation of additional factors. There is *no* thermal contribution to the material softening—thermal softening can be induced by high strain and/or high strain rate cycling but experimental conditions were carefully controlled to preclude this occurrence. Cyclic softening is *not* a manifestation of slow crack growth, as evidenced by the approximate symmetry of both the tensile and compressive stresses. The transition (softening) region thus represents a true mechanical instability. In all ductile polymers, this transition region extends over only a few cycles, constituting a negligible fraction of the fatigue life. However, the changes that are induced in both mechanical response and molecular packing in the transition region have a profound effect on the fatigue resistance of ductile polymers. These phenomena are revealed in studies of the stress–strain behavior of ductile polymers in the cyclic steady state (see later).

The processes of large scale flow in ductile polymers probably involve the formation, growth, multiplication, and motion of mobile defects, the particular types of defect depending upon the polymeric system. The formation and multiplication of mobile defects takes place heterogeneously at adventitious sites of strain localization (compositional or morphological variations, flaws, inclusions, etc.). The process is most likely autocatalytic, in the sense of the Robertson theory of polymer yield (Robertson, 1966) in which the formation of one defect makes the formation of neighboring defects easier. Thus, softening occurs in an avalanche mode in which the rate of defect production and accumulation is a sensitive function of the test limitations.

In structurally inhomogeneous ductile polymers, regions of high local stress exist and defect production in these regions can occur at an accelerated rate. In addition, by virtue of structure or thermomechanical history, a steady state population of defect multiplication centers may exist in the virgin state. The combination of these effects allows immediate defect generation and multiplication, leading to immediate cyclic softening. Thus in ductile polymers with inhomogeneous microstructures, namely semicrystalline polymers and ductile composite polymers, the transition region or softening is the immediate response to cyclic deformation.

In contrast, in microstructurally homogeneous ductile polymers such as amorphous polymers an incubation period occurs prior to the transition region, Fig. 10. This sort of incubation period is not unique to cyclic deformation. In constant stress (creep) experiments on discontinuously yielding metals and ductile polymers, an incubation period of very low creep rate (delay time) is often encountered between the immediate elastic response and the onset of primary creep or neck formation. These similarities suggest that phenomenologically the deformation mechanisms can be described in an analogous manner (in creep and fatigue, in metals and polymers). The incubation period (or delay time) then constitutes an early stage in the deformation when the population of mobile defects is too low to account for a significant nonelastic-strain rate, but is increasing slowly. When the mobile defect population attains a critical value, softening (or primary creep) occurs.

The rate of defect production and accumulation, and hence the extent of the incubation period, varies with the state of the material and the test configuration. For example, the integrated time-under-load prior to softening in fatigue is always less than the delay time in constant stress deformation at the same initial peak stress. Since the time at *peak* stress is only a small fraction of the total incubation period in the cyclic experiments, and since the rate of defect production is very stress sensitive, the localization of plastic strain that is promoted in reversed deformation must provide a potent driving force for defect production. Also, if a sufficient mobile population exists on (or immediately after) the first fatigue cycle, incubation is markedly reduced or even eliminated. Thus, as noted earlier, the incubation period is absent when polycarbonate is cycled above a threshold strain (or stress) level (cf. Fig. 11a); the production of mobile defects on the first tensile stroke satisfies the nonelastic-strain rate requirements for cyclic softening.

In general, multiphase microstructures show immediate softening with no incubation period whereas single phase, homogeneous microstructures tend to exhibit an incubation period. It is not meant to suggest that the character of the mobile defects, nor the mechanisms of their formation, multiplication, and interaction are similar within these material groupings; only the phenomenology of cyclic deformation is being described. In spite of the commonality of cyclic softening in all ductile polymers, it is noteworthy that the opposite response, cyclic hardening (common in many metals displaying a variety of yield behaviors), is never manifest in polymers. Metals cyclically harden for the same basic reasons that they work harden in monotonic deformation—mobile defect exhaustion, annihilation, mutual interference, etc. Work hardening in polymers has not been studied simply because the molecular orientation hardening that invariably accompanies large scale bulk flow overwhelms and obscures any defect-hardening mechanism. However, the fact that polymers do not cyclically harden would indicate that such defect work hardening is small.

In the fully softened condition, a ductile polymer maintains a new, steady state stress–strain relation as cyclic deformation proceeds. The energy dissipated in driving the specimen through a complete cycle remains constant from one cycle to the next, and dynamic strain recovery processes balance the cyclically applied strain to give a constant cyclic nonelastic strain range. The large defect population and molecular rearrangement (see Fig. 21) produced in the softening process are sustained in the continuous deformation of the cyclic experiment. In a broader sense, however, the softened material can be considered a new material state, the properties of which can often be examined outside the fatigue situation that created it. For example, when a sample of PC is cycled into the steady state response and then removed from the test equipment after returning the stress on the sample to zero, the new softened state persists indefinitely at room temperature. We

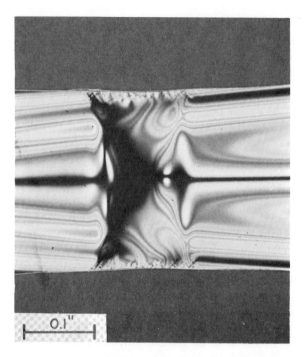

Fig. 21. Thin section through softened region in PC specimen.

designate this the softened *cyclic stable state.* It must be emphasized, however, that cyclic softening and the attainment of a cyclic steady state do not imply such static stability. For example, ductile PMMA (above about 40° C and below T_g) manifests a softened steady state response during cycling, but recovers rapidly to the initial condition on removal from the fatigue test; the recovery is so rapid that attempts to quench PMMA to room temperature to preserve the softened state at zero applied stress have thus far proved unsuccessful. On a qualitative assessment based on the ductile polymers included in this study (all but PMMA tested at room temperature only), the attainment of cyclically stable behavior depends on the location of the temperature range of ductile response relative to the temperature of strong molecular relaxations. In other words, independent of the fatigue considerations necessary to produce the softened state, stability depends on the rate of thermally activated recovery processes.

The discussion of the mechanics of cyclic softening has been general to the ductile polymer class. Observations have been accumulated at room temperature on PC, nylon, nylon with 13% glass fibers, ABS, polyoxymethylene, polypropylene, and a partially cured epoxy, and on PMMA at 40° C and

60° C. While the phenomenology of deformation is uniform within the ductile class, the specific mechanisms of plastic flow vary from one polymer to the next. For example, marked structural changes accompany cyclic softening in PC. Figure 21 is a thin longitudinal section taken through the center of a PC sample after cycling to the softened stable state condition. The softened material is clearly delineated as a region of marked molecular reorientation. Little dimensional change can be detected in the softened region on examination at low magnification—this is in marked contrast to neck formation in monotonic deformation. However, density measurements on softened PC indicate a 1% *increase* in density in the cyclic stable state. Similar observations have been made on softened samples of nylon. In nylon, the softened zone was delineated by a hardness trace along the specimen gage length and by X-ray diffraction studies. The X-ray studies indicate a decrease in lattice d-spacing (compared to the initial material) and an increase in crystallite perfection for the softened material; again, softening produces an increase in material density.

Note that the densification of ductile polymers in cyclic deformation results in a decrease in resistance to nonelastic deformation. It is difficult to account for this behavior on any free volume model of nonelastic deformation in rigid plastics. On the other hand, it is interesting to speculate on the possible correlation between these mechanical observations and the existence of ordered regions (presumably of higher density than the bulk sample) in amorphous thermoplastics (Siegmann and Geil, 1970; Yeh, 1972). On the assumption that the order regions have a lower flow stress than the bulk polymer—resulting, for example, from molecular packing efficiencies and a concomitant diminution in inter/intramolecular constraints (entanglements, configurational contortions, etc.) on large scale nonelastic flow—cyclic softening may reflect the stress induced growth and coalesence of these regions. Whatever the mechanism, however, the softened stable state provides a unique material condition in which to study the process(es) of large scale nonelastic flow in ductile polymers. In conventional constant stress and constant strain rate experiments, the transition into large flow is always accompanied by extreme strain localization (necking) and significant molecular orientation. The change in polymer properties with molecular orientation so overwhelms any other mechanical response, that observation of the underlying deformation mode may be totally obscured. In the softened stable state, a significant population of mobile defects can be established with no concomitant orientation; on redeformation of the softened material, the "defect" structure and properties can be studied in isolation.

It was pointed out earlier that some ductile composite polymers do not achieve a true *cyclic stable state* in that the stress decreases slowly throughout the "steady" state portion of life. Both ABS and glass fiber reinforced nylon (33%) exhibit this phenomenon, Figs. 15 and 20. If it is possible to generalize from these limited data, it appears that ductile composite polymers with large volume fractions of second phase exhibit this response. Note that these two types of composites are fundamentally different in the manner in which the ductility is produced. In ABS, a brittle matrix of styrene-acrylonitrile is made ductile by the introduction of soft, rubber particles. In glass fiber reinforced nylon, a very ductile matrix (nylon) is stiffened, and the overall ductility reduced, by the introduction of hard, brittle fibers. Thus, specific deformation mechanisms cannot be responsible for the continuing decrease in the "steady" state stress level in both types of composite. It is likely that the decrease reflects a continuous change in deformation modes between two competitive mechanisms as cycling continues. Some support for this hypothesis can be obtained from the shape of the hysteresis loops for ABS, Fig. 16. Crazing is generally believed to be the major mode of compliance in this material; the bulk strain is accommodated in that volume fraction of normal polymer which has been converted to craze. However, it has been suggested (Rabinowitz and Beardmore, 1972) that the bulk flow processes characteristic of the unmodified polymer matrix may be induced to operate by the complex local stress fields and local constraints associated with the particle/craze structure. The ultimate ductility may then result from a combination of crazing and bulk flow.

The hysteresis loops for ABS, Fig. 16, suggest that this combination of mechanisms is indeed operative. Furthermore, the change in shape with increasing cycles is in agreement with a continuing change in mechanism. Specifically, the stress–strain curve of a craze, Fig. 22, is comparable to the later stages of fatigue life (the tensile portion of hysteresis loop). The initially softened state hysteresis loop is more symmetric, characteristic of predom-

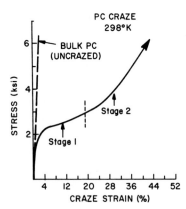

Fig. 22. Typical tensile stress–strain behavior of a single craze—see Rabinowitz and Beardmore (1972) and Kambour (1973) for details.

inantly bulk flow. Thus, the hysteresis loops suggest that the initial soften-ing is primarily accomplished by bulk flow with some (small) component of strain accommodated by crazing. As cycling proceeds, the balance of strain accommodation shifts—from bulk flow to crazing until a significant volume fraction of the matrix is converted to craze structure accommodating the bulk of the strain. Concomitantly, the tensile half of the hysteresis loop assumes the shape characteristic of craze deformation. This interplay between competitive deformation modes is presumably possible in other composite polymers—in 33% glass fiber reinforced nylon, it is probably the balance between bulk flow and matrix/fiber interface deformation which changes with cycles.

The usefulness of the hysteresis loop shape in depicting changing deforma-tion mechanisms with cycles suggests that such a shape analysis may be indicative of the different deformation mechanisms operative in different polymers. While such a detailed analysis has not been carried out in the work reported here, there are obvious trends which can be readily ap-preciated. For example, ductile amorphous polymers which accommodate strain by homogeneous bulk flow exhibit essentially symmetric hysteresis loops in the softened state. This can be seen in Fig. 11b, a typical example of this class. On the other hand, many semicrystalline polymers show distinctly asymmetric loops, Fig. 8. This asymmetry is usually characteristic of a ten-sile cavitation type deformation mechanism (cf. crazing) which is obviated under compressive stress conditions, and compressive deformation thus is achieved by bulk flow. In semicrystalline polymers, tensile deformation can occur by fibril extension, chain unfolding, etc., a process which is associated with a density decrease (microvoid formation (Peterlin, 1967). In a sense, this is very reminiscent of crazing in amorphous polymers, and in direct analogy, the voids are closed by compressive stresses and the deformation mechanism changes. The qualitative degree to which the fibril extension mechanism operates may be measured by the hysteresis loop shape. Thus, it is a much greater component of the total strain in polyoxymethylene and polypropylene than it is in nylon (cf. Figs. 7 and 8).

It is clear from the above discussion that, regardless of the particular defect mechanisms involved in the cyclic plastic deformation of ductile poly-mers, overall cyclic softening is the phenomenological result. Polymers ex-hibiting a variety of deformation mechanisms validate this conclusion. The specific response of any particular mechanism may be detected in the hyster-esis loop shape, and it may be that further analysis of this type of data could lead to a better understanding of the particular deformation phenomena involved.

In contrast to ductile polymers, in polymers of limited ductility only a small amount of continuous softening occurs throughout the fatigue life,

Fig. 13. Essentially, the entire fatigue life is spent in the incubation period. Consistent with the small amount of softening, little change occurs in the hysteresis loop prior to fracture. Two competitive strain accommodation processes are operative on the tensile half-cycle, namely craze formation and bulk flow, whereas only bulk flow occurs in compression. In PMMA at 298° K, crazes are the inherent flaw development mechanism which preempts extensive bulk flow. Thus, in cyclic deformation just as in monotonic tensile deformation, crack formation by craze breakdown occurs rapidly and failure occurs within the incubation period.

In brittle polymers, deformation is macroscopically elastic throughout the fatigue life. As in other brittle materials, plastic deformation occurs only in very localized regions where it is promoted by stress raisers such as inclusions. The localized nature of the plastic flow is not detectable as gross strain, so no softening occurs and the hysteresis loop stays completely linear.

4. Anomalous Cyclic Softening Effects

There are several phenomena which can occur in the cyclic testing of polymers which, at first sight, might be interpreted as a special type of mechanical softening akin to that described in the previous section. Some of these are given below as examples, and to familiarize the reader with anomalous stress changes which can occur during cyclic deformation.

Slow, incremental fatigue crack growth is accompanied by a corresponding decreasing peak tensile stress caused by a cycle-by-cycle increase in crack area, Fig. 10. The peak compressive stress is maintained approximately constant. The type of hysteresis loop pattern characteristic of incremental fatigue crack growth is shown in Fig. 23. This is directly analogous to fatigue

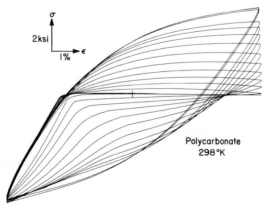

Fig. 23. Typical nested hysteresis loops for the crack propagation stage of fatigue failure in PC (see Fig. 10).

crack growth in ductile metals. Many polymers do not exhibit this slow growth and fracture occurs abruptly. Examination of the fracture surface usually provides immediate evidence that this tensile stress decrease is caused by slow crack growth.

A second type of stress instability not to be confused with bulk softening, again manifest as "tensile-only" softening, can occur in the cyclic stress–strain behavior of *semiductile* polymers when craze growth proceeds to unusually large craze area prior to crack formation. Figure 24 shows the

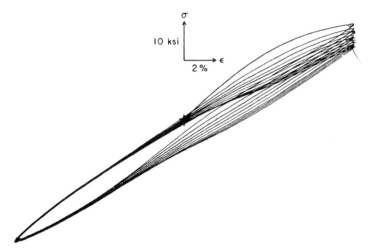

Fig. 24. Complete set of hysteresis loops for a sample of PC tested at 77° K in liquid nitrogen. $2N_f = 23$ reversals.

complete set of hysteresis loops for a sample of PC tested at 77° K, in liquid nitrogen, at $\Delta\varepsilon_T = 0.22$ and $2N_f = 23$ reversals to failure. At this high peak tensile strain in cyclic deformation, PC crazes grow completely through the specimen cross section prior to fracture in liquid nitrogen. The crazes actually grow incrementally on each tensile half-cycle. On the initial cycle, the principal mode of strain accommodation is bulk flow in both tension and compression—note the large strength differential on the first cycle, characteristic of ductile type bulk flow. As craze area and craze thickness increase on each successive cycle, more and more of the (fixed) total tensile strain is accommodated in the crazes, at the expense of bulk flow. Since the crazes are more compliant than the uncrazed polymer (Rabinowitz and Beardmore, 1972), a progressive decrease in the resistance to tensile deformation occurs. On the compression half of each cycle there is no craze-strain contribution to the total applied strain, and the stress–strain relation remains unchanged. Note, however, that the point at which the current stress–strain relation merges with the initial cycle response shifts to ever-increasing compressive stresses as deformation proceeds. This effect is probably caused by the increased difficulty of craze closure as craze thickness increases.

In many cases, confirmation of the incremental nature of craze growth, and the causal association with the stress-strain behavior shown in Fig. 24,

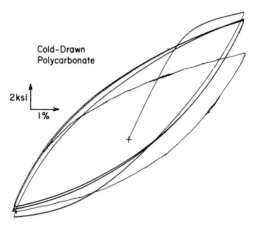

Fig. 25. Hysteresis loops of the initial three cycles and a cycle close to failure in the cyclic deformation of oriented PC at 298° K.

can be obtained directly from the fatigue fracture surface (see Section V). Fatigue fracture occurs abruptly, as in other semiductile polymers; the sudden drop in load to zero is shown in Fig. 24. Crack nucleation within the craze proceeds in the conventional manner. However, with a large expanse of well-formed craze ahead of the crack as it begins to grow, the fatigue crack is channeled in the craze and a very smooth, planar fracture surface results (see Section V).

In all the previous discussions of cyclic softening, the general trends have been delineated for polymers essentially isotropic in mechanical behavior and generally free of residual stresses. One example of the unique type of cyclic response that can result from biasing the molecular orientation of a polymer is provided in the following experiment. A sample of polycarbonate was extended in uniaxial tension at room temperature until cold drawing was complete. The resulting specimen had a high degree of molecular orientation paralled to the tensile axis; on further deformation parallel to the orientation direction the compressive flow stress was about 0.65 of the tensile flow stress. The specimen was allowed to relax at zero stress for several days; then, the specimen was cycled, parallel to the orientation direction, between fully reversed strain limits.

Figure 25 shows several hysteresis loops in the cyclic deformation—the first three cycles, and a cycle closer to the point of failure. A large reduction in the resistance to tensile deformation occurs (i.e. softening) whereas the compressive flow stress remains approximately constant. This is shown more clearly in Fig. 26 in which the peak tensile and compressive stress on

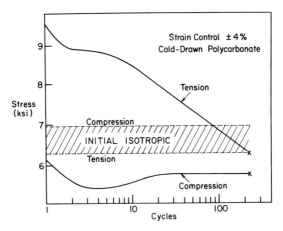

Fig. 26. Peak tensile and compressive stress against accumulated cycles for the experiment shown in Fig. 25.

each cycle is plotted against the accumulated cycles. The compressive peak stress remains essentially constant throughout the fatigue life. In marked contrast, the peak tensile stress decreases rapidly and continuously to the approximate isotropic strength. This readjustment in deformation resistance occurs with no significant change in specimen birefringence, indicating that a high degree of molecular orientation is maintained. Presumably, the tensile softening results from residual stress relaxation and a possible intrinsic ductile softening; the compressive stress "stability" probably reflects a competition between intrinsic ductile softening and residual stress relaxation.

Note that cycling this sample has produced a unique structure/property combination—a highly oriented molecular structure, with little strength differential between tension and compression. The example cited above is used to illustrate the manner in which structural and/or internal stress variations in a material can produce unique cyclic responses. It is important to have some information about structural and mechanical isotropy of a material before conclusions regarding the origin of the cyclic response can be drawn.

The three examples above are given to illustrate the complexities that can occur, and possibly be misinterpreted, in the stress response of polymers. However, there are many unique characteristics to the individual responses and the basic mechanism can usually be resolved with additional simple tests.

B. Cyclic Stress–Strain Behavior

1. CONCEPT OF CYCLIC STRESS–STRAIN CURVE

The most convenient and meaningful way to describe mechanical behavior in the cyclic steady state is with a graphical representation of the steady state stress–strain relation—this is termed the cyclic stress–strain curve (Morrow, 1965; Landgraf, 1970). The applied peak strain and the steady state peak tensile stress give, for example, one data point on the cyclic tensile stress–strain curve. By repeating the experiments shown in Figs. 5, 6, and 10 at several strain levels, the entire cyclic stress–strain relation can be generated. In Figs. 27–32, the cyclic stress–strain curves for several homopolymers and composites at 298° K are given. The monotonic stress–strain curves are included for comparison. In comparison of the monotonic and cyclic stress–strain curves, it is clear that the two stress–strain relations are identical in the initial region. Since the deformation in this low strain region is primarily elastic, the conclusion may be drawn that cyclic softening is a nonelastic strain process, involving primarily the anelastic and viscoplastic components of the strain. The amount of softening, i.e. the percentage stress

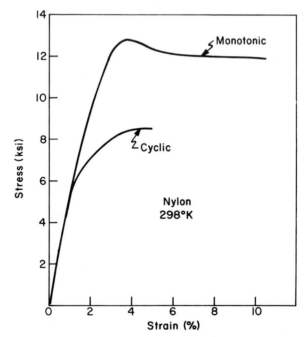

Fig. 27. Tensile cyclic and monotonic stress–strain curves for nylon at 298° K.

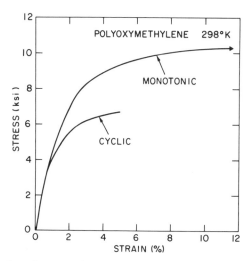

Fig. 28. Tensile cyclic and monotonic stress–strain curves for polyoxymethylene at 298° K.

difference between the monotonic and cyclic curves at constant strain, in-creases with increasing strain, to a maximum of the order of 40% at the upper yield point. These general comments are equally true in tension and compression, and a compressive cyclic stress–strain curve may be defined analogously, e.g. Fig. 33. The magnitude of the strength differential between tensile and compressive flow stresses (S-D effect) manifest in the initial (precycled) condition is approximately the same as is obtained in the cyclic steady state. Recalling that the S-D effect can be quantitatively related to the

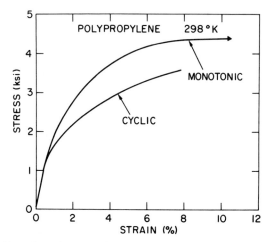

Fig. 29. Tensile cyclic and monotonic stress–strain curves for polypropylene at 298° K.

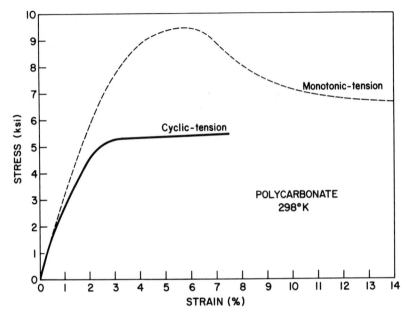

Fig. 30. Tensile cyclic and monotonic stress–strain curves for PC at 298° K.

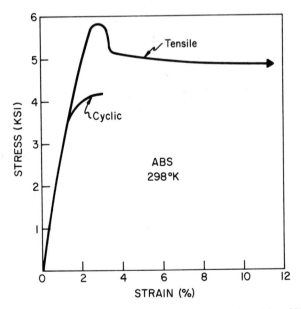

Fig. 31. Tensile cyclic and monotonic stress–strain curves for ABS at 298° K.

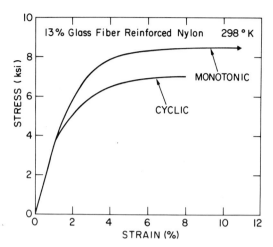

Fig. 32. Tensile cyclic and monotonic stress–strain curves for 13% glass fiber reinforced nylon at 298° K.

hydrostatic pressure coefficient of nonelastic flow in ductile polymers (Duckett *et al.*, 1970), this result (Fig. 33) supports the hypothesis that cyclic softening does not involve a change in the fundamental character of non-elastic deformation, but rather reflects a significant change in the relative rates of elastic and nonelastic strain processes, all of which are active throughout the fatigue life.

For a ductile polymer in the softened stable state, the cyclic stress–strain relation defined above indicates the general mechanical behavior of the softened material. Since the largest fraction of the fatigue life is spent in the

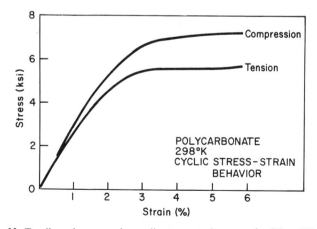

Fig. 33. Tensile and compressive cyclic stress–strain curves for PC at 298 K.

softened steady state, it is clearly the mechanical response depicted in the cyclic stress–strain curve that ultimately controls fatigue failure.

The validity of the cyclic stress–strain curve can be illustrated by removing a softened specimen from the fatigue apparatus and then retesting in monotonic tension. The stress–strain relation thus obtained would trace the cyclic stress–strain curve. One obvious and important conclusion to be drawn from the results in Figs. 27–32, then, is that the yield discontinuity exhibited by many polymers in monotonic deformation is eliminated in the cyclic stable state. The cyclic stress–strain curve is useful for determining the cyclic response in different control modes. In strain limited tests (as discussed here), the transition to the softened state is always controlled and results in a steady state condition. However, in load control tests the transition may not be controlled, depending on the peak stress level and specimen configuration. For example, at an initial stress level above the cyclic yield plateau (see Figs. 27–32) runaway creep failure would occur; at lower stress levels, a new steady state would be attained as defined by the same cyclic stress–strain curve obtained in strain control.

2. Determination of Cyclic Stress–Strain Curve

As mentioned above, the standard procedure for determining the cyclic stress–strain curve is to obtain one data point from each specimen tested in fatigue. In cases such as ABS, where the "steady" state stress is not constant, an average curve can be used by measuring the peak stress at half the specimen life, Fig. 31. It is frequently desirable to obtain the information inherent in the cyclic stress–strain curve but the tedious experiments to generate it are a dissuading factor. This problem has been overcome in metals by the development of special tests which allow the cyclic stress–strain curve to be generated from a specialized test on a single specimen (Landgraf et al., 1969). This rapid procedure has been evaluated for polymers and is outlined below.

The so-called incremented step test (Landgraf et al., 1969) involves subjecting a specimen to blocks of gradually increasing and then decreasing strain amplitudes as shown in the upper half of Fig. 34. After only a few of these blocks, the stress response stabilizes. A typical test record of the imposed strain amplitude blocks and the corresponding stress response is shown in Fig. 34 for PC. The stress envelope is then the trace of the cyclic stress–strain curve. The overall cyclic softening effect is immediately obvious in such a test. A similar test for polypropylene is shown in Fig. 35. The corresponding plot of the hysteresis loops for PC throughout a stabilized

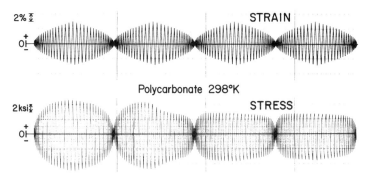

Fig. 34. Incremental step test for PC at 298° K. The imposed blocks of strain cycles are shown together with the corresponding change in the stress response.

block is shown in Fig. 36, in which the locus of the loop tips corresponds to the cyclic stress strain curve.

The incremental step test provides a convenient, fast method for assessing the degree of softening exhibited by ductile polymers. So far, it has only been evaluated on the limited number of polymers noted here. The test appears to be generally useful for homopolymers and composite polymers with small amounts of second phase, e.g. 13% glass fiber reinforced nylon. However, application of the procedure to ABS did not induce the appropriate degree of softening, so the test may not prove to be as valid for highly complex, multiphase microstructures which show a change in deformation mechanism as a function of accumulated cycles (see earlier).

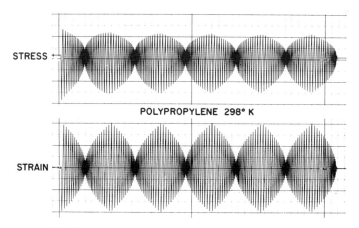

Fig. 35. Incremental step test for polypropylene at 298° K.

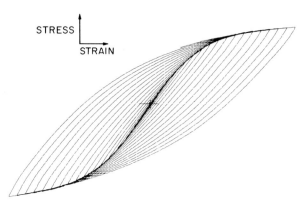

Fig. 36. Typical block of hysteresis loops corresponding to a stabilized block of cycles in the incremental test of PC at 298° K (see Fig. 34).

C. Patterns of Response

One of the most useful derivatives of a study of a large number of materials is the development of characteristic patterns of behavior. Such generalization of response characteristics allow, for example, the immediate identification of the cyclic response of any polymer (at least qualitatively) from a simple tensile test. As mentioned in various sections above, there are clear patterns of cyclic behavior of polymers based on their phenomenological behavior and, to a lesser extent, on their microstructure.

The following general conclusion may be drawn:

(a) Ductile polymers exhibit *marked* cyclic softening irrespective of microstructure and deformation mechanisms.

(b) Polymers of limited ductility are either cyclically stable or soften *slightly*.

(c) Polymers do *not* cyclically harden.

(d) Amorphous polymers, semicrystalline polymers, and composites of limited second phase content achieve an essentially stable, cyclically softened state.

(e) Composite polymers of high volume fraction second phase tend to show a dynamically unstable, cyclically softened state due to competitive deformation mechanisms.

These general conclusions may be used to estimate the type of cyclic response of a given polymer by a simple classification procedure. It is not suggested that the studies upon which these conclusions are based are complete or sufficiently comprehensive. However, the data are sufficient for at least qualitative assessments and provide the best current background for the phenomenological cyclic response of rigid polymers.

IV. Fatigue Behavior

The standard technique for cataloging the fatigue response of materials is in terms of log $(\Delta \varepsilon_T)$ vs log $2N_f$ or σ_A vs log $2N_f$. Such data are extremely valuable to the design engineer who has to design components to withstand certain stress of strain levels. However, the accummulation of these data does not provide insight into the mechanisms of fatigue. Since we are primarily concerned here with the deformation aspects of fatigue, it is the micromechanistic observations which usually accompany such data accummulation which are of most interest. In this section, therefore, only a summation of the S-N type data are given, mainly for reference purposes in order to compare different polymers in overall fatigue resistance.

Patterns of Behavior

There are numerous S-N curves for individual polymers throughout the literature. Rather than document a large number of these data, here we are primarily concerned with developing patterns of behavior and classifying the polymers on the basis of these patterns with respect to their mechanisms of deformation.

Generally, the overall fatigue response of any polymer can be described schematically as shown in Fig. 37 (Rabinowitz *et al.*, 1973; Beardmore and Rabinowitz, 1974). The life can be divided into three regions, as indicated in Fig. 37; regions II and III are common to all polymers irrespective of their deformation and failure mechanisms and irrespective of the degree of

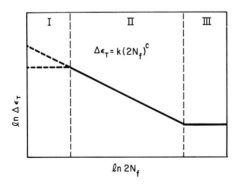

Fig. 37. Schematic representation of the general fatigue response of polymers in terms of strain range versus number of reversals to failure.

ductility. (The relative positions of the fatigue life curve in regions II and III are, of course, very dependent on the properties of the particular polymer.) The fatigue response in region I is, however, very dependent on the deformation mode. In polymers which do not craze, region I is essentially an extension of region II to shorter lives. Polymers which craze, however, show a distinct change in slope of the fatigue curve in the transition from region I to region II. Essentially, region I corresponds to the test conditions under which crazes form on the initial tensile quarter cycle (see later). In region II, the strain (stress) level is insufficient to generate crazes on the initial cycle.

From Fig. 37, it is clear that the major difference in patterns of fatigue response is that of crazing as a deformation mode. In addition, this is basically only evident in the short life region (region I). Examples are given in

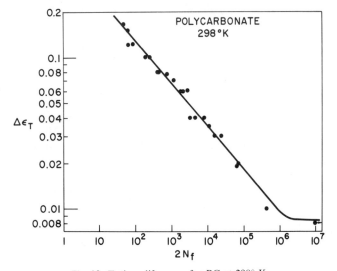

Fig. 38. Fatigue life curve for PC at 298° K.

Figs. 38, 39, and 40 to support this viewpoint. In Fig. 38, the fatigue response of PC at room temperature is shown; since PC does not craze at room temperature, region I is absent. In contrast, PS crazes very heavily at room temperature and the corresponding fatigue data (Rabinowitz et al., 1973) are shown in Fig. 39. Note the very clear delineation of the three stages (cf. Fig. 37). The mechanisms involved in the fatigue of PS (Rabinowitz et al., 1973) are detailed in Section V. An intermediate example is shown in Fig. 40, namely PMMA fatigued at room temperature. At this temperature, PMMA forms only small surface crazes but the effect of even this limited amount of crazing is immediately evident from the appearance of region I, Fig. 40.

The generality of these two distinct patterns of response, Fig. 37, can be

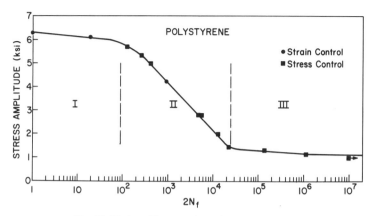

Fig. 39. Fatigue life curve for polystyrene at 298° K.

clearly illustrated by testing a polymer under two different conditions, one in which crazing occurs and one in which crazing is absent. The results of such a test can be seen by comparison of Figs. 41 and 38. In Fig. 41, data for PC cycled at 77° K (in liquid nitrogen) are given; under these testing conditions, large crazes are formed. Note the immediate introduction of region I due to the crazing (stage III is not included merely because the data were not developed to sufficiently long lives).

The individual S-N curves for particular polymers are only useful in the present context as a reference scale for comparison between polymers. More importantly, the generalized curve of Fig. 37 can be used as a schematic

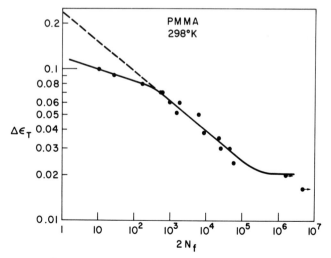

Fig. 40. Fatigue life curve for PMMA at 298° K.

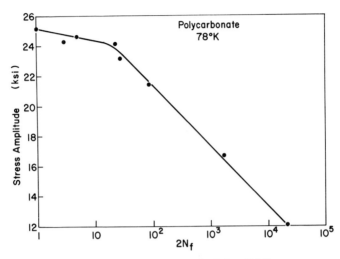

Fig. 41. Fatigue life curve for PC at 78° K.

representation. The deformation mechanisms involved in the separate stages are of more concern here, and while the data on the micromechanisms of deformation are far from complete, the current understanding is presented in the following section by detailing studies of typical polymers where such studies have been carried out.

V. Mechanisms of Fatigue

Fatigue failure in all materials occurs in basically the same way. The three stages of fatigue consist of (a) reversed plastic deformation leading to strain localization, (b) crack initiation resulting from the strain localization, and (c) crack propagation. The specific atomic (molecular) mechanisms by which the three stages occur vary, of course, depending on the material. The detail in which any of the three stages has been characterized varies widely between materials and is by far the most advanced for metals. The rudimentary nature of the understanding of molecular deformation mechanisms in polymers relative to (say) that of dislocations in metals is reflected in the current lack of detailed mechanistic models in the fatigue processes in polymers. Indeed, there are very few studies in rigid polymers which have attempted even a phenomenological interpretation of the mechanisms of fatigue. In the present section, the current understanding of the mechanisms of fatigue in rigid polymers is summarized by detailing specific examples.

In the outline above of the three stages of fatigue, the implicit assumption was the absence of a preexisting crack. Certain brittle materials (e.g. ceramics) are always characterized as containing inherent microcracks. Such materials will go immediately into the third stage of fatigue, i.e. crack propagation, under cyclic testing. The same effect can be achieved by deliberately introducing a notch or crack prior to cyclic evaluation. In these cases, the effect of preempting the first two stages of fatigue on the total fatigue life will be strongly dependent upon the relative influences of the three stages on the total life. For example, if an uncracked specimen spends 90% of the total fatigue life in crack propagation (generally characteristic of high strain, low cycle fatigue) then the effect of a preexisting crack will only be to eliminate the initial 10% of life spent in the crack initiation stage; thus, the effect on the overall life will not be severe. Conversely, in long life (low strain) fatigue, crack initiation generally occupies the predominant portion of the total fatigue life, so the introduction of a preexisting crack will have a drastic effect on the overall fatigue resistance.

The effects of extraneous variables such as notches must always be borne in mind in any assessment of fatigue behavior. For the present purpose, we are concerned solely with the fatigue mechanisms when all three stages of fatigue occur. For a fuller discussion of the fatigue of precracked specimens readers are referred to Manson and Hertzberg (1973).

Crazing and Fatigue

With reference to Fig. 37 in Section IV, polymers which deform by crazing tend to exhibit three distinct stages to the fatigue life. It is with reference to the existence of these three discrete life ranges that the following description of a specific example of the fatigue of polymers which craze is given.

1. FATIGUE OF POLYSTYRENE

Polystyrene is an ideal material in which to study crazing and the details of craze breakdown, because of the high propensity for craze formation and the large areal size of the crazes formed. The lack of significant bulk flow prior to fracture leads to the classification of polystyrene as a semibrittle material (Rabinowitz and Beardmore, 1972), although on a local scale plastic deformation in the crazes is very extensive.

An appreciation of tensile deformation and fracture behavior is necessary to provide a reference for interpretation of the cyclic deformation and fracture behavior (Rabinowitz et al., 1973). Figure 42 shows the tensile stress–strain relation. The flow stress for uniform nonelastic deformation is almost equal to the fracture strength, and the bulk response is largely elastic below

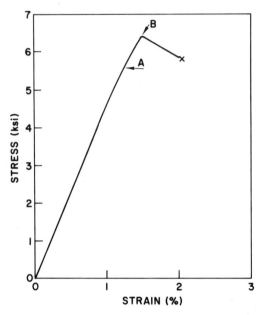

Fig. 42. Tensile stress–strain curve for polystyrene at 298° K. Point A marks the craze appearance stress (visual examination).

about 5 ksi. The predominant mode of nonelastic deformation is in the form of crazes, which appear (visual examination) at about 6 ksi, and grow rapidly across the specimen cross section. Craze development is so extensive that the rate of nonelastic strain (predominantly craze strain) quickly increases beyond the constant applied strain rate at point B (Fig. 42), and the load begins to decrease slowly. Catastrophic failure follows soon after the load maximum. It is emphasized that the load maximum does not relate to the onset of extensive bulk ductility (as in, say, the cold-drawing of thermoplastics), but rather results from intense localized straining in the form of crazing.

The large (areal) crazes, in most cases traversing the entire specimen cross section (Rabinowitz et al., 1973), provide a low energy path for crack propagation and thus act as a preferential fracture path (Hull, 1970; Rabinowitz and Beardmore, 1972). This can be seen on the tensile fracture surface, Fig. 43, where the striking mica-like appearance of the fracture surface is a direct consequence of crack propagation through preexisting craze material (Rabinowitz et al., 1973).

The polystyrene fatigue data (Rabinowitz et al., 1973) are summarized in Fig. 39 in Section IV. The data point at one reversal is tensile fracture. These data are of prime interest here as a scaling reference for the discussion to follow.

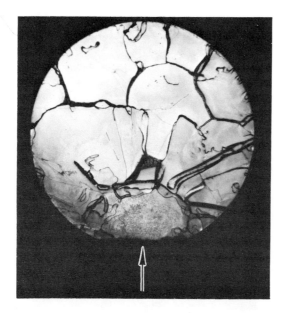

Fig. 43. Polystyrene fracture surface in simple tension. The arrow indicates the origin of fracture.

Comparison of the data in Figs. 42 and 39 shows that the transition from region I to region II coincides with the constant-strain rate craze initiation stress (visual observation) in unidirectional testing. Thus, a clear distinction between region I and regions II and III can be drawn on the following basis: in region I crazes form on the first tensile quarter cycle—the fatigue behavior in the region is then essentially that of a precrazed sample; in regions II and III, on the other hand, crazes form as a consequence of cyclic deformation, so that the fatigue life reflects both craze initiation and crack nucleation and propagation. There are fractographic features typical of fatigue failure in each life region, and these provide a convenient characterization of the fatigue process. The cyclic stress–strain response in regions II and III is common to a wide variety of polymers, and will be discussed first. Region I is peculiar to polymers that have a high propensity for crazing and will be considered later.

a. Fatigue Failure in Region II. Region II fatigue fracture in polystyrene follows the well characterized pattern of craze induced fracture described in Rabinowitz and Beardmore (1972): craze nucleation, followed by craze growth and the deterioration of craze integrity, followed by crack formation, terminated by catastrophic crack propagation. The primary effect of reversals in the cyclic deformation is to accentuate certain stages in this process.

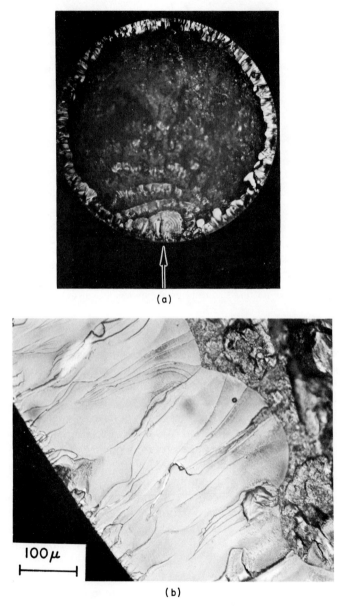

(a)

(b)

Fig. 44. (a) Typical polystyrene fatigue fracture surface in region II. (b) Higher magnification image of a portion of the annular ring.

Figure 44a is a low magnification microphotograph of a representative fatigue fracture surface in the intermediate life region (II). This particular sample failed after about 900 reversals at ± 4.2 ksi. A portion of the prominent annular ring on the fracture surface is shown in greater detail in Fig. 44b. Comparison of Figs. 44a and 44b with the tensile fracture surfaces shown in Fig. 43 allows several conclusions to be drawn. First, the annular ring marks the extent of areal craze growth—craze penetration—prior to catastrophic crack propagation; the mica-like appearance of the surface within the annulus is directly comparable to the fracture surface in Fig. 43. Second, the rough, circular central region in Fig. 44a constitutes the portion of the fracture during which the crack was moving catastrophically through previously uncrazed bulk polymer (Rabinowitz et al., 1973). In particular, the coarse striae in this central region are not a consequence of incremental fatigue crack growth, but rather derive from a tensile mechanism (Rabinowitz et al., 1973). Finally, the extent of true fatigue crack growth is limited to the small semicircular region at the bottom of Fig. 44a. This region is shown in greater detail in Figs. 45a and 45b, where a series of fatigue striae is clearly discernible.

Consider first the *craze* nucleation stage in the failure process. As noted above, crazes do not appear on the first tensile quarter-cycle in region II, but develop with continued cycling. In a qualitative assessment, the number of cycles required to nucleate a craze (visual observation) decreases as the peak stress (or strain) increases. At any stress, crazes appear much earlier in the fatigue life than would be predicted on a simple extrapolation of time-under-load from creep experiments (Beardmore and Rabinowitz, 1974), even computing the cyclic time-under-load as the sum of the total times for each tension half-cycle. In this regard, an important effect of cyclic deformation is to localize nonelastic strain relative to the more homogeneous deformation characteristic of unidirectional testing at the same stress levels. This is analagous to the well-documented localization of plastic strain in fatigued metals (Avery and Backofen, 1963). It is suggested that in these more intensively strained regions, craze formation is easier than in the normal polymer. In support of this hypothesis, recent experiments (Wang et al., 1971) indicate that craze formation is governed by a critical value of the local strain or strain energy density. In addition, we have observed that cycling specimens of poly(methylmethacrylate) and polycarbonate under only compressive stresses results in a decrease in the craze appearance stress in subsequent tensile testing (Beardmore and Rabinowitz, 1974). Thus, in the first stage of fatigue failure a true cyclic effect is manifest as an acceleration of craze initiation—a necessary prerequisite to fracture in polystyrene in this stress range.

The effects of reversed straining are also evident in the crack nucleation stage. As in constant strain rate and creep failure, crack nucleation in fatigue

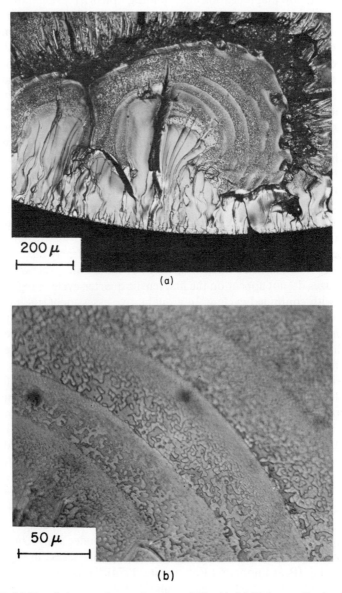

Fig. 45. (a) Slow fatigue crack growth region of Fig. 44. (b) High magnification image of slow growth region showing detail of craze breakdown.

occurs via a process of craze breakdown closely tied to the rate and extent of craze thickness growth. The process of craze breakdown is one of strain localization in a small region of the craze and this occurs relatively independently of the areal growth of the remainder of the craze (Rabinowitz and Beardmore, 1972). In the characteristic unidirectional fracture process in polystyrene, the rate of areal craze growth so exceeds the rate of crack formation within the craze that crazes traverse the entire specimen cross section prior to crack propagation (see, for example, Fig. 43). In fatigue, this strain localization within the craze is enhanced, resulting in a more rapid rate of crack nucleation relative to tensile fracture. In addition, creep tests under the range of stresses in region II produce total-cross section crazing prior to crack nucleation (Fig. 48a) in much the same manner as in constant-strain rate deformation. It is clear that compared to tensile and tensile creep failure, the process of crack nucleation within the growing craze occurs more rapidly in cyclic deformation.

The crack nucleation event occurs in a submicroscopic volume of craze. The next stage in the failure process, be it unidirectional tension, tensile creep, or fatigue, involves the slow growth of the crack nucleus to a size (determined by some modified Griffith criterion) at which unstable crack propagation occurs. It is this slow crack growth process that is currently the best understood aspect of fracture in polystyrene, largely because of the comprehensive work of Murray and Hull (1969, 1970) on tensile failures. Examination of the slow fatigue crack growth region in Fig. 45 reveals that the mechanism of slow crack growth in cyclic deformation is identical to that observed in unidirectional testing. The characteristic patchy texture on the surface has been attributed by Murray and Hull (1969, 1970) to the meandering course of the growing crack, meeting first one craze/normal polymer interface, and then another. It is important to point out that the number of fatigue striae is very small compared to the number of cycles to failure. Thus, the slow crack growth stage of fatigue failure constitutes a small fraction of the total fatigue life in polystyrene. The majority of the fatigue life of polystyrene is spent in crack nucleation.

Within region II there are several effects manifest with increasing fatigue life. As the peak stress amplitude decreases (increasing fatigue life) the extent of areal craze growth, i.e. the width of the annular ring, decreases. Conversely, the extent of slow crack growth prior to catastrophic crack propagation increases with increasing life. The areal growth of crazes in fatigue appears to exert no direct influence on the extent of slow fatigue crack growth. Rather, the size of the slow growth region is determined by the applied peak stress amplitude, in accord with previous observations relating the size of the slow fatigue crack growth region to a Griffith type criterion (Feltner, 1967; Havlicek and Zilvar, 1971). Finally, within the slow crack growth region the number of striae increases, and the interstriation spacing decreases, as the life increases. This demonstrates the true fatigue nature of the slow crack growth region. No attempt has been made to assess whether slow crack growth occupies a greater or lesser fraction of the total fatigue life as the stress amplitude decreases through region II.

The final stage of fracture in region II is the rapid propagation of the crack across the remaining specimen cross section. This aspect of fatigue fracture is no different than in tensile fracture. The transition in crack propagation mode from a crack moving through pre-uncrazed material is very distinct (Rabinowitz *et al.*, 1973).

b. Fatigue Failure in Region III. The long life region of fatigue (III) constitutes the so-called endurance limit range, where the slope of the fatigue–life relation is extremely shallow. Figure 46 shows a typical region III fatigue fracture surface. Fractographically, the region II to region III transition is a gradual one. In region III, however, several distinct changes in the fracture process can be noted. First, the craze nucleation event is extremely rare, and requires a long cyclic incubation; ultimately, craze nucleation probably depends on a fortuitous perturbation in stress distribution.

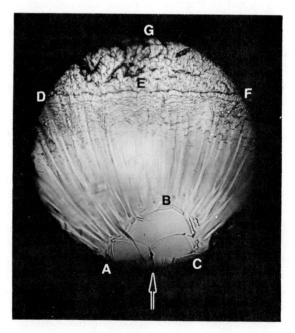

Fig. 46. Typical polystyrene fatigue fracture surface in region III (long life). The area ADEFCA is slow fatigue crack growth.

Secondly, craze areal growth, which appears to be directly related to peak stress amplitude, is severely restricted and there is no region corresponding to preexisting (with respect to crack propagation) craze on the fracture surface. Third, most of the fatigue fracture in region III occurs in the slow crack growth mode. Unlike slow fatigue crack growth in region II, however, slow growth in region III (ADEFCA in Fig. 46) occurs in two stages: the first stage appears as the very smooth, glassy region ABCA in Fig. 46, and is peculiar to region III fracture; the second stage, ADEFCBA, is conventional slow fatigue crack growth as seen in region II. The remainder of the fracture surface, DGFED, comprises the ultimate tensile separation of the specimen.

The earliest stage of fracture (ABCA) is not well understood. One important difference between fracture in long life fatigue and previously reported observations in tension and tensile creep is the extremely low stress level at which long life fracture occurs. The only comparable data have been reported by Havlicek and Zilvar (1971), who noted a region similar to ABCA in their long life, low stress tests. The region ABCA is featureless at the highest magnification available in visible light microscopy, with the exception of a faint but definite array of crack arrest striae characteristic (presumably) of fatigue crack growth, Fig. 47a. The interstriation spacing is on the order of 8 μ, evidence of the slow progress of the crack through this region. There is no evidence of crazing on the fracture surface. This does not obviate the possibility, however, that crack growth still proceeded via the usual mechanism of craze breakdown. For example, if crack nucleation and growth occurred in a very thin craze, topographical relief on the fracture surface would be minimal. The postulate of low stress fracture in a thin craze goes counter to the relationship between increasing fracture stress and decreasing craze thickness noted by Murray and Hull (1970). However, the potency of reversed deformation in accelerating the craze breakdown process becomes increasingly strong as the peak stress level decreases; crack nucleation may thus occur within the craze at an early stage in its thickness growth. It is also possible that region ABCA derived from crack propagation exclusively along one craze/normal polymer interface—at the low peak stresses in region III, the tendency for the crack to bifurcate or skip from one craze plane to another would be minimized. However, in this fracture mode it would be expected that one of the matching fracture surfaces would show some evidence of residual craze structure and no such features were detected. Finally, it must also be considered that the early stages of crack propagation in long life fatigue occur with no craze participation, i.e. crack nucleation and propagation with no associated craze-type plastic deformation. Since crazing involves the dissipation of strain energy over some finite volume of material, the extreme localization of stress concentrations at the low fatigue stresses may preclude craze formation. In any case, the sharp transition along the boundary ABC is difficult to account for.

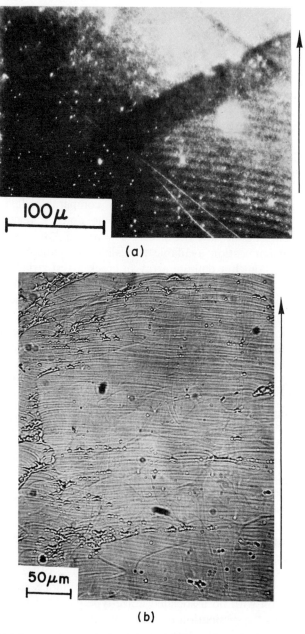

Fig. 47. (a) Fatigue crack arrest striae in region ABCA in Fig. 46. The arrow indicates the crack growth direction. (b) Fatigue crack arrest striae in region ADEFCBA.

Region ADEFCBA results from a fatigue crack growth mechanism similar to that in region II type fracture—sequential craze formation, craze growth, and craze breakdown in a volume of material directly ahead of the slowly growing crack. This is particularly evident toward the latter stage of slow crack growth (approaching DEF in Fig. 46) where a patchy texture of "mackerel" markings, as noted by Hull and co-workers, can be seen. These markings are characteristic of "meandering" crack growth within a craze. Striations, assumed to be fatigue crack arrest markings, are discernible throughout the region ADEFCBA; Fig. 47b is a representative example, showing an interstriation spacing on the order of 10 μ.

c. Creep Effects in Regions II and III. In the discussion above it has been emphasized that there is a true fatigue effect in the fracture process in regions II and III. The events tied to the strain localization associated with reversed deformation appear to control the fracture sequence, and are manifested primarily as an acceleration of processes that would otherwise similarly have been operative in constant load or constant-strain rate experiments. There is also some creep-type component, however small, to the deformation in regions II and III. Two additional experiments were performed to vary the amount of creep time at peak load, and assess the relative influence of this creep component on region II fracture. First, a specimen was allowed to creep at a stress of 5 ksi; the time to fracture was 16 min. In this failure mode, crazes grew completely across the specimen cross section prior to crack propagation, resulting in the fracture surface shown in Fig. 48a. This creep fracture surface is identical to that obtained in constant-strain rate tensile fracture (Fig. 43) with the exception of the size of the slow growth region.

A second experiment involved the imposition of a low frequency tensile square wave function, i.e. a tensile creep stress of 5 ksi was interrupted every 2.5 sec by a rapid excursion to zero stress and back. The time to failure at 5 ksi in this mode was reduced to 3.6 min, that is, 88 cycles. The resulting fracture surface is shown in Fig. 48b. Several important points are noteworthy. The dramatic reduction in creep time in square wave loading demonstrates the potent effect of reversed deformation on the craze breakdown process at constant stress, viz. 16 min vs 3.6 min. On the other hand, the number of reversals to failure in the square wave experiment (176) is considerably less than the 420 reversals to failure in the conventional fatigue tests (Fig. 39); this illustrates the potential of the creep component in reducing fatigue life. Comparison of Figs. 48a and 48b with Fig. 44a shows that the extent of areal craze growth prior to crack propagation increases as the creep component in the failure process increases. Areal craze growth may, in fact, provide a sensitive indication of the relative amounts of creep and

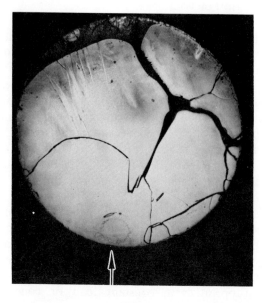

(a)

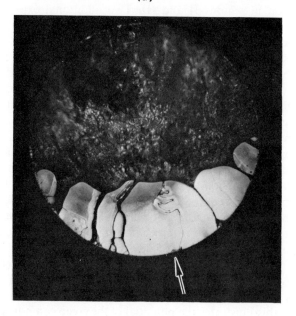

(b)

Fig. 48. (a) Typical tensile creep fracture surface of polystyrene in the stress range defined by region II in Fig. 39. (b) Fracture surface resulting from combined creep/fatigue in stress range of region II.

fatigue in any individual failure. On this basis, the creep component in the conventional fatigue tests (cf. Fig. 44) is small, and decreases with increasing fatigue life through regions II and III. In contrast, with increasing peak stress amplitude, i.e. moving into region I, the creep component becomes increasingly important.

d. Failure in Region I. As the peak cyclic stress amplitude is increased through region II, polystyrene experiences a rather abrupt deterioration in fatigue resistance as reflected by a sharp reduction in the slope of the fatigue life plot, Fig. 39. This abrupt transition occurs because the applied peak stress in region I exceeds the constant-strain rate craze initiation stress; crazes are generated on the first tensile quarter-cycle, and the entire fatigue life consists of craze growth and craze breakdown. The sharpness of the transition gives testimony to the potency of crazes as the sites and vehicles for crack initiation in thermoplastic polymers. With the elimination of the craze initiation part of fatigue in polystyrene, the fatigue life is drastically reduced from what would be predicted by an extrapolation to shorter lives of the cyclic response in region II. This abrupt transition, coincident with the critical crazing stress, is characteristic of all plastics which fail via craze nucleated fracture (see Section IV).

The fatigue fracture surface in region I reflects the large areal craze development and rapid crack propagation characteristic of high stress deformation. A typical region I fractograph is shown in Fig. 49. One conclusion that can be drawn from a comparison of Figs. 43, 48, and 49 is that creep plays a very significant part in the failure process in region I; areal craze growth prior to fracture is quite extensive. Unquestionably, reversed deformation also plays an important role in region I by accelerating the breakdown process so that the failure mechanism in region I is most properly described as combined creep/fatigue. Experimentally, region I failure is difficult to define because of the extreme stress sensitivity, i.e. the shallowness of the stress–life relation in Fig. 39, but small changes in peak stress amplitude can make marked changes in the appearance of the fatigue fracture.

The extent of areal craze growth prior to crack propagation decreases with decreasing stress. Under the high stress creep conditions that prevail on the portion of each tensile half-cycle above the constant-strain rate craze appearance stress, the rate of areal craze growth is a sensitive function of peak stress amplitude. In all but the unidirectional failure (1 reversal), there is insufficient stress/time to permit craze growth completely across the cross section in one cycle. There is clear evidence in the fractograph in Fig. 49 that areal craze growth proceeds incrementally on each tensile half-cycle; craze arrest markings, quite analogous to the conventional fatigue crack arrest

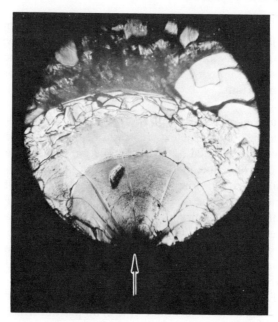

Fig. 49. Typical polystyrene fatigue fracture surface in region I.

striae, are prominent. Such incremental craze growth is a characteristic feature of region I fatigue in thermoplastic polymers.

Virtually all of the fracture surface in Fig. 49 derived from catastrophic crack propagation, first through preexisting craze and finally through previously uncrazed polymer. However, small patches of craze can be seen growing in from the periphery in the final fracture region, giving a combination rough/smooth texture. It is only toward the lower stress end of region I that visible manifestation of slow fatigue crack growth is resolvable prior to rapid crack propagation. Throughout region I the part of the fracture surface near the crack nucleus shows the patchy texture characteristic of craze induced fracture.

2. ADDITIONAL STUDIES ON CRAZING POLYMERS

Both PC and PMMA form large (areal) crazes when deformed at 77° K in liquid nitrogen. In the case of PC, the large crazes are relatively thick and the craze formation is also accompanied by a significant amount of homogeneous bulk flow (Beardmore and Rabinowitz, 1971; Rabinowitz and Beardmore, 1972). For PMMA, the large (areal) crazes are extremely thin and no bulk flow is evident, resulting in essentially linear fracture behavior (Beardmore and Rabinowitz, 1971; Rabinowitz and Beardmore, 1972). Under these testing conditions, however, both polymers can be classified as having a high propensity for areal craze growth and thus have a commonality with polystyrene tested at room temperature. There is some data available on PC and PMMA tested in fatigue at 77° K in liquid nitrogen (Beardmore and Rabinowitz, 1974), and while the data are far from comprehensive in nature, they provide a good comparison for judging the generality of fatigue mechanisms in polymers with a high propensity for crazing.

At 77° K, both PC and PMMA conform to the general pattern of response shown in Fig. 37. Using PC as an example (see Fig. 41), typical fracture faces are shown in Fig. 50. In tension, catastrophic failure occurs after the crazes have grown a considerable depth into the specimen but not completely across the cross section—this is clear in Fig. 50a where the wide annulus represents the extent of craze existing prior to crack propagation. The small, circular center portion represents material where the crack has propagated through unprecrazed material. By contrast, in fatigue failure in region I, the effect of cyclic straining is to promote additional craze growth prior to crack propagation, Fig. 50b, resulting in craze growth completely through the cross section prior to craze breakdown and crack initiation and propagation. The fracture face is smooth and planar and shows the high reflectivity typical of a surface created by crack propagation through preexisting craze. Transition into region II and increasing the life in this region is characterized by craze growth becoming limited to an ever narrowing annulus at the surface of the specimen, Figs. 50c and 50d, and a transition to no annulus at all, Fig. 50e. These features are directly analogous to the change in fatigue fracture characteristics through region II as illustrated for PS in Figs. 49, 44, and 46. The overall similarity of the fracture faces confirms the mechanistic conclusions drawn from the PS observations as being generally representative of fatigue failure in heavily crazing polymeric systems.

There is an additional striking feature obvious on fracture surfaces of PC characteristic of region I, Fig. 50b. As shown in greater detail in Fig. 51, a series of concentric ring-shaped markings cover the fracture surface (Beardmore and Rabinowitz, 1972). A longitudinal section through such a sample [by a technique described in Beardmore and Rabinowitz (1971)] is shown in Fig. 52, and detailed examination established unequivocally that the shorter longitudinal markings (parallel to the stress axis) are in fact crazes. The transverse markings are the traces of conventional crazes formed perpendicular to the stress axis. The concentric rings in Figs. 50b and 51 mark the sites of intersection of annular, longitudinal crazes with the fracture surface. It is emphasized that the orientation of these crazes is *parallel* to the applied stress axis.

Three characteristics of the longitudinal crazes are noteworthy. First, they are much shorter than the "conventional" transverse crazes, and appear for

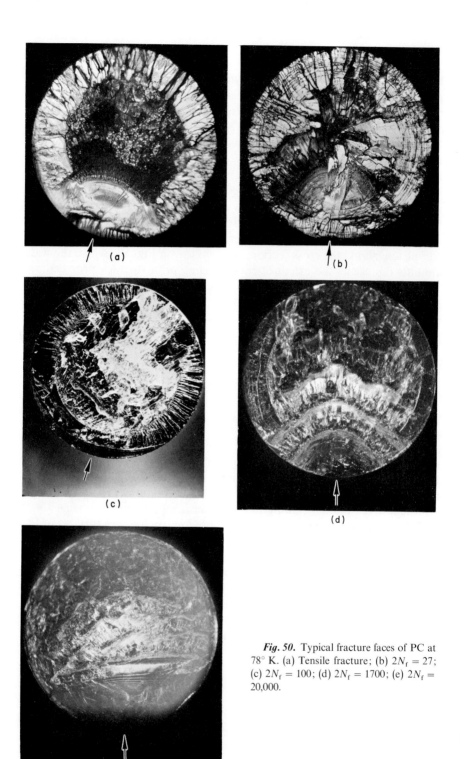

Fig. 50. Typical fracture faces of PC at 78° K. (a) Tensile fracture; (b) $2N_f = 27$; (c) $2N_f = 100$; (d) $2N_f = 1700$; (e) $2N_f = 20,000$.

0.5 mm (a)

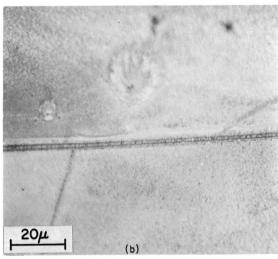

20μ (b)

Fig. 51. Higher magnification photographs of the concentric rings of Fig. 50b. (a) Two of the rings are indicated by C and one ring is emphasized by the dashed arc. (b) High magnification of part of one concentric ring.

Polycarbonate 78°K

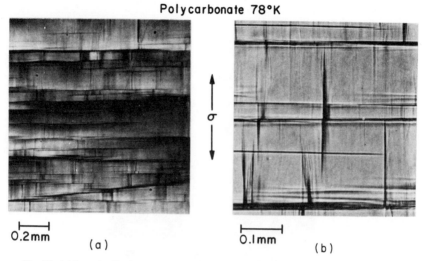

σ

0.2mm
(a)

0.1mm
(b)

Fig. 52. (a) Longitudinal cross section through the sample of PC shown in Fig. 50b. Stress axis is vertical. (b) Higher magnification showing detail of craze intersections.

the most part to originate and terminate at conventional crazes. Second, the lateral spacing between longitudinal crazes appears to be fairly uniform. Third, as seen clearly in the center of Fig. 52b, when the two craze networks intersect there is a distinct displacement of the longitudinal craze plane. The occurrence of longitudinal crazes, and their general characteristics, are not specific to polycarbonate; similar observations have been made in PMMA at 77° K, and the phenomenon is believed common to all crazing polymers in cyclic deformation.

One effect of cyclic deformation is to localize nonelastic deformation relative to the more homogeneous deformation obtained in unidirectional testing at the same stress levels. This is analogous to the well-documented localization of plastic strain in fatigued metals. The experiments by Wang *et al.* (1971) indicate that craze formation in glassy polymers is governed by a critical value of the local principal strain or local strain energy density. Thus, the strain localization associated with cyclic deformation should enhance the propensity for crazing relative to virgin material. In addition, the stress field ahead of a growing craze may be likened, in a qualitative sense, to that of a similarly oriented crack. In particular, as has been shown by Sneddon (1946), there is a component of tensile stress in the plane of the craze (crack). We suggest that the tensile component of the stress in the plane of the craze, although well below the normal crazing stress, is sufficient to nucleate and grow longitudinal crazes in the intensively strained region at the tip of the growing transverse craze. Figure 53 is a schematic drawing of the sequence of events believed to lead to the development of the crazes imaged in Figs. 51 and 52. It is known that the conventional transverse crazes grow incrementally through the specimen cross section on each tensile half-cycle (Rabinowitz *et al.*, 1973). The longitudinal crazes develop in advance of the growing transverse craze on each tensile half-cycle. The driving force for longitudinal crazing—the tensile stress component parallel to the transverse craze—diminishes rapidly with distance from the transverse craze plane. Thus, the extent of longitudinal craze growth is restricted. In support of this

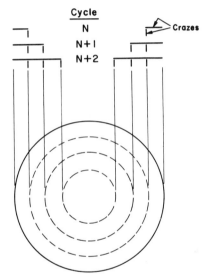

Fig. 53. Schematic illustration of the sequence of events leading to the formation of longitudinal crazes.

hypothesis, the number of concentric rings on the fracture surface (i.e. the traces of the longitudinal crazes) is in one-to-one correlation with the number of cycles for the transverse crazes to grow completely across the cross section; the spacing between longitudinal crazes is then dictated by the size of the incremental growth step of the transverse craze.

The data developed on PMMA at 77° K in liquid nitrogen confirm the generality of the observation made on PC. Thus, the patterns of behavior in fatigue of polymers with a high crazing propensity will follow the characteristics of the particular polymeric structure, but the general behavior will be similar.

VI. Summary

The current state of understanding of the fatigue deformation of polymers indicates that the development of detailed deformation mechanisms in polymers generally is still in its infancy. The bulk of the studies so far have been macroscopic and phenomenological in nature. However, as the specifics of compliance modes available in polymers are unraveled on a molecular scale, the sophistication of such fatigue studies will increase rapidly. The deformation processes occurring during cyclic straining are inherently the same as those occurring in other types of testing. The unique character of fatigue deformation accrues from the continued reversal of the plastic flow processes and the resulting damage accumulation. As summarized here, the present development of the fatigue of polymers imparts two significant indicators; (1) cyclic deformation can contribute to the development of understanding of basic deformation mechanisms and therefore should be conducted concurrently with the more standard types of deformation studies, and (2) the usefulness of fatigue studies is greatly enhanced by concurrent examination of both phenomenological and mechanistic properties. Hopefully, such an approach will ultimately lead to the systematic development of understanding of polymer fatigue behavior without the surfeit of peripheral data which accrues from the more haphazard approach previously adopted in other materials. Certainly, the understanding of the cyclic deformation of polymers presents an interesting and stimulating area of research with the promise of a practical payoff in terms of sensible applications in engineering structures.

References

Andrews, E. H. (1969). In "Testing of Polymers" (J. V. Schmitz, ed.), Vol. IV, p. 237. Wiley (Interscience), New York.
Avery, D. H., and Backofen, W. A. (1963). In "Fracture of Solids" (D. C. Drucker and J. J. Gilman, eds.), p. 339. Wiley (Interscience), New York.
Beardmore, P. (1969). Phil. Mag. [8] 19, 389.
Beardmore, P., and Rabinowitz, S. (1971). J. Mater. Sci. 6, 80.
Beardmore, P., and Rabinowitz, S. (1972). J. Mater. Sci. 7, 720.
Beardmore, P., and Rabinowitz, S. (1974). To be published.
Dillon, J. H. (1950). Advan. Colloid Sci. 3, 219.
Duckett, R. A., Rabinowitz, S., and Ward, I. M. (1970). J. Mater. Sci. 5, 909.
Feltner, C. E. (1967). J. Appl. Phys. 38, 3576.
Havlicek, V., and Zilvar, V. (1971). J. Macromol. Sci., Phys. 5, 317.
Higuchi, M., and Imai, Y. (1970). J. Appl. Polym. Sci. 14, 2377.

Hull, D. (1970). *J. Mater. Sci.* **5**, 357.

Kambour, R. P. (1973). *J. Polym. Sci., Macromolec. Rev.* **7**, 1.

Kambour, R. P., and Robertson, R. E. (1972). *In* " Polymer Science—Materials Science Handbook" (A. D. Jenkins, ed.), Vol. I, p. 687. North-Holland Publ., Amsterdam.

Landgraf, R. W. (1970). *Amer. Soc. Test. Mater., Spec. Tech. Publ.* **467**, 3.

Landgraf, R. W., Morrow, J., and Endo, T. (1969). *J. Mater. Sci.* **4**, 176.

Manson, J. A., and Hertzberg, R. W. (1973). *CRC Crit. Rev. Macromol. Sci.* **1**, 433.

Morrow, J. (1965). *Amer. Soc. Test. Mater., Spec. Tech. Publ.* **378**, 45.

Murray, J., and Hull, D. (1969). *Polymer* **10**, 451.

Murray, J., and Hull, D. (1970). *J. Polym. Sci., Part A-2* **8**, 1521.

Peterlin, A. (1967). *J. Polym. Sci., Part C* **18**, 123.

Rabinowitz, S., and Beardmore, P. (1972). *CRC Crit. Rev. Macromol. Sci.* **1**, 1.

Rabinowitz, S., and Beardmore, P. (1974). *J. Mater. Sci.* **9**, 81.

Rabinowitz, S., Krause, A. R., and Beardmore, P. (1973). *J. Mater. Sci.* **8**, 11.

Robertson, R. E. (1966). *J. Chem. Phys.* **44**, 3950.

Siegmann, A., and Geil, P. H. (1970). *J. Macromol. Sci., Phys.* **4**, 557.

Sneddon, I. N. (1946). *Proc. Roy. Soc., Ser. A* **187**, 229.

Tomkins, B. (1969). *J. Mater. Sci.* **4**, 532.

Wang, T. T., Matsuo, M., and Kwei, T. K. (1971). *J. Appl. Phys.* **42**, 4188.

Yeh, G. S. Y. (1972). *J. Macromol. Sci., Phys.* **6**, 465.

Low Temperature Deformation
of Crystalline Nonmetals

R. G. WOLFSON

Thayer School of Engineering
Dartmouth College
Hanover, New Hampshire

I. Introduction

Most crystalline solids are nonmetallic, and their low-temperature mechanical response ranges from the extreme brittleness of the very stable inorganic ceramics to the partially viscoelastic behavior of well-crystallized thermoplastic polymers. The ceramics are of marginal interest here, since they exhibit appreciable plasticity only at high temperatures, and their deformation is discussed in the article by Wilcox. The technologically important polymers cannot be treated adequately in the present, general review because of their incomplete crystallinity and the resultant complexity and variability of their microstructures; one significant aspect of their deformation is examined in the article by Beardmore and Rabinowitz. The myriad nonpolymeric organic crystals have received increasing attention recently, but, with the possible exception of the aromatic hydrocarbons, too little is known of their mechanical properties to warrant inclusion; the available information on dislocations in crystalline organic compounds has been summarized by Thomas and Williams (1971). Thus, we shall be concerned with the behavior of those inorganic nonmetals that can deform by crystallographic slip in the vicinity of room temperature.

As contrasted with the metals, these nonmetallic crystals are characterized by a pronounced tendency toward brittle fracture, particularly at low homologous temperatures and high strain rates. There are degrees of brittleness among them, and they have been classified by Stokes (1971) as completely brittle, semibrittle, and ductile:

a. In completely brittle crystals, deformation is elastic up to fracture; if dislocations are present, they are totally immobile.

b. In semibrittle crystals, the dislocations are mobile, but high lattice friction stresses and restricted slip systems limit the amount of plastic deformation, resulting in accommodation problems that initiate brittle fracture.

c. In ductile crystals, dislocation motion is sufficiently easy to permit extensive plasticity terminated by shear fracture.

This classification is far from rigid, for dislocation mobility increases with temperature. Completely brittle materials become semibrittle, and semibrittle materials tend to become more nearly ductile, as the dislocation velocity rises and secondary slip systems are activated. However, true ductility is very rare below the recrystallization temperature, so that the mechanical behavior of plastically deformable nonmetals is typically semibrittle and strongly temperature-dependent.

Plastic deformation in the semibrittle nonmetals does not differ fundamentally from that in the metals. On the contrary, it very much resembles that in the body-centered cubic metals, and those differences which do exist, however distinctive, are of degree rather than of kind. Therefore, the general treatments of the mechanical strength and plastic flow, worked out in detail for the metals, can be applied without significant modification. These are discussed in Section II. The theoretical strength and the Peierls stress are reviewed briefly in Sections II,A,1 and II,A,2, respectively, and the several mechanisms proposed for thermally activated flow are compared in Section II,A,3. The phenomenological description of plastic deformation is then summarized—the microdynamical rate equations in Section II,B,1 and the relationship between slip morphology and ductility in Section II,B,2.

Low-temperature deformation has been observed in a large number of crystalline nonmetals, far too many, in fact, to be included here. A sampling of the literature is given in Table I. The present discussion has been limited to those crystals which meet the following criteria:

a. Deformation must occur by translational glide rather than by twinning.

b. Dislocation motion must have been observed in the vicinity of room temperature.

TABLE I

SOME NONMETALS KNOWN TO DEFORM BY GLIDE NEAR ROOM TEMPERATURE

Structure type	Example	Slip systems	Reference
Arsenic	As	$\langle 11\bar{2}0 \rangle (0001)$	
		$\{\bar{1}100\}$	Shetty and Taylor (1968)
Tellurium	Te	$\langle 11\bar{2}0 \rangle \{\bar{1}100\}$	Stokes et al. (1961)
Graphite	C	$\langle 11\bar{2}0 \rangle (0001)$	Soule and Nezbeda (1968);
			Gillin and Kelly (1969)
Tungsten carbide	WC	$\langle 11\bar{2}3 \rangle \{\bar{1}100\}$	Luyckx (1970)
Chalcopyrite	$CuFeS_2$	$\langle \bar{1}10 \rangle \{111\}$	Buerger (1928)
Montroydite	HgO	$[001] (010)$	Palache et al. (1944)
Spinel	Fe_3O_4 [a]	$\langle \bar{1}10 \rangle \{100\}$	
		$\{111\}$	Charpentier et al. (1968)
Corundum	Al_2O_3 [b,c]	$\langle 11\bar{2}0 \rangle (0001)$	Hsu et al. (1967)
Paralaurionite	Pb(OH)Cl	$[001] (100)$	Palache et al. (1951)
Calcite	$CaCO_3$	$\langle \bar{1}102 \rangle \{20\bar{2}1\}$	
		$\{1\bar{1}01\}$	Keith and Gilman (1960)
Anhydrite	$CaSO_4$	$[100] (010)$	Buerger and Washken (1947)
Gypsum	$CaSO_4 \cdot 2H_2O$	$[001] (010)$	Palache et al. (1951)
Low quartz	SiO_2 [b,d]	$\langle 11\bar{2}0 \rangle (0001)$	Hartley and Wilshaw (1973)
Naphthalene	Naphthalene	$[010] (001)$	
		$[110] (001)$	Robinson and Scott (1967)
	Anthracene	$[010] (001)$	
		(100)	
		$(20\bar{1})$	Williams et al. (1969)
	p-terphenyl	$[110] (1\bar{1}0)$	
		$[010] (201)$	Williams et al. (1969)

[a] Slip direction inferred from observed slip planes.

[b] Room temperature glide indicated by formation of microhardness indent.

[c] Slip system not reported but assumed to be that determined by Conrad (1965).

[d] Slip system not reported but assumed to be that determined by Baeta and Ashbee (1969).

c. Sufficient data must be available to correlate the deformation with the stress required for sensible dislocation glide.

d. The active slip modes must be capable of yielding appreciable plasticity.

These four criteria insure that meaningful comparisons can be made with the low-temperature behavior of metals, and they effectively restrict the description of individual crystals to a few simple structure types of high symmetry and homodesmic ionic or covalent bonding. The deformation of the ionic compounds is reviewed in Section III and that of the covalent crystals in Section IV. Within each section, the specialization of the dislocations by the chemical bonding is discussed first. Then, the several most prominent structure types are considered separately; for each, the distinction is made between those crystals in which the bonding approaches the ideal type and those in which the bonding is of demonstrably mixed character.

Finally, the possibility of modifying the intrinsic behavior of the crystalline nonmetals is treated in Section V. Although these characteristically semibrittle materials have many specific low-temperature applications, their general usefulness is limited by their notch sensitivity and susceptibility to brittle fracture. This is particularly true of applications where their high strength-to-weight ratios, thermal and electrical insulating capabilities, and chemical and biological inertness would otherwise result in extensive utilization. However, it is feasible to enhance the plasticity and to suppress the brittle fracture of the crystalline nonmetals by solute additions. The flow stress can be lowered significantly by solid-solution softening, which is discussed in Section V,A, while the fracture strength can be raised by a factor of two or more by precipitation hardening, which is described in Section V,B.

II. Semibrittle Behavior

The temperature dependence of the flow stress is, perhaps, the most salient general feature of the low-temperature deformation of the crystalline nonmetals. As is well known from studies on the body-centered cubic metals, which exhibit similar behavior, these stresses can be resolved into approximately athermal and thermal components. The athermal component τ_μ, which is proportional to the shear modulus μ, is usually identified with the value of the applied shear stress τ on the comparatively flat, high-temperature portion of the τ versus temperature curve. The thermal component τ^* is obtained as the difference between τ and τ_μ, the latter corrected for the temperature dependence of the shear modulus.

$$\tau^* = \tau - \tau_\mu \tag{1}$$

The athermal component, which is also known as the internal stress, is due to the interaction of mobile dislocations with the long-range internal stress field and corresponds, apparently, with its maximum value (Conrad, 1970); as such, τ_μ is relatively insensitive to temperature and strain rate. The thermal component, frequently called the effective stress, is associated with the thermally activated surmounting of short-range barriers to dislocation motion and is, therefore, a function of both temperature and strain rate. The central problem in explaining the low-temperature deformation of these crystals is to account for the effective stress.

A. Mechanical Strength

1. THEORETICAL (IDEAL) STRENGTH

Prior to the development of the dislocation model, all attempts to explain the mechanical strength of crystalline materials were based upon the behavior of the perfect crystal. Frenkel (1926) estimated the limiting theoretical shear strength τ_{max} to be

$$\tau_{max} = \mu b/2\pi a \simeq \mu/2\pi \tag{2}$$

where b is the atomic spacing in the direction of shear, and a is the spacing of the shear planes. The similarly derived estimate of the limiting theoretical fracture strength σ_{max} is, after Kelly (1966, p. 1ff),

$$\sigma_{max} = E/\pi = \pi\gamma/a \tag{3}$$

where E is Young's modulus, γ is the specific surface energy of the cleavage plane, and a is now the spacing of the cleavage planes. These expressions are simplistic in that they neglect the actual form of the atomic interaction and the effects of the crystal structure. On the other hand, by virtue of the same excessive simplicity, they apply to any crystalline solid, and they correlate the strength with a single, easily measured bulk parameter. Surprisingly, they are in good agreement with the results of more exact treatments, which generally predict values somewhat higher for τ_{max} and lower for σ_{max}, but by a factor of not more than 2 (Kelly et al., 1967). In view of the very high strengths exhibited by certain crystals (Table II) it is possible to conclude, along with Macmillan (1972), that the theoretical strength has probably been observed experimentally in one or two cases.

TABLE II

Observed Very High Strengths of Nonmetallic Crystals

Crystal	Shear strength	Tensile strength
Ge	$0.32\ \mu^a$	$0.076E^a$
MgO	$0.15\ \mu^a$	$0.035E^a$
Al_2O_3	$0.086\ \mu^b$	$0.048E^c$
SiC	$0.072\ \mu^b$	$0.043E^c$
BeO	$0.068\ \mu^b$	$0.038E^c$
NaCl	$0.030\ \mu^d$	$0.026E^d$

[a] From microdeformation of single-crystal surfaces (Macmillan, 1972).
[b] Calculated from tensile strength of whisker crystals by Eisenstadt (1971).
[c] From measurements on whisker crystals (Rauch et al., 1968).
[d] Shear strength calculated from tensile strength of whisker crystals (Mehan and Herzog, 1970); see also Macmillan (1972).

Unfortunately, the extant treatments of the theoretical strength are not directly applicable to the complex displacements within the core of a dislocation or at the tip of a microcrack. All assume that the deformation is a very slowly varying function on the atomic scale, and, in most, the calculations are referred to the structure in static equilibrium. Thus, the essentially inhomogeneous, dynamic nature of the deformation process is ignored, and the temperature dependence of the theoretical strength cannot be estimated reliably.

2. Peierls Stress

The relative weakness of crystalline solids in shear is explained by the dislocation model. According to the series of treatments known collectively as the Peierls–Nabarro theory, the energy of the dislocation varies with its crystallographic location; it is a minimum at the positions of symmetry, which occur at intervals determined by the lattice period in the direction of glide, and it is higher in between. The energy fluctuates during glide, and the maximum value of the resultant restoring stress, known as the Peierls stress, is identified as the intrinsic resistance of the crystal to dislocation motion. In the original form of the theory (Peierls, 1940; Nabarro, 1947), the equations for an isotropic elastic continuum are assumed to be valid everywhere within the crystal except across the glide plane, where a sinusoidal force law is operative. The model implies that the total relative displacement of the two half-crystals (the Burgers vector **b**) spreads out over many interatomic distances, so that the effective width of the dislocation on the glide plane is large. Thus, the magnitude of the width can be used as a qualitative internal test of the validity of the model. Employing several analytic techniques subsequently devised for the complete solution (Eshelby, 1949; Nabarro, 1952; Cottrell, 1953, p. 98), the Peierls stress τ_p can be derived:

$$\tau_p = (2\mu b/cK) \exp\left[-4\pi(a/2K)/c\right] \tag{4}$$

per unit length, where, as before, a is the spacing of the shear planes, and b is the magnitude of the Burgers vector; c is the lattice period in the direction of glide, and the constant $K = 1$ for the screw orientation and $K = 1 - v$ for the edge orientation (v is Poisson's ratio).

The factor $(a/2K)$ in the exponent is the distance from the center of the dislocation at which the relative displacement across the glide plane reaches half of its extreme value. It has been generally accepted as a convenient measure of the effective width of the dislocation:

$$\text{Width} = 2(a/2K) \tag{5}$$

For the original Peierls supposition of a sinusoidal force law, this width is small, on the order of an atomic spacing, which is contrary to the requirements of the basic model. As would be expected, the corresponding values of the Peierls stress are generally too high, although they are two to three orders of magnitude smaller than the limiting theoretical shear strength of Eq. (2).

This discrepancy is due, in part, to the assumed force law, which uniquely determines the atomic displacements on the glide plane. Using the formalism of Foreman *et al.* (1951), each force law, expressed as $\tau = \text{fctn(displacement)}$, implies both a corresponding displacement function, displacement = fctn (distance from dislocation), and an internal stress function maintaining the displacements, $\tau = \text{fctn(distance from dislocation)}$. The former specifies a core structure that can be represented by the effective width, defined as the interval within which the displacement exceeds half its extreme value; the latter, upon suitable manipulation, yields the Peierls stress. Foreman and his associates investigated the implicit functional relationships by progressively modifying the force law to correspond to a weaker atomic interaction with a more realistic, anharmonic form. As the maximum of the assumed force function is lowered and skewed toward zero relative displacement, i.e. as the crystal becomes less capable of opposing the shear displacements of the dislocation, the width increases slowly, and the Peierls stress decreases very rapidly. For a modified force law that gives a width only 1.27 times larger than the Peierls–Nabarro value of Eq. (5), the Peierls stress is smaller by a factor of approximately 2×10^{-3}; when the width is twice the Peierls–Nabarro value, the Peierls stress is vanishingly small. Here, again, the width of the dislocation on the glide plane, on the order of perhaps two atomic spacings, is too narrow to justify the use of classical elasticity within the two contiguous half-crystals. Clearly, the magnitude of the Peierls stress is determined by the atomic bonding in a manner more complex than can be shown by a parametric relationship such as Eq. (4).

It appears that the incapability of classical elasticity to treat large atomic displacements cannot be circumvented by assuming a periodic force law operative only across the glide plane. Rather, the Peierls stress must be formulated in terms of a discrete atomic model. This has been done for several crystal types, e.g. the ionic rocksalt structure (Granzer et al., 1968) and the covalent diamond cubic structure (Teichler, 1967), but the calculations are extremely difficult to complete without introducing simplifying assumptions of questionable validity. Nevertheless, the Peierls stress has been shown to increase both with increasing bond strength and with increasing deviation from the central, additive, nonpolar interactions of pure metallic bonding. Large values have been associated with pronounced directionality of the interatomic bond (Gridneva et al., 1969), and the Peierls stress is generally believed to be highest in the Group IV elements C, Si, and Ge; however, Starodubtsev et al. (1966) have obtained internal friction data indicating that the ratio τ_p/μ in Si is comparable to that in the alkali halides.

3. FLOW STRESS

When dislocations tend to become straight and to lie in low-energy valleys parallel to close-packed directions on their glide planes, their motion can be reasonably interpreted on the basis of the Peierls stress. [See, for example, the reviews of Conrad and Hayes (1963) and Guyot and Dorn (1967), where the primary references to the earlier literature may be found.] This is the usual situation in the crystalline nonmetals at low temperatures, and the double-kink model of thermally activated glide, originally proposed by Seeger (1956), has been widely identified as the rate-controlling deformation mechanism. However, the intrinsic Peierls resistance to glide is not the only source of dynamic frictional stresses acting on mobile dislocations, and the origin of the effective stress is still the subject of controversy. It has been studied most extensively in the body-centered cubic transition metals, where it has been attributed to thermally activated mechanisms for overcoming: (a) the resistance due to the nonconservative motion of jogs in screw dislocations; (b) the spontaneous dissociation of screw dislocations into sessile configurations; (c) generalized Peierls effects; and (d) the interaction of gliding dislocations with residual interstitial impurity atoms. These mechanisms have been applied to the crystalline nonmetals, of course, but the systems studied are too diverse in character to permit a concise exposition of the several models. Therefore, the proposed rate-controlling mechanisms will be introduced here largely in the context of the original and far more voluminous literature on the body-centered cubic metals.

a. Jogs in Screw Dislocations. Jogs in screw dislocations are incapable of conservative motion on any plane other than the plane of the jog. Thus, they can keep up with the main dislocations during glide only by the creation of point defects, and they act as mobile pinning points requiring thermal activation. Schoeck (1961) proposed that the presence of very high jog densities can account for the strongly temperature-dependent yield stress in the body-centered cubic metals; this model was supported by the results of Rose et al. (1962) and Lawley and Gaigher (1964). However, jogs alone cannot explain all of the observed features of the deformation. To do so, it would be necessary to assume either that the jogs are extended, see Section II,A,3,b below, which is almost certainly precluded by the very high stacking fault energies in these metals (Sherwood et al., 1967), or that the jog density is unrealistically high at the start of the deformation (Bowen et al., 1967). In ionic crystals, where jogs exist in thermal equilibrium, very high densities are possible. Furthermore, the jogs are charged, and electrostatic attraction may overcome the elastic repulsion between jogs of the same sense, permitting coalescence into superjogs, although this seems unlikely (Braekhus and Lothe, 1972). On the other hand, pressures up to 4.3 kbar have been found to have no effect upon either the dislocation velocity or the macroscopic deformation in LiF and KCl, which indicates that plastic flow is not controlled by point defect generation (Haworth and Gordon, 1970).

b. Dissociated Screw Dislocations. The temperature dependence of cross slip in the body-centered cubic metals has been attributed to the spontaneous dissociation of screw dislocations into sessile configurations of partials (Nabarro and Duncan, 1967; Kroupa and Vitek, 1969). In order for the extended screws to cross slip, they must first recombine by some thermally activated process, perhaps analogous to the mechanism of Friedel (1964); even for glide on a single slip plane, some such thermally activated transformation into a higher-energy glissile configuration must occur. This model can obviously be used to explain the temperature dependence of the yield and flow stresses (Brown and Ekvell, 1962; Mitchell et al., 1963; Vitek and Kroupa, 1966; Takeuchi, 1969). However, there is no direct evidence of pronounced dislocation dissociation in the body-centered cubic metals, although its occurrence has been inferred by Sleeswyk (1971) from activation-energy analyses. On the contrary, because of the high stacking fault energies, the equilibrium separation of the partials must be so small that they can have little physical reality. Much the same situation exists in the simple ionic crystals. Even in compounds with the cesium chloride structure, where electrostatic contributions to the stacking fault energy are minimized by the high coordination and radius ratio, the energy of the fault boundary opposing dissociation is very large, e.g. 1990 erg/cm^2 in CsBr (Potter, 1969/1970). Of course, a low stacking fault energy does not necessarily mean that flow is controlled by thermally activated recombination. In the adamantine-type covalent crystals, the deformation is dominated by Peierls effects, despite low stacking fault energies; e.g. on the order of 50–70 erg/cm^2 in Si and Ge and about half as large in InAs and InSb (Alexander and Haasen, 1972), ranging down to 1.9 erg/cm^2 in SiC (Stevens, 1972). Recently, however, Mendelson (1972) criticized the Peierls barrier model usually assumed for these crystals, proposing instead a model based upon the nonplanar dissociation of the glissile dislocations.

c. Generalized Peierls Effects. Kroupa and Vitek (1967) interpreted the putative extended configuration of screw dislocations in the body-centered cubic metals as an approximate description of the core structure. This approach was also adopted by Dorn and Mukherjee (1969), who generalized the Peierls mechanism to include the effects of an asymmetric core. They developed an alternative explanation for thermally activated cross slip, but their psuedo-Peierls model cannot be differentiated from the original recombination model (Duesbery and Hirsch, 1968). Indeed, both are now treated as competitive variations on the double-kink model. Much recent work on the low-temperature deformation of the body-centered cubic metals has been devoted to separating Peierls effects from impurity effects. Some investigators have found that interstitial solute atoms have little or no influence on the temperature dependence of the yield and flow stresses (Keh and Nakada, 1967; Lau and Dorn, 1968; Spitzig, 1970; Sastry and Vasu, 1972); in all these studies, the activation parameters were found to agree with a Peierls mechanism.

d. Residual Impurities. There is, nevertheless, considerable evidence to support the view that the rate-controlling deformation mechanism in these metals is the thermal activation of mobile dislocations past residual interstitial atoms, as suggested by Cochardt *et al.* (1955), Fleischer (1962a), and Kelly (1969). Specifically, the removal of interstitial carbon, nitrogen, and oxygen has been shown to significantly decrease the thermal component (Christ and Smith, 1967; Fleischer, 1967; Stein, 1967; Ravi and Gibala, 1969, 1971; Kamenetskaya *et al.*, 1970; Smialek *et al.*, 1970). Furthermore, the analysis of the deformation kinetics often yields activation parameters appropriate to interstitial hardening, as in the recent investigation of Tanaka and Conrad (1972). On the other hand, Leslie and his colleagues (Leslie and Sober, 1967; Leslie et al., 1969; Solomon et al., 1969), working with titanium-scavenged iron, have shown that the strong temperature dependence persists as the interstitial concentration approaches zero; Lahiri and Fine (1970) obtained similar results. This same difficulty in differentiating between the dynamic frictional stress due to the Peierls barrier and that due to residual impurities occurs in the crystalline nonmetals, where it is aggravated by electrostatic interactions. In an ionic crystal, a substitutional impurity with a valence different from that of the host ion associates with a vacancy to produce severe lattice distortion and rapid hardening (Fleischer, 1962b). The presence of such impurities in concentrations on the order of 1 ppm can obscure the effects of the moderately high Peierls energy, and it appears that residual impurities control the deformation of so-called high purity crystals of the alkali halides (Frank, 1968), MgO (Srinivasan and Stoebe, 1970), and CaF_2 (Keig and Coble, 1968). The problem is less serious in the covalently bonded semiconductors, particularly Si and Ge, for residual impurity concentrations can be reduced to extraordinarily low levels and the Peierls barrier is very high. Even here, however, the double-kink model fails to yield quantitative agreement with experiment, and more complex models involving impurities have been invoked; Celli et al. (1963), in developing their Peierls mechanism for dislocation glide in these semiconductors, found it necessary to assume an interstitial impurity content of about 2 ppm. Impurity effects are discussed at greater length in Section III,A,3 for ionic crystals and in Section IV,A,4 for covalent crystals.

It has become increasingly evident that no one model can explain the temperature-dependent plastic flow of the body-centered cubic metals. For each investigation which succeeds in identifying a single thermally activated mechanism, there is generally an equally convincing study, using comparable experimental methods and the accepted techniques for data analysis, which demonstrates that a different mechanism is solely responsible for the deformation of the same material in the same range of temperature and strain rate. There is also considerable evidence that two mechanisms can jointly control the deformation (Soo and Galligan, 1969; Turner and Vreeland, 1970), in which case the two can be consecutively rate-controlling with changing temperature, as observed by Orava (1967) and Smidt (1969). Obviously, simultaneously operative mechanisms call for a combined model such as the "extended-core-impurity" model of Frank and Sestak (1970). The situation is further complicated by the interaction of the possible mechanisms and by the similarity of their effects. Ono and Sommer (1970) and, more recently, Lachenmann and Schultz (1972) have shown that intrinsic core effects and impurity effects are not simply additive. Even the validity of the conventional values of the effective stress must be questioned; among others, Takeuchi (1970) has pointed out that the long-range stress field in a dilute solid solution should assist the dislocation in surmounting short-range barriers. Finally, as stated by Seeger and his colleagues (1968), because the proposed models for thermally activated flow in the body-centered cubic metals often have very similar functional dependences upon temperature and strain rate, the unequivocal identification of rate-controlling mechanisms is unlikely in the near future. This judgement would appear to have remained valid up to the present time.

The controversy over the origin of thermally activated flow has extended to the recently recognized phenomenon of solid-solution softening. This effect, which occurs characteristically in systems with strongly temperature-dependent flow stresses, is now known to be a regular feature of the body-centered cubic metals, and it has been thoroughly documented for the close-packed hexagonal magnesium–lithium system. It has also been reported in a small number of typically ionic and covalent systems as well as in ice. The discussion of the flow stress will be resumed in connection with solid-solution softening in Section V,A.

B. Plastic Deformation

1. MICRODYNAMICAL RATE EQUATIONS

Whatever the nature of the resistance to flow, low-temperature deformation can be correlated with the stress required to produce sensible dislocation glide, and the flow rate (the shear strain rate $\dot{\gamma}$) is directly proportional to the magnitude of the Burgers vector b, the density of mobile dislocations N, and the mean dislocation velocity v:

$$\dot{\gamma} = bNv \tag{6}$$

where N is a function of strain and v is a function of stress at constant temperature. Although extremely small initial densities of mobile dislocations often produce yield-point drops, the basic features of plastic flow are determined by the dynamical behavior of gliding dislocations. The flow rate equation was used by Gilman (1968) as the basis of a phenomenological theory which successfully relates the microdynamics of dislocation motion with the conventional variables of macroscopic deformation; namely, stress, strain, and time. By introducing the empirical dependences of N upon strain and of v upon stress into the equation, he was able to construct stress-strain curves which closely approximate to experience. With the use of additional, largely statistical relationships, the prediction of more complex processes becomes possible (Gilman, 1969).

This microdynamical approach is particularly useful in dealing with the low-temperature semibrittle behavior of the crystalline nonmetals. Since the extent of plastic flow preceding fracture is usually limited, macroscopic stress-strain parameters are difficult to measure; on the other hand, individual dislocations are generally far easier to observe in these materials than in metals, and there are many data on the temperature and stress dependences of their mean velocity. Furthermore, the flow rate equation permits the available information to be organized in abstraction from interpretation, facilitating comparison of the several theories of temperature-dependent flow. In the simplest case, the rate-controlling process is a single thermally activated mechanism for glide, so that the mean dislocation velocity of Eq. (6) is an exponential function of the absolute temperature T.

$$v = v_0 \exp \{-\Delta G/kT\} \tag{7}$$

where the preexponential factor can be interpreted as the area which would be swept out per unit time by unit length of dislocation line if all attempts to surmount the barrier were successful, where ΔG is the Gibbs free energy of activation, and where k is the Boltzmann constant. Provided that the density of mobile dislocations is independent of temperature and strain rate, Eq. (6) can be rewritten as

$$\dot{\gamma} = \dot{\gamma}_0 \exp (-\Delta G/kT) \tag{8}$$

Equations (7) and (8) are rate equations for the flow produced by thermally activated glide, and the thermodynamic parameters usually employed for analyzing the activation mechanism are the free energy of activation ΔG, the activation enthalpy ΔH,

$$\Delta H = \Delta G + T\Delta S \tag{9}$$

where ΔS is the activation entropy, and the stress-related activation volume V^*, which is a measure of the work done during the activation event.

$$V^* = -(\partial \Delta H/\partial \tau^*)_T \tag{10}$$

Since the parameters ΔG, ΔH, ΔS, and V^* are all unique functions of the effective stress τ^*, several dislocation velocity–stress relationships have been proposed, of which two have gained wide acceptance: the power law of Johnston and Gilman (1959),

$$v = B(\tau^*)^{m^*} \tag{11}$$

where B is the average dislocation velocity at unit effective stress and m^* is the velocity–stress exponent, which is usually assumed to be independent of stress and strain rate but is not necessarily independent of strain; and the inverse exponential law, recently restated by Gilman (1968) as

$$v = v_0 \exp (-D/\tau^*) \tag{12}$$

where v_0 is the terminal velocity and D is the characteristic drag coefficient (stress). Li (1968) has pointed out that the power law of Eq. (11) implies an inverse proportionality between V^* and τ^*, while the exponential law of Eq. (12) implies an inverse proportionality between V^* and the square of τ^*.

The fact that neither of these empirical velocity–stress relationships is generally applicable, as detailed for the individual crystallochemical types described in Sections III and IV, has prompted Krausz (1968, 1970) to consider the mean dislocation velocity in terms of absolute rate theory, assuming that the apparent activation energy of the single rate-controlling process is a linear function of stress; i.e.

$$\Delta G(\text{apparent}) = \Delta G - V^*\tau^* \tag{13}$$

In his treatment, the energy barrier opposing motion is permitted to be asymmetric, and the possibility of a backward transition, with different values of the preexponential factor v_0, the activation energy ΔG, and the activation volume V^*, is explicitly included.

$$v = v_{0,\,f} \exp\left[-(\Delta G_f - V_f^*\tau^*)/kT\right] - v_{0,\,b} \exp\left[-(\Delta G_b + V_b^*\tau^*)/kT\right] \tag{14}$$

where the subscripts f and b refer to forward and backward motion, respectively. This equation was found to provide a good description of the stress dependence of the dislocation velocity in NaCl, LiF, ice, Si, Ge, InSb, and GaSb. Furthermore, the temperature dependence specified by this expression, with ΔG and V^* assumed to be temperature-independent, is in good agreement with the thermally activated dislocation motion observed in Ge, Si, and CaF_2, as well as in several metals, over the entire temperature range.

2. SLIP MORPHOLOGY

The microdynamical rate equations presented above in Section II,B,1 take implicit account of the active slip system through the parameters b, N, and v; however, they give no information on the capability of the crystal to accommodate an arbitrary stress by an arbitrary change of shape. This is determined by the slip morphology, which is of the utmost importance, since it controls the notch sensitivity and the extent to which the crystal can deform plastically without fracture.

In order for a material to be fully ductile in both single-crystal and polycrystalline form, it must possess five independent slip systems which become operative at comparable shear stresses and which communicate by easy cross slip (Kelly, 1966, p. 77ff). These five systems may be provided either by one family of high multiplicity or by several families of lower multiplicities. For example, AgCl and the isomorphous rocksalt-type alkali halides have comparable room temperature values of the critical resolved shear stress on the primary $\{110\}\langle\bar{1}10\rangle$ family, yet only AgCl is ductile. The difference arises from the slip morphology. AgCl deforms with equal facility on $\{110\}$ and $\{100\}$, which together provide five independent slip systems, subject to the reduction in multiplicity due to pencil glide; the calculations of Chin (1973) have demonstrated that there are 84 stress states which can activate slip on five such systems without exceeding the yield stress on the remaining systems. On the other hand, the rocksalt-type alkali halides deform preferentially on $\{110\}$, which yields only two independent systems, and, although secondary $\{100\}$ slip is common at room temperature, the critical resolved shear stress is much higher. As a result, the notch sensitivity of these crystals remains appreciable, and they are semibrittle. The geometrical features of the slip morphology, i.e. the number of independent systems, can be determined by the method of Groves and Kelly (1963), but the occurrence of a particular slip mode cannot always be accounted for in simple terms.

Observed slip systems are often rationalized as follows: shear occurs preferentially in the direction of the smallest lattice translation vector, identified as the Burgers vector **b**, and on the planes of maximum spacing in the zone of **b**. This rule of thumb follows directly from the proportionality between the limiting theoretical shear strength and the ratio b/a, Eq. (2). It also corresponds with the minimization of the total self-energy of the dislocation, both the elastic energy, which is proportional to b^2, and the core energy, which is very sensitive to the value of a. There are exceptions to this rule, of course; for example, the preferred slip direction in mercury does not coincide with the minimum repeat vector (Singleton and Crocker, 1971), and the primary slip planes of the rocksalt-type and fluorite-type ionic compounds are not the most widely spaced (see Sections III,B,1 and III,D). Its shortcoming lies, first of all, in the neglect of elastic anisotropy. Foreman (1955) demonstrated that although the anisotropic energy factor does not generally affect the choice of Burgers vector, it can alter the plane of lowest elastic energy, which is the shear plane predicted by the above criterion. Furthermore, the inclusion of elastic anisotropy is not sufficient, for a dislocation with higher total self-energy can nevertheless have a higher glide mobility, permitting it to multiply more rapidly and to dominate deformation. This may be the factor favoring $\{111\}$ slip in the diamond cubic elements (Brar and Tyson, 1972). In addition, dissociation reactions and peculiarities in the core structure itself are capable of determining the operative slip system. All these complications are important among the crystalline nonmetals, where the slip morphology can often be explained only in terms of the chemical bonding.

Thus, the semibrittle behavior typical of the crystalline nonmetals may be due either to high lattice friction stresses or to the incapability of the primary

TABLE III

SLIP MORPHOLOGIES OF CRYSTAL TYPES DISCUSSED IN SECTIONS III AND IV

Structure Bonding	Slip systems (Primary given first)	Number of independent systems[a]	See Section
Rocksalt			
Predominantly ionic	$\langle\bar{1}10\rangle\{110\}$	2	III,B,1
	$\{001\}$	3^b	
	$\{111\}$	5	
Partially covalent/metallic	$\langle\bar{1}10\rangle\{001\}$	3	III,B,2
	$\{110\}$	2^b	
Predominantly covalent/metallic	$\langle\bar{1}10\rangle\{111\}$	5^c	III,B,3
	$\{110\}$	2	
	$\{001\}$	3	
Cesium chloride			
Predominantly ionic	$\langle100\rangle\{011\}$	3	III,C
Predominantly covalent/metallic	$\langle\bar{1}11\rangle\{110\}$	5	III,C
Fluorite			
Predominantly ionic	$\langle\bar{1}10\rangle\{001\}$	3	III,D
	$\{110\}$	2^b	
Partially covalent/metallic	$\langle\bar{1}10\rangle\{111\}$	5	III,D
Ice I			
Ionic: hydrogen bond	$\langle11\bar{2}0\rangle(0001)$	2	III,E
	$\{hki0\}$	?	
Diamond cubic			
Nonpolar covalent	$\langle\bar{1}10\rangle\{111\}$	5	IV,B,1
Zinc blende			
Predominantly covalent	$\langle\bar{1}10\rangle\{111\}$	5	IV,B,2
Wurtzite			
Partially ionic	$\langle11\bar{2}0\rangle(0001)$	2	IV,B,3
	$\{\bar{1}100\}$	2^d	

[a] After Groves and Kelly (1963).

[b] The independent systems provided by this secondary family have no crystallographic connection with those provided by the primary family; subject to the restrictions imposed by the occurrence of pencil glide, the total number of independent systems is 5.

[c] The maximum number of independent systems is 5.

[d] Because the primary and secondary families are not connected crystallographically, the total number of independent systems is 4.

slip mode to provide a sufficient number of independent systems. This two-fold problem is treated in the discussion of individual crystallochemical types which follows in Sections III and IV, with special emphasis on the relationship between the morphology and the character of the interatomic bonding. The crystallographic features of the deformation are summarized in Table III.

III. Ionic Crystals

A. Dislocations in Ionic Structures

1. BONDING AND STRUCTURE TYPES

Most inorganic crystals are ionic and can be visualized, according to the Born theory, as electrically neutral arrays of positive and negative ions. This model presupposes certain structural characteristics. First, each ionic species is treated as if it has a definite radius for a given coordination, with the interionic spacing equal to the sum of the cation radius and the anion radius; i.e. nearest-neighbor ions are in contact. Second, since the interaction is considered to be both central and additive, it exists between each ion and all other ions in the crystal, so that there are neither numerical nor spatial limitations on the bond. Lastly, only those structures which permit strict electrical neutrality can occur. Consequently, ionic structures are determined largely by the geometrical requirement that each ion be coordinated by the maximum number of oppositely charged nearest-neighbor ions. In an ideal ionic compound, this reduces to the packing of hard spheres of characteristic sizes, as expressed by the radius ratio, with each ionic charge completely or very nearly neutralized by its nearest neighbors.

This simplification is best realized among the alkali halides, and their two structure types, the rocksalt and the cesium chloride arrangements, are commonly regarded as archetypical ionic structures. Even here, however, the degree of ionic character varies considerably, as the electronegativity-difference scale of Pauling (1960, p. 97ff) makes clear. The bonding approaches most closely to the ideal type in LiF, being on the order of 90% ionic and 10% covalent; in NaCl and the cesium halides, it is about 75% ionic, and in the other alkali halides somewhat less. The rocksalt-type alkali hydrides and alkaline-earth chalcogenides are also predominantly ionic. Among the latter, the oxides are comparable in ionicity to the alkali halides, but there is a steady decrease in ionic character in passing from the oxide to the sulfide to the selenide and, finally, to the telluride. The deviation is even more marked in the isomorphous silver halides and lead chalcogenides, because the covalent component is decidedly nonlocalized, i.e. metallic. In addition, there are at least 75 compounds of the transition metals, the lanthanides, and the actinides which can be described as partially ionic, with the electrostatic interactions contributing from nearly one-half to considerably less than one-third of the total cohesive energy (Straumanis et al., 1967; Sole et al., 1968; Ramqvist, 1971). At the other end of the scale, the rocksalt and the cesium chloride structures are known to occur in well over 125 intermetallic phases (Taylor and Kagle, 1963); the ionicity is very small, but it can affect the mechanical behavior.

Experimental investigations of low-temperature deformation in ionic crystals have been largely restricted to these two structure types. This emphasis, which is necessarily reflected here, is justifiable on the basis of structural simplicity, ease of single-crystal growth, and frequent occurrence. However, the same can be said of many other typically ionic crystals whose deformability has been noted but not yet adequately documented. A recent exception is the fluorite type; although the slip morphology of the fluorite-group halides has been known for more than half a century, room temperature deformation by dislocation glide was not conclusively demonstrated until the work of Burn and Murray (1962) on CaF_2 and Katz and Coble (1970) on BaF_2 and SrF_2. Nevertheless, the extensive research on compounds of these few structure types has succeeded in revealing the basic features of dislocation behavior in ionically bonded crystals, in part because of the physical simplicity of the bond itself.

2. DISLOCATION MORPHOLOGY

So long as the bonding is at least partially ionic, the need for local electrical neutrality in the structure dominates the dislocation morphology. Although the familiar criterion of least elastic self-energy is generally valid for the Burgers vector, electrostatic contributions to the core energy frequently favor slip planes which are neither the most widely spaced nor the most densely packed; this is true of the rocksalt and the fluorite structures. Consider, for example, the nearest-neighbor coordination across the plane of densest packing in rocksalt, $\{001\}$. When the structure is in registry, each ion below (001) has a single oppositely charged nearest neighbor above the plane at $(a_0/2)[001]$. After a glide of half the Burgers vector, $\frac{1}{2}(a_0/2)[110]$, each ion below the plane would have four nearest neighbors above at $(a_0/4) \times \langle 112 \rangle$. Since two of these would be negatively charged and two positively charged, the nearest-neighbor binding energy across the plane would vanish; in other words, $\{001\}$ shear is precluded by electrostatic faulting. Similarly, the dissociation of a total dislocation into partials is opposed by the mutual repulsion of ions of like charge, since Burgers vectors smaller than the minimum repeat distance of the structure shift the relative positions of the cation and anion sublattices, which raises the surface energy of the fault ribbon.

The violation of local charge neutrality at jogs causes further electrostatic effects. An edge dislocation in an ionic crystal must generally be associated with a double extra half-plane normal to the Burgers vector. For the rocksalt structure, two types of unit jog can occur in the edge. If the offset is formed by the passage of a Burgers vector oblique to that of the main dislocation, the resultant jog is charged by an uncompensated ion on the step. If the jog is due to a Burgers vector normal to that of the main dislocation, it is neutral, but the step itself constitutes an unbalanced dipole. It is not certain that jogs can be formed in just this way, but dislocation intersection is not necessary. Jogs exist in thermal equilibrium, and, in the alkali halides at least, they spontaneously develop net charges. This is due to the smaller free energy of formation of cation vacancies at low temperatures, which favors positive jogs. Thus, the dislocation singular line has a net positive charge at equilibrium. This charge can be built up more rapidly if the dislocation is moving; Rueda and Dekeyser (1963) concluded that fresh dislocations created by indentation in NaCl acquire net charges during glide by taking up the vacancies in the volume swept by the core. The positively charged singular lines induce surrounding negative space-charge regions made up mostly of cation vacancies, and one immediate consequence is the static pinning of the dislocations. As pointed out by Van Bueren (1961, p. 576), the strong interaction between the dislocations and their point defect atmospheres can produce a yield-point drop in even the purest of crystals. Furthermore, the charge on the dislocation, and hence the strength of the electrostatic pinning, is a function of temperature. According to Eshelby *et al.* (1958), when the charge passes through zero, at a temperature they identified with the isoelectric temperature, the yield stress passes through the anomalous minimum which had been observed in NaCl. Ahlquist and Carlsson (1972) corroborated this explanation by demonstrating that the minimum in the critical resolved shear stress of LiF does indeed occur at the isoelectric temperature. The effect is large; the minimum critical resolved shear stress of LiF, which lies slightly below 200° C, is about one-third its room temperature value.

3. Interaction with Impurities

The need for local charge neutrality also dominates the effect of solutes on the deformation properties; differences in valence are far more important than ionic misfit. In NaCl, monovalent cation impurities are some two orders of magnitude less effective than divalent cations in increasing the yield stress, regardless of misfit (Gilman, 1961). Similarly, in MgO, Fe^{2+}

produces negligible hardening compared to Fe^{3+}, which approximately doubles the flow stress at a concentration of 1 cation % (Groves and Fine, 1964). In this case, the magnitude of the misfit parameter is greater than 0.2 for the ineffective Fe^{2+} cation but less than 0.02 for Fe^{3+}. The addition of NiO to MgO has been reported to produce considerable hardening (Liu *et al.*, 1964; Day and Stokes, 1967), which would seem to contradict the generally observed valence effect. However, more recent work by Srinivasan and Stoebe (1970) on very high purity MgO indicates that Ni^{2+} is not a potent hardening agent; indeed, their results suggest that many of the earlier studies on MgO may have been invalidated by the presence of perhaps 0.01% of undetected impurities. The importance of such small residual concentrations can hardly be overestimated. Frank (1968) demonstrated that additions on the order of 0.0001 cation % of divalent cations to the alkali halides can increase the critical resolved shear stress appreciably.

This rapid hardening was treated by Fleischer (1962a,b) with specific reference to Mg^{2+} in LiF. In order to maintain electrical neutrality, the introduction of the divalent cation is accompanied by the creation of a positive-ion vacancy to which it is bound, producing a severe tetragonal lattice distortion. Fleischer represented these defect pairs by elastic dipoles, and he assumed that low-temperature dislocation glide proceeds by thermally activated motion through their stress fields. He succeeded in predicting the observed temperature dependence of the flow stress, the variation of the dislocation velocity–stress relationship with temperature and defect concentration, and the room temperature activation energy for dislocation motion.

In view of the overriding importance of electrostatic effects in the low-temperature deformation of predominantly ionic crystals, it is somewhat surprising that their flow curves so strongly resemble those of the face-centered cubic metals. They even show the three-stage deformation recognized in the metals by Seeger (1957): (I) easy glide, (II) linear work hardening, (III) dynamical recovery. However, as demonstrated by the work hardening theory of Frank (1970), the physical causes are fundamentally different in the ionics, despite the similarity of the macroscopic plastic behavior. Central to this theory, which is based upon LiF and NaCl, is the distinctive role played by divalent cations as foreign ion–cation vacancy dipoles, dipole aggregates, and precipitates. The theory successfully predicts, quantitatively and in great detail, the Stage I and Stage II linear hardening. It is in general agreement with the more qualitative treatment by Davis and Gordon (1969) of work hardening in LiF, NaCl, KCl, and KI; in this latter study, Stage III dynamical recovery was also considered and was tentatively attributed to the thermally activated cross slip of screw dislocations, as in the face-centered cubic metals.

Subsequent work has corroborated the ideas of both Fleischer and Frank. Reppich (1972) measured the critical resolved shear stress of pure and Mg-doped LiF single crystals over the entire range from room temperature to the melting point. He concluded that the Fleischer interaction controls the low-temperature deformation, below 200° C; that an induced Snoek interaction between the dipoles and the dislocations dominates the temperature-independent plateau lying above 200° C; and that the dipoles begin to dissociate near the end of the plateau, around 500° C. In both the low-temperature and the plateau regions, he found quantitative agreement with the theory of Frank. It is significant that Frank and Reppich accept the presence of appreciable concentrations of residual divalent cations in supposedly pure alkali halides.

4. Surface Effects

No discussion of special effects peculiar to the deformation of crystalline nonmetals is complete without mention of the surface condition. The yield stress, the ductility, and the fracture stress may all be changed dramatically by altering the procedures of specimen preparation and handling. This dependence of the mechanical behavior upon the variable surface condition is pronounced among ionic crystals, especially the rocksalt-type alkali halides, where the inherent ductility can be suppressed by surface damage (Lad et al., 1958) and by surficial reactions with the atmosphere (Gorum et al., 1958; Bassett, 1959; Class et al., 1961). It had been thought that surface effects do not influence the dislocation mobility; rather, the embrittlement had been attributed to the creation of microcracks which alter the density and distribution of the dislocations and which nucleate premature cleavage. Although these flaws are important, the clarification of the Rebinder effect, whose various manifestations are general among all but predominantly metallic compounds, has led to the recognition of a more fundamental interaction: adsorbed ions, atoms, and molecules can affect the dislocation mobility by inducing electrostatic changes in near-surface layers. For example, adsorbed water produces anomalous indentation creep in such diverse compounds as LiF, MgO, α-Al_2O_3, and TiC (Westbrook and Jorgensen, 1965), and many surface-active species produce surficial hardening by decreasing the mean dislocation velocity (Westwood and Goldheim, 1968). Sashital and Vedam (1972) demonstrated that atmospheric oxygen is responsible for the sizable hardening of CaF_2 surfaces at room temperature, as had also been shown with LiF. They attributed the observed decrease in the mobility of the dislocations to the rapid diffusion of O^{2-} ions along the singular lines, which introduces electrostatically bound O^{2-}–anion vacancy dipole pairs; these, of course, act as centers of strong tetragonal distortion and give rapid Fleischer-type hardening.

B. The Rocksalt Structure

The rocksalt structure (cubic, space group $F\ 4/m\ \bar{3}\ 2/m$) can be visualized in three general ways: as a pair of interpenetrating face-centered cubic anion and cation sublattices with a relative displacement of $(a_0/2)[100]$; as an array of coordinating anion octahedra, each with all edges shared; or, approximately, as a cubic closest packing of anions with all octahedrally coordinated interstices occupied by cations. The structure is adopted by a large number of AX compounds with radius ratios suitable for 6 : 6 coordination, and it is a typically ionic arrangement. However, the bonding always contains a covalent component which is often metallic in nature. Rocksalt-type crystals are here classified into three groups on the basis of the character of the bond: predominantly ionic, partially covalent/metallic, and predominantly covalent/metallic.

1. PREDOMINANTLY IONIC BONDING

The best known of the predominantly ionic rocksalt-type compounds are the alkali halides (with the exception of CsCl, CsBr, and CsI, which have the cesium chloride structure), and the alkaline-earth oxides, sulfides, selenides, and tellurides (with the exception of the beryllium compounds and MgTe, which have either the zinc blende or the wurtzite structures). At low temperatures, all these crystals deform preferentially on $\{110\}\langle\bar{1}10\rangle$ slip systems and fracture on $\{100\}$. Thus, as would be predicted geometrically, the slip direction coincides with the minimum repeat distance of the structure, $(a_0/2)\langle110\rangle$, which is the Burgers vector, and cleavage occurs parallel to the most densely packed plane. The primary slip plane, however, is $\{110\}$, since unit shear on $\{100\}$ would involve electrostatic faulting. The preference for $\{110\}$ slip is a distinctive feature of the ionic nature. Secondary slip on $\{100\}\langle011\rangle$ becomes easy only at elevated temperatures, where the character of the bonding is modified by increasing polarization; in LiF, for example, the critical resolved shear stress on $\{100\}$ always tends to be greater than that on $\{110\}$, although it decreases more rapidly with increasing temperature (Budworth and Pask, 1963). High-temperature $\{111\}\langle\bar{1}10\rangle$ slip is also observed.

The geometry of the primary $\{110\}\langle\bar{1}10\rangle$ slip system would severely limit the plasticity of these crystals, if it were not for secondary slip, which normally occurs on $\{100\}$ and is common at room temperature (Johnston and Gilman, 1960; Washburn et al., 1960; Argon, 1962), despite the much higher critical resolved shear stress on this plane. Dislocation reactions also hinder primary glide, and, as shown by Frank (1970), they contribute to the work hardening. There are three types of product dislocation resulting from the reaction between total dislocations on intersecting $\{110\}$ planes, but only one of these is definitely stable: a pure edge dislocation with the usual $(a_0/2)\langle 110 \rangle$ Burgers vector lying along $\langle 111 \rangle$. [Amelinckx (1958), however, obtained evidence for the stability of dislocation segments with $\mathbf{b} = a_0 \langle 100 \rangle$ in nets in KCl.] Such a dislocation has a unique $\{211\}$ glide plane and is sessile; it blocks the pair of slip planes in much the same manner as the analogous Lomer lock in the face-centered cubic metals. Dislocation intersection poses further difficulties because of the electrostatic effects accompanying jogging.

Nevertheless, the flow curves of rocksalt-type ionic crystals generally show pronounced Stage I easy glide. This requires comment, for there are never less than two primary slip systems in equivalent orientations with respect to any deformation axis, and multiple slip near the yield point has been reported in NaCl, KCl, and LiF (Davis and Gordon, 1968). This seeming paradox has been resolved for NaCl, at least, by Hesse and Matucha (1972); they demonstrated that the initial multiple slip is restricted to the very narrow transition regions between blocks with single slip on different $\{110\}$ planes.

The fundamental microdynamical aspects of the deformation have been reviewed by Gilman (1961), who used LiF, MgO, and NaCl as prototypes. He established that the macroscopic yield stress is determined primarily by the frictional resistance of the crystal to dislocation glide and that dislocation multiplication takes place by multiple cross glide. Using these two empirical facts, he was able to correlate macroscopic plastic flow with the average dislocation velocity, which he found to be an inverse exponential function of both temperature, as in Eq. (7), and stress, as in Eq. (12). Decreasing the temperature displaces the dislocation velocity–stress curve uniformly to higher stresses, and the primary effect of impurities and point defects is qualitatively the same. That is, the behavior of the dislocation is affected over the entire range of velocities, which implies that the observed hardening is due to an increase in the frictional resistance of the crystal to glide. Gilman suggested that the dislocation mobility and the yield stress are controlled either by the Peierls–Nabarro stress or by size–misfit interactions with isolated impurity ions. He also noted that the mobility of edge segments is much greater than that of screw segments, by a factor of about 50 at room temperature; this disparity seems to be characteristic of the rocksalt-type ionics.

More recent work has shown that the dependence of the dislocation velocity upon stress is more complicated and less clear-cut, as would be expected in view of the complex stress-strain behavior discussed in Section III,A,3. For example, Gupta and Li (1970) found that the velocity–stress relationship in high purity LiF and NaCl follows the power law of Eq. (11) rather than the inverse exponential law. The velocity–stress exponent m^* remains constant for LiF up to 6% plastic strain, but it increases with plastic strain for NaCl. This was attributed to the fact that the deformation of NaCl was largely Stage II, where short-range obstacles to glide are generated. Lee *et al.* (1969/1970) measured the velocity of screw dislocations as a function of stress and temperature in LiF containing 1 ppm and 80 ppm Mg. They concluded that the velocity–stress relationship, which is approximately linear in the range of stresses studied, can be described by either the inverse exponential law or the power law. The activation energy for glide, determined over the low-temperature range from 100° C to 200° C, is independent of the stress but varies with the doping; it rises from 0.4 eV with 1 ppm of Mg to 0.5 eV with 80 ppm of Mg, which indicates that the Fleischer dipole interaction controls the microdynamics.

Although electrostatic interactions dominate the low-temperature deformation of all of these crystals, the need to maintain local charge neutrality becomes less critical as the minor covalent component of the bond increases. This decrease in the ionicity can be brought about either by raising the temperature or by the isomorphous substitution of ions of greater polarizability. The effect of increased temperature on the slip morphology was noted previously. In LiF, the flow stress on the preferred {110} slip planes decreases slowly with increasing temperature, while that on {100} drops sharply until 400° C and less rapidly thereafter (Budworth and Pask, 1963); their values become comparable around 400–500° C, with that on {100} remaining larger. In CaO, the critical resolved shear stress on {110} drops from 2000 psi at 26° C to 1300 at 1500° C, while that on {100} goes from about 86,000 psi to 1400 psi within the same temperature interval; the two slip systems become approximately equivalent at 1000° C (Hulse, 1966). A similar effect is observed when these isomorphous compounds are arranged according to the relative polarizabilities of the constituent ions. As pointed out by Hulse (1966), glide is much easier in CaO than in MgO at all temperatures, and the equivalence of {110} and {100} slip occurs at a much lower temperature. CaO is thus softer than MgO, and it is more difficult to cleave as well; the difference is due to the greater ionic polarizability of Ca^{2+}, which is 5 times that of Mg^{2+} (Rice, 1969). Among the potassium halides, the relative ease of {100} slip increases in the order $KCl \rightarrow KBr \rightarrow KI$, as the anionic polarizability increases (Gilman, 1959). A more extreme example is provided by comparing predominantly ionic NaCl with partially covalent/metallic AgCl. NaCl is semibrittle, glides preferentially on {110} with straight, fine slip lines, and retains its easy cleavage almost up to the melting temperature; AgCl is ductile, shows little preference for a definite glide plane, has wavy, coarse slip bands, and is notch sensitive only below room temperature (Stokes and Li, 1962).

2. PARTIALLY COVALENT/METALLIC BONDING

With regard to low-temperature deformation, the most interesting transition of bond type is in the direction of increasing metallic character. Among the rocksalt-type crystals with partially covalent/metallic bonding, the metallic component is most pronounced in the silver halides (with the exception of AgI, which has the zinc blende structure), the lead chalcogenides (with the exception of PbO, which exists in two rather complex polymorphic forms), and the manganous chalcogenides (with the exception of MnTe, which has the nickel arsenide structure).

Because of the peculiarities of the lead atom, whose stereochemistry is still not fully understood, the partially covalent/metallic bonding in PbS, PbSe, and PbTe is markedly metallic, even more so than in the ductile isomorphous silver halides. The deformation of these three compounds has not been extensively studied, and only that of PbS is known in detail. The primary slip system at room temperature is $\{100\}\langle011\rangle$, with secondary $\{110\}\langle\bar{1}10\rangle$ slip occasionally observed. The occurrence of $\{100\}\langle001\rangle$ slip was reported in PbTe by Rachinger (1956) but was disputed by Gilman (1959). However, Matthews and Isebeck (1963) have since identified presumably sessile dislocations with $\mathbf{b} = a_0\langle001\rangle$ in $\{100\}$ planes in PbS; these may be similar in origin to the $a_0\langle001\rangle$ dislocation segments in nets in KCl (Amelinckx, 1958). Lyall and Paterson (1966), who investigated the deformation of PbS under hydrostatic confining pressure upon which a uniaxial stress was superimposed, found no evidence in any orientation for $\{100\}\langle001\rangle$ slip. They were able to induce yielding, at higher stresses, on secondary $\{110\}\langle\bar{1}10\rangle$ systems, but the later stages of deformation are invariably dominated by single slip on the primary system of initial maximum resolved shear stress; when two such systems are equally favored by the initial stress, the one with minimum glide path length takes over. In almost all orientations, the crystals show some degree of easy glide before the onset of linear work hardening, which they attributed either to jogging or to the formation of sessile dislocations in $\{110\}$.

The low-temperature plastic deformation of the manganous chalcogenides clearly shows the transition from partially covalent/metallic bonding to predominantly covalent/metallic bonding. MnO, MnS, and MnSe have the rocksalt structure, with the degree of ionic character decreasing in the order given. MnTe, with minimal ionicity, has the hexagonal nickel arsenide structure, which is a common type among intermetallic compounds. Here, the bonding is primarily covalent, as indicated by the large axial ratio of MnTe, but metallic Mn–Mn bonds of appreciable strength exist along [0001]. MnO deforms by $\{110\}\langle\bar{1}10\rangle$ glide, MnS by primary $\{110\}\langle\bar{1}10\rangle$ and secondary $\{111\}\langle\bar{1}10\rangle$ glide (Chao et al., 1964), and MnSe by primary $\{111\}\langle\bar{1}10\rangle$ and secondary $\{110\}\langle\bar{1}10\rangle$ glide (Riewald and Van Vlack, 1969). In all three of these rocksalt-type compounds, primary cleavage occurs readily on $\{100\}$, with secondary $\{110\}$ cleavage. Riewald and Van Vlack (1969, 1970) investigated the slip behavior in MnS–MnSe and MnSe–MnTe solid solutions. The progressive replacement of S by Se produces a gradual and continuous change in the preferred slip plane from $\{110\}$ to $\{111\}$; the temperature dependence of primary glide decreases considerably in going from pure MnS to pure MnSe. Within the MnSe-rich single-phase region of the MnSe–MnTe system, up to 13 mole % MnTe, $\{110\}$ slip is increasingly favored with respect to the primary $\{111\}$ planes. At the other end of the system, MnTe and its single-phase solid solutions, up to at least 19 mole % MnSe, deform readily on (0001) and show pencil glide with $\langle11\bar{2}0\rangle$ zone axes. Pure MnTe is soft and is characterized by easy dislocation glide.

A similar but less striking transition occurs in the series US, UN, UC, where the bonding also goes from partially to predominantly covalent/metallic. The degree of ionic character is greater in the sulfide than in the other two compounds, and US deforms by primary $\{110\}$ and secondary $\{111\}$ slip (Sole et al., 1968). In UN, the primary slip plane is $\{110\}$ and in UC, $\{111\}$; however, since both are predominantly covalent/metallic with about the same ionicity, the occurrence of $\{110\}$ slip in the nitride is probably a geometrical effect of the atomic radius ratio (van der Walt and Sole, 1967).

3. PREDOMINANTLY COVALENT/METALLIC BONDING

The rocksalt structure is also adopted by the interstitial monocarbides of the Group IV and Group V transition metals and of at least some of the lanthanides and actinides; there are a number of less well studied isomorphous mononitrides and monoxides, as well. In his review of the mechanical behavior of the carbides, Hollox (1968/1969) pointed out that referring these crystals to the rocksalt structure is somewhat misleading. The equivalent description as a cubic closest packing of the metal atoms, with the carbon atoms occupying the octahedrally coordinated interstices, is to be preferred, for their high-temperature deformation is very similar to that of the face-centered cubic metals. Above the ductile–brittle transition, which lies in the range from 800° C to 1200° C for these compounds, extensive plastic deformation occurs readily on $\{111\}\langle\bar{1}10\rangle$, with some evidence of secondary $\{110\}\langle\bar{1}10\rangle$ and $\{100\}\langle011\rangle$ slip (Hollox et al., 1971).

Since all of these interstitial compounds are predominantly covalent/metallic in nature and are among the hardest and most refractory of crystals, they would not ordinarily be discussed in this chapter. However, Ramqvist (1971) has noted that the ionic contribution to the bonding in these compounds can approach one-third of the total cohesive energy, although it is probably much less, in general. Furthermore, recent work by Rowcliffe and Hollox (1971) has shown that limited room temperature plastic flow can be produced in the carbides by indentation. The low-temperature deformation of the Group IV and the substoichiometric Group V transition metal carbides occurs preferentially on $\{110\}\langle\bar{1}10\rangle$, while only the stoichiometric Group V carbides deform by $\{111\}$ slip as at high temperatures; $\{110\}$ slip is characterized by lesser ductility, regardless of the relative plasticity of the compound at high temperatures. The preference for $\{110\}$ slip can be correlated with the increased importance of the covalent metal–carbon bonding; the lesser ductility, on the other hand, appears to arise either from the smaller multiplicity of $\{110\}\langle\bar{1}10\rangle$ slip systems or from the increased difficulty of dislocation dissociation.

Ramqvist (1971) also suggested that ionic contributions may be of comparable importance to the predominantly covalent/metallic bonding in the isomorphous refractory borides, nitrides, and oxides. [A theoretical study of the electronic structure in rocksalt-type transition metal oxides is currently in progress (Honig and Wahnsiedler, 1972), with the results for TiO published.] However, there appears to be no evidence of general low-temperature deformation in these compounds.

C. The Cesium Chloride Structure

The cesium chloride structure (cubic, space group $P\ 4/m\ \bar{3}\ 2/m$) is the arrangement predicted for ionic AX compounds with radius ratios greater than 0.732; i.e. for 8 : 8 coordination. It is conveniently described either as a pair of interpenetrating primitive anion and cation sublattices with a relative displacement of $(a_0/2)[111]$ or, preferably, as an array of coordinating anion cubes, each with all six faces shared. This latter representation emphasizes the fundamental instability of the structure, for the cations are relatively close and not well screened by the anions. Consequently, the structure is avoided by compounds in which the cations are small or multivalent, even when the radius ratio equals unity. Among ionic compounds, it appears to be limited to a very few halides, cyanides, and hydrosulfides of cesium, thallium, and ammonium. On the other hand, it is common among intermetallic compounds, where it can often be visualized as the result of ordering in the body-centered cubic structure.

Low-temperature deformation has been observed in the cesium halides CsCl, CsBr, and CsI; in the thallium halides TlCl, TlBr, and TlI; and in the ammonium halides NH_4Cl and NH_4Br. The occurrence of these highly symmetric, ionic ammonium halides is due to the free rotation of the NH_4^+ group, which acts like a spherically symmetric cation; in NH_4F, where rotation is prevented by the formation of hydrogen bonds between the nitrogen and fluorine atoms, the ion is locked into the fourfold coordination of the wurtzite structure.

These predominantly ionic compounds deform on $\{110\}\langle001\rangle$ slip systems; $\{100\}$ slip is occasionally reported in compilations of the older literature, but recent work has failed to corroborate its occurrence. The Burgers vector is the minimum repeat distance of the lattice, $\mathbf{b} = a_0\langle100\rangle$, and, unlike the rocksalt structure, glide in this direction can proceed without electrostatic faulting or excessive like-ion repulsion on the most densely packed planes, $\{110\}$. Rachinger and Cottrell (1956) compared the slip morphology in the mixed thallium halides Tl(Cl,Br) and Tl(Br,I) with that in cesium chloride-type intermetallic compounds, some of which exhibit $\langle111\rangle$ glide. They pointed out that a total dislocation with $\mathbf{b} = a_0[111]$ can dissociate into a pair of partial dislocations, each with $\mathbf{b} = (a_0/2)[111]$, with a reduction in the elastic energy; the partials, however, are connected by a ribbon of fault boundary whose surface energy controls the extent of dissociation. In the ionic compounds, this stacking fault energy is so large that dissociation cannot occur, and dislocations with $a_0\langle111\rangle$ Burgers vectors are unstable relative to those with $a_0\langle100\rangle$ Burgers vectors. In the isomorphous intermetallics, the energy of the antiphase boundary is a function of the degree of order and, hence, the ordering energy; as the ordering energy decreases, $\langle111\rangle$ slip becomes more likely. In the limit of the totally disordered alloy, the cesium chloride structure reduces to the body-centered cubic, and $(a_0/2)\langle111\rangle$ dislocations are simply the stable unit dislocations. Thus, $\langle100\rangle$ slip is characteristic of ionic bonding, while $\langle111\rangle$ slip is distinctively metallic.

This theoretical rationalization of the slip morphology was extended by Ball and Smallman (1966), who calculated the anisotropic elastic self-energy and the glide parameter of the four most probable slip dislocations; they concluded that the $\{110\}\langle001\rangle$ system is favored in CsBr, although the choice of slip plane appears to depend upon the glide parameter rather than the self-energy. Potter (1969/1970) refined these calculations and demonstrated that the observed slip system is predicted by both the self-energy and the glide parameter. In other words, in ionically bonded crystals with the cesium chloride structure, $a_0\langle100\rangle$ dislocations have the smallest elastic self-energy for all orientations, which favors their preponderance; they also have the maximum mobility, on $\{110\}$, which favors their rapid multiplication and motion.

Dislocation motion is far less restricted by the slip geometry than is the case for the rocksalt structure. There are two mutually perpendicular $\{110\}$ slip planes for each $a_0\langle100\rangle$ Burgers vector, so that cross slip is easy; the dislocation reactions that would produce sessile dislocations are energetically unlikely; and, although jogs can be charged, electrostatic effects are less pronounced because of the higher coordination.

The cesium chloride-type ionic compounds are unusual in that they lack cleavage planes. Johnson and Pask (1964) found that single-crystal CsBr fails by ductile fracture which nucleates at slip-band intersections and which propagates in the direction of the maximum resolved shear stress. In specimens loaded to yield a Schmid factor of zero for the active slip systems, little plastic deformation precedes fracture, which occurs on $\{110\}$ in $\langle\bar{1}10\rangle$ for tension and on $\{112\}$ in $\langle11\bar{1}\rangle$ for compression. In specimens in optimum orientation for shear, with a Schmid factor of approximately $\frac{1}{2}$, up to 10% strain is observed before fracture; cross slip occurs readily at low stresses, and strain hardening, attributed to jog formation, is associated with slip on two sets of $\{110\}$ planes intersecting at $60°$.

Although there are no known compounds with transitional bonding of moderate ionicity, the persistence of a small degree of ionic character is observed among the intermetallics with the cesium chloride structure. Here, the bonding is predominantly covalent/metallic, and there is little evidence of low-temperature deformation. AuCd, AuZn, LiTl, MgTl, and NiAl all deform on the characteristically ionic $\{110\}\langle001\rangle$ slip system; CuZn deforms on the distinctively metallic $\{110\}\langle\bar{1}11\rangle$ system, and AgMg, which may be a borderline case, deforms on $\{321\}\langle\bar{1}11\rangle$ (Rachinger and Cottrell, 1956; Potter, 1969/1970). The effect of increasing metallic character in compounds with the same, extremely small ionicity is shown by the series TiFe → TiCo → TiNi. The bond goes from strongly covalent in TiFe to almost completely metallic in TiNi, and the ductility increases in the order given (Scholl *et al.*, 1968).

D. The Fluorite Structure

The fluorite structure (cubic, space group $F\ 4/m\ \bar{3}\ 2/m$) is the typically ionic arrangement for AX_2 compounds with radius ratios larger than 0.732. It has 8 : 4 coordination and can be derived from the cesium chloride structure by removing every second cation. Thus, it is an array of coordinating anion cubes which share edges, rather than faces, and the concomitant reduction in the cation–cation repulsion accounts for its great stability. The fluorite structure is adopted by the fluorides and chlorides of the alkaline-earth metals (with the exception of the beryllium compounds); the dioxides of the actinide elements; and a number of diborides, dicarbides, dihydrides, dinitrides, and other dioxides. The antifluorite structure, with 4 : 8 coordination, is adopted by the chalcogenides of the alkali metals lithium, sodium, and potassium, plus rubidium oxide and rubidium sulfide. The fluorite and antifluorite structures are also known to occur in perhaps two dozen metal alloys.

Predominantly ionic compounds with the fluorite structure deform preferentially on $\{100\}\langle011\rangle$ slip systems. Secondary slip becomes common on $\{110\}\langle\bar{1}10\rangle$ at moderate temperatures, and high-temperature $\{111\}\langle\bar{1}10\rangle$ slip is observed in partially covalent/metallic compounds such as ThO_2 and UO_2; $\{111\}$ slip is strongly opposed by electrostatic faulting involving the tetrahedrally coordinated anions, and its occurrence is indicative of decreased ionic character. In all cases, the Burgers vector is $(a_0/2)\langle110\rangle$. Cleavage is perfect and easy on $\{111\}$.

Little attention has been given in the past to the low-temperature deformation of predominantly ionic fluorite-type crystals. Phillips (1961) measured the critical resolved shear stress of CaF_2 in compression between $25°$ C and $1000°$ C. He found that plastic flow is thermally activated, that the dislocation density is a linear function of strain, and that appreciable ductility appears above $400°$ C; he also noted that secondary cleavage occurs on $\{110\}$ and $\{100\}$. Using lower strain rates, on the order of 10^{-4} per minute, Burn and Murray (1962) plastically deformed CaF_2 at room temperature up to 0.007% strain; they also demonstrated that considerable plastic flow and even creep to fracture occur at temperatures as low as $175°$ C.

As higher purity crystals became available, the ductile-to-brittle transition in CaF_2 was moved to lower temperatures. It was bracketed within the range $25–100°$ C by Keig and Coble (1968). These investigators measured the steady-state dislocation velocity as a function of stress over this temperature range, following the edge dislocations only, since the screws moved too rapidly at the applied stress levels of 0.1 to 1.0 kg/mm^2. The velocity can be represented by the power law, but the inverse exponential law of Eq. (12) gives a better fit except at the lowest stresses. The terminal velocity v_0 is a function of temperature, increasing from about 10^{-4} cm/sec at $25°$ C to 10^{-2} cm/sec at $100°$ C; the drag coefficient has a constant value of 0.89 kg/mm^2 independent of temperature. Keig and Coble noted that the combination of temperature-dependent terminal velocity and temperature-independent drag coefficient can be explained neither by an intrinsic Peierls mechanism nor by vacancy-controlled dissipative drag. Furthermore, the dislocation velocity is far less sensitive to stress than is the case for LiF at the same temperatures, and the activation energy for glide, 0.54 eV, is close to that for anion vacancy migration. They concluded that the dislocation motion is controlled by impurities, perhaps oxygen, and is consistent with the relaxation of Fleischer-type divalent anion–anion vacancy dipoles.

Room temperature dislocation motion on the primary slip planes has also been observed in BaF_2 and SrF_2 (Katz and Coble, 1970). As in CaF_2, the screw dislocations move more rapidly than the edge dislocations.

The bonding in the fluorite-type hydrides of the transition metals is partially covalent/metallic. In zirconium hydride and titanium hydride, where low-temperature deformation has been observed in polycrystalline specimens, the metallic component is pronounced. Barraclough and Beevers (1969) studied the delta form of zirconium hydride over the composition range $ZrH_{1.27}$–$ZrH_{1.66}$ between room temperature and about 500° C. They found that polycrystalline specimens are completely brittle in tension but show a ductile-to-brittle transition around 100° C in compression; by 120° C, plastic strains up to 3% can be accommodated without fracture. The yield stress decreases rapidly until 200° C and very little thereafter. The active slip planes were identified as {111}, and there was no indication of slip on either {100} or {110} at any temperature. They concluded that the composition dependence of the deformation is consistent with a lattice friction mechanism controlled by interactions with anion vacancies. {111}⟨$\bar{1}$10⟩ slip was also found by Irving and Beevers (1972) in their detailed investigation of polycrystalline titanium hydride, $TiH_{1.53}$–$TiH_{1.99}$, between $-35°$ C and 200° C. With the exception of near-stoichiometric $TiH_{1.99}$, all compositions show some plasticity above 0° C. For example, $TiH_{1.75}$ deforms to 2% strain at $-35°$ C. The yield stress decreases with increasing temperature and with decreasing anion vacancy concentration. It also decreases with decreasing strain rate, as does the temperature of the ductile-to-brittle transition. Irving and Beevers demonstrated that these several dependences can be correlated with the hydrogen ion mobility and that the ductile-to-brittle transition occurs at a specific value of the mobility for a constant strain rate. They attributed the hardening produced by deviation from stoichiometry to long-range interactions between the dislocations and aggregates of anion vacancies, a conclusion supported by observed activation volumes on the order of several hundred b^3. They also surmised that electrostatic faulting during {111} glide is avoided, in this largely ionic compound, by hydrogen ion diffusion out of the path of the dislocation core.

E. Ice: The Hydrogen Bond

Because of its small size and simple extranuclear structure, hydrogen often plays a distinctive crystallochemical role. Under certain circumstances, it can be coordinated by two atoms to which it is more or less equally attracted, forming the hydrogen bond. This bond has occasionally been described in terms of covalent bonding or van der Waals bonding, as an extreme form of molecular polarization, but Pauling (1960, p. 449ff) has shown that it is largely ionic in origin. Its strength is perhaps 10% that of the typical ionic bond, and, when it is solely responsible for the cohesion of a crystalline solid, the stability is small. Most hydrogen-bonded crystals melt below room temperature, and the deformation characteristics of only one are known in great detail: ice.

Ice is an exceptional substance in many ways, and its stable form under ordinary conditions, ice I, is anomalous with respect to the other hydrogen chalcogenides. The structure superficially resembles the wurtzite type, with water molecules occupying all of the positions within the unit cell. Since the water molecule is both neutral and asymmetric, ice can be pictured, incorrectly of course, as a van der Waals solid with a very large dipole moment. However, since the intermolecular distance and the tetrahedral coordination are anomalously small and the melting temperature and mechanical strength anomalously high for such bonding, it is clear that ice is hydrogen-bonded. On the model of Bernal and Fowler (1933), the oxygen atoms are tetrahedrally coordinated and are linked by hydrogens located between them, but each oxygen has only two nearest-neighbor hydrogens; the other two are next-nearest neighbors. In other words, each hydrogen must be associated with one of the two oxygen atoms that it coordinates. This bond asymmetry results in a great number of possible configurations, among which a random distribution is maintained (Pauling, 1935).

The resultant proton disorder complicates dislocation motion, and it accounts for the distinctive mechanical behavior of ice. According to the so-called Bernal–Fowler rules, (1) there are two hydrogens near each oxygen, and (2) there is one hydrogen on each oxygen–oxygen link. Violation of the first rule gives rise to ionic defects; it corresponds to the self-dissociation of H_2O into H_3O^+ and OH^-. Violation of the second rule gives rise to the rotational defects known as Bjerrum defects, which also exist in thermodynamic equilibrium; they comprise the vacant bond, or L-defect, and the double bond, or D-defect. These two types of electrically active point imperfections dominate the deformation process.

Glen (1968) pointed out that the proton disorder in ice would cause a moving dislocation to emit either ionic or Bjerrum defects, and he estimated that such nonconservative motion would require an applied stress on the order of 3400 kg/cm². Since the observed flow stress is only about 1 kg/cm², dislocation glide must involve some process of proton rearrangement. In other words, glide must proceed by the lateral motion of kinks along the singular line, with new kinks forming wherever the randomly distributed hydrogen bonds ahead of the line become favorably oriented; the number of bonds which must be reoriented is a function of the applied stress. Glen proposed that the bond reorientation is accomplished by the passage of Bjerrum defects, and the general experience on ice tends to support this glide mechanism. In particular, it explains the very low dislocation velocities calculated from stress–strain curves obtained at high homologous temperatures, e.g. 50×10^{-8} cm/sec at $-50°$ C, and the strong dependence of the velocity on the concentration of bonding defects (Jones and Glen, 1969). It also explains the dependence of the dielectric properties on the mechanical state of the crystal, e.g. the variation of dielectric relaxation with shear strain (Ackley and Itagaki, 1969).

With regard to the dielectric properties of deformed ice, Itagaki (1969) has suggested that the motion of charged dislocations may account for part of the observed effects. He subsequently obtained direct evidence for the existence of the charges (Itagaki, 1970). Using X-ray diffraction topography, he showed that dislocations vibrate under an alternating electric field; the charge on the singular line is positive, and the charge density is comparable to the values observed in the alkali halides. Itagaki also reported that straight dislocations lying along $\langle 11\bar{2}0 \rangle$ do not vibrate, at least for fields up to 160 V/cm, which he attributed to trapping in the low-energy Peierls valleys. However, since these dislocations are in the screw orientation, it is equally plausible that they are uncharged.

Although the mechanism of glide seems reasonably well established, experimental difficulties have hampered the investigation of the dislocation microdynamics. The existence of charged dislocations is disputed, and even the glide morphology is vigorously debated. The latter is summarized below.

Ice deforms preferentially on the basal (0001) plane by thermally activated dislocation glide (Readey and Kingery, 1964), and evidence has been presented for cross slip on prismatic planes (Muguruma and Higashi, 1963). The Burgers vectors of total glide dislocations lie along $\langle 11\bar{2}0 \rangle$ (Webb and Hayes, 1967); the observation that slip occurs in the direction of the shear-stress vector, with little preference for a particular crystallographic direction, can be explained by the lack of resolution associated with the nonlinear flow law at high homologous temperatures (S. F. Ackley, private communication, 1971). It has been suggested that screw dislocations with $\mathbf{b} = (a_0/3)\langle 11\bar{2}0 \rangle$, which predominate in ice, may be able to dissociate into partials on the basal plane (Unwin and Muguruma, 1971), but further proof is required. On occasion, prismatic dislocation loops normal to [0001] have been reported (Jones, 1970).

There is an enormous literature on the deformation of pure ice, particularly as it relates to glacier flow, and on the mechanical behavior of lake and sea ice. Unfortunately, with the exception of the information already presented, it has little relevance to basic glide processes. On the other hand, recent studies on doped single-crystal ice have revealed the occurrence of pronounced solid-solution softening; these are discussed in Section V,A,2.

IV. Covalent Crystals

A. Dislocations in Covalent Structures

1. BONDING AND STRUCTURE TYPES

Covalent structures are determined largely by the spin valence, which must generally be referred to hybridized orbitals; e.g. for elemental carbon, the coplanar trigonal sp^2 bonds in graphite and the tetrahedral sp^3 bonds in diamond. Thus, the covalent bond has two distinctive structural characteristics. First, the number of bonds per atom is limited to the maximum number of paired electrons that can be accommodated in the valence subshell. Second, the bonds have a definite configuration in space, since they form along, or close to, the directions of maximum electronic charge density. Because of the stringency of these two limitations, covalent bonding usually gives rise to discrete molecular units. In graphite, for example, the atoms are covalently bonded into sheets, but the sheets themselves are held together by residual forces; graphite is thus classed with the van der Waals crystals. Covalent crystals such as diamond, in which there is a three-dimensional network of directed bonds, are uncommon, and the important structure types are both few and simple. The structures tend to be open, with low coordination and a comparatively small atomic density; unlike predominantly ionic, metallic, or van der Waals crystals, geometrical packing is not a major consideration.

Tetrahedral sp^3 bonds are formed by most of the B-subgroup elements, and the resultant structure types are typically covalent. The simplest are the diamond cubic structure of the Group IV elements C (diamond), Si, Ge, and α-Sn; the zinc blende and the wurtzite structures of most III–V, II–VI, and I–VII compounds; and the numerous carborundum structures. The more complex types include the chalcopyrite, the enargite, and the famatinite structures. Since all have a fourfold tetrahedral coordination of the atoms, they are very similar crystallochemically, as is indicated by the frequent occurrence of polymorphic transformations among the compounds. Newman (1963) demonstrated that their structures can be generated by the stacking of nearly plane stoichiometric nets of hexagonal symmetry, and he termed such tetrahedrally coordinated covalent crystals as adamantine.

This structural approach is not artificial, for distinguishable planar units exist in the adamantine crystals, despite the homodesmic bonding. In the simple structure types, these planes occur in pairs, each component containing only atoms of one kind. Every atom is bonded to three nearest neighbors on the adjacent component of the same double layer and to one atom on the next double layer. As a result, the two components are tightly bonded into a composite "puckered" plane with a spacing much smaller than that between successive double layers. In the diamond cubic and the zinc blende structures, the fundamental stoichiometric nets are the {111} double layers, and the stacking sequence can be represented as aB–bC–cA–aB–bC–cA ⋯; in the wurtzite structure, the nets are the (0001) double layers and the stacking is aB–bA–aB–bA–aB–bA ⋯; in the various carborundum polytypes, the stacking is mixed but repetitive. Within a double layer such as aB, the arrangement is geometrically similar to closest packing, while layers such as B–b lie directly above one another. These simple adamantine types are the preeminent representatives of covalent structures, and their mechanical behavior will be used to epitomize that of covalently bonded crystals.

2. DISLOCATION MORPHOLOGY

The nature of dislocations in covalent structures is dominated by the directionality of the bonding. In the adamantine crystals, the Burgers vector is the edge of the bonding tetrahedron, and the preferred slip plane is the fundamental stoichiometric double layer. This morphology conforms to the qualitative rule that shear occurs in the direction of the minimum lattice translation vector and on the planes of maximum spacing. However, although the observed Burgers vector is predicted by all criteria, the double layer is not necessarily the plane of lowest elastic energy, so that its operation must be attributed to core effects (Brar and Tyson, 1972). The influence of the bonding within the core is also seen in the alignment of the dislocations parallel to close-packed directions. In Si and Ge deformed at relatively low temperatures, the preferred singular-line directions are $\langle 110 \rangle$ (Dash, 1956) and $\langle 112 \rangle$ (Geach et al., 1957; Patel, 1958), although deviations of several degrees from $\langle 110 \rangle$ have been reported in Frank–Read sources (Authier and Lang, 1964). Diffusion-induced dislocations also favor these orientations (Schwuttke and Queisser, 1962, 1964; Jaccodine, 1964; Joshi and Wilhelm, 1965). The only other strongly preferred singular-line direction in the elements is [001], which has been observed for sessile edge dislocations in low-angle grain boundaries in [001]-grown Ge (Vogel, 1955; Pfann and Lovell, 1955; Pfann and Vogel, 1957).

When the singular line lies along $\langle 110 \rangle$, the dislocation can be either screw, 60°, or edge. The first two are the familiar slip dislocations; the edge dislocation, which is sessile, can be formed by the reaction of the glissile types (Washburn et al., 1964), and it has been shown to decompose into an extended Lomer–Cottrell dislocation in Si (Wolfson et al., 1966a) and GaAs (Schwuttke and Rupprecht, 1966). When the singular-line orientation is $\langle 112 \rangle$, the dislocation can be either 30° or edge, the latter being the type introduced during plastic deformation. The structures of the slip dislocations have been treated by Hornstra (1958). In the most probable form of the screw dislocation, the covalent bonds are rotated away from their normal tetrahedral configuration, but they remain unbroken; dissociation into partials is unlikely, since it would require the breaking of bonds. In the $60°$ dislocation, the edge character predominates, and the extra half-plane ends in a line of dangling bonds; dissociation into a pair of Shockley partials on the glide plane seems favored by energy considerations. In the edge dislocation along $\langle 112 \rangle$, the number of dangling bonds per unit length is about 15% greater than that for the $60°$ dislocation, and it too should dissociate into an extended Shockley pair.

However, as pointed out by Alexander and Haasen (1972), there is considerable uncertainty over the dissociation of these dislocations. Since the preferred glide planes are composite double layers, two types of dislocation are possible. The first exists between double layers, e.g. between B and b, and produces a relative displacement of the composite stoichiometric planes. It breaks only one bond per atom and is presumably the type responsible for plastic flow, but it cannot dissociate on a single plane without violating the tetrahedral coordination. The second type exists between the components of the double layer, e.g. between a and B, and shifts the two closely spaced layers. It breaks three bonds per atom and is presumably sessile, but it can dissociate into Shockley partials like the analogous $\{111\}\langle \bar{1}10 \rangle$ and $(0001)\langle \bar{2}110 \rangle$ slip dislocations in the face-centered cubic and hexagonal close-packed metals, respectively. These authors concluded that the contradictory reports of dislocations in both total and extended forms cannot be explained, unless the two types of dislocation coexist in deformed crystals. The same problem was considered by Mendelson (1972), who maintained that the nonplanar dissociation of glissile dislocations does occur. In particular, he showed that such dissociation can account for the existence of the stable, glissile edge dislocations along $\langle 112 \rangle$ and for the fact that $60°$ dislocations are less mobile than screws at low temperatures and strain rates in Ge. He argued that neither of these observations is consistent with the generally accepted model of a very high Peierls barrier which must be surmounted by double-kink nucleation.

A related question, also unresolved at the present time, is that of the elimination of broken bonds by rearrangements within the dislocation core. This was originally proposed for the diamond cubic elements (Hornstra, 1958), and it seems equally likely for the compounds (Holt, 1962), but experimental verification is lacking, at least for dislocations lying in the primary (double layer) slip planes.

To the contrary, these dislocations have been shown capable of trapping electrons in n-type material. This presumably occurs at the dangling bonds associated with edge components (Shockley, 1953), although it may also take place near the severely distorted bonds characteristic of screw components (Heine, 1966). Consequently, the singular line acquires a negative charge which induces a surrounding positive space-charge region (Read, 1954). As in ionic crystals, the dislocation charge gives rise to strong pinning interactions with charged defects (Van Bueren, 1961, p. 193), Rebinder effects (Westbrook and Jorgensen, 1965), and a striking sensitivity to the valence of substitutional impurities (Patel and Chaudhuri, 1966). Furthermore, because the charge is electronic rather than ionic in origin, the Coulomb interactions are functions of the carrier density, which can be altered by photogeneration as well as by doping. Indeed, light-induced changes in the low-temperature mechanical properties appear to be of general occurrence among the simple adamantine crystals.

3. LIGHT-SENSITIVE MECHANICAL EFFECTS

Photoplasticity is especially pronounced in II–VI compounds, where the absorption of light photons with energies near the bandgap energy produces large, reversible increases in the flow stress. The effect has been attributed to the pinning of negatively charged dislocations by a Coulomb-type interaction with photoionized point defects (Carlsson and Svensson, 1968). This model was dramatically confirmed by Carlsson (1971); he demonstrated that photoplasticity produces a 100% increase in the flow stress of single-crystal ZnO (wurtzite-type) oriented for primary slip, but that it does not even appear when the crystals deform by pure secondary slip, which proceeds by the glide of uncharged dislocations. In ZnO, the charged defects were identified as Zn^+ interstitials which become doubly ionized by the capture of photogenerated holes within the positive space-charge regions surrounding the negatively charged dislocations (Carlsson and Svensson, 1970). In CdTe (zinc blende-type), where the yield stress is increased by as much as 70% by illumination, the defects are photoionized vacancies or self-interstitials (Carlsson and Ahlquist, 1972).

The surficial photomechanical effect, which is the light-induced change in the indentation microhardness, is a widespread, variable, and incompletely understood phenomenon. In Ge, it has been associated with the photoionization of trapped electrons from dislocation acceptor levels (Kuczynski et al., 1972). Here, the effect appears below 300° C, and its magnitude increases with decreasing temperature. In the vicinity of room temperature, microhardness drops of about 20% are produced by illumination with wavelengths less than 0.4 μ and greater than 1.7 μ plus a band around 0.57 μ (Iyer and Kuczynski, 1971); illumination also increases the room temperature fracture strength by 10–20% (Iyer et al., 1971).

4. Interaction with Impurities

The carrier density and the charge on the dislocation are also altered by electrically active substitutional impurities, and doping with minute quantities of acceptor and donor atoms produces very large changes in the dislocation mobility. Patel and Chaudhuri (1966) found that the dislocation velocity in Ge remains constant as the dopant concentration increases from $10^{13}/cm^3$ to the vicinity of $10^{18}/cm^3$, which corresponds to the intrinsic carrier concentration at the test temperature. When this is exceeded, the velocity in p-type Ga-doped specimens decreases while that in n-type As-doped specimens increases. At 515° C and the maximum doping level of $2 \times 10^{19}/cm^3$, roughly 0.04 at. %, dislocations in the n-type Ge move about 8 times faster than those in the p-type Ge; this is reflected in the macroscopic stress–strain behavior, the n-type Ge being 30% softer and the p-type Ge 70% harder than the undoped intrinsic Ge. Patel and Chaudhuri correlated these changes with the activation energy for glide, which falls to 1.2 eV in n-type and rises to 1.75 eV in p-type for the maximum dopant concentration, compared to 1.57 eV in pure intrinsic Ge. Finally, they demonstrated that the electrically neutral Group IV impurities Si and Sn have no effect upon the dislocation microdynamics, at least in concentrations up to 10^{20} per cm^3. Patel and Freeland (1967) showed that As and Ga have much the same effects upon the dislocation velocity in Si at 600° C as they do in Ge around 500° C.

This doping phenomenon cannot be explained by misfit-induced strain. Among other obstacles to such an explanation is the fact that the normal covalent tetrahedral radii of As, Ga, and Ge are all very nearly the same, not too far different from that of Si, but much smaller than that of Sn (Dickinson, 1970). Rather, it must be referred to some modification of the double-kink mechanism which is sensitive to the valence of the solute. Kulkarni and Williams (1972) have suggested that the trapping of electrons at kinks nucleating in screw dislocations reduces the formation energy of stable double kinks in n-type material. However, the activation energy for dislocation glide can also be lowered by p-type dopants, e.g. B in Si (Erofeev et al., 1970). These and other, more complex doping effects are discussed in connection with solid-solution softening in Section V,A,3.

B. The Simple Adamantine Structures

The crystallochemical similarity of the adamantine structures is particularly pronounced for the diamond cubic, the zinc blende, and the wurtzite types. Each of these three simple structures can be visualized as a pair of identical interpenetrating sublattices with the atomic sites distributed as in closest packing. Each sublattice is occupied by atoms of one kind; every atom has four nearest neighbors on the other sublattice and twelve next-nearest neighbors on its own sublattice. The bonding in the diamond cubic elements is completely nonpolar, but it always has some ionic character in the zinc blende-type and wurtzite-type compounds. Estimating the covalency of the bond from the electronegativity difference, as was done for the ionic compounds, it is found that the predominantly covalent III–V and II–VI compounds adopt the zinc blende structure in preference to the wurtzite structure; the more rigorous calculations quoted by Vukcevich (1972) confirm this observation. The corresponding wurtzite-type compounds are partially ionic. Therefore, despite the close structural relationships, these simple adamantine crystals can be differentiated on the basis of bonding characteristics: the nonpolar diamond cubic elements, the predominantly covalent zinc blende-type compounds, and the partially ionic wurtzite-type compounds.

The distinction between the diamond cubic elements and the zinc blende-type compounds appears clearly in their mechanical behavior. Indeed, Gridneva et al. (1969) have proposed a criterion for bond directionality which is directly related to the ease of dislocation motion: the parameter $\Delta G/kT_m$, which is the product of the homologous temperature T/T_m and the exponent of the rate equation for thermally activated glide, Eq. (7). ΔG, the activation energy, is a measure of the Peierls barrier, and kT_m, where T_m is the melting temperature, is characteristic of the bond strength. These investigators have correlated the magnitude of the parameter with the degree of directional bonding, the critical resolved shear stress, and the upper temperature limit to the ductile-to-brittle transition. Its value is approximately 15 for the diamond cubic elements, 10 for the zinc blende-type III–V compounds, and 1 for the body-centered cubic transition metals.

1. Nonpolar Covalent Bonding (The Diamond Cubic Structure)

In the diamond cubic structure (space group $F\,4_1/d\,\bar{3}\,2/m$), the two component sublattices are face-centered cubic with a relative displacent of $(a_0/12)[111]$. The slip system is $\{111\}\langle\bar{1}10\rangle$, with secondary slip not observed; the preferred cleavage plane is also $\{111\}$, and, although it can be rationalized in terms of the bonding, it has been explained more satisfactorily by microcrack nucleation due to the pileup and coalescence of glissile dislocations (Abrahams and Ekstrom, 1960). The diamond cubic elements do not show perceptible plasticity much below an homologous temperature of 0.6. For example, in Ge, there is no evidence for room temperature dislocation glide, and the extrapolation of high-temperature data to 25° C indicates that the dislocation velocity would be on the order of 10^{-16} cm/sec at about one-fifth the theoretical shear stress (Johnson, 1965, 1966). Nevertheless, their mechanical behavior is of importance here. First, it exhibits the extreme covalent characteristics which are moderated by ionic and/or metallic bond components in the compounds. Second, Si and Ge are available as macroscopic single crystals of extraordinary perfection and purity: completely free of dislocations, stacking faults, and twins, and with the total impurity content limited to the level of parts per billion or better; as a result, dislocation processes can be observed in extreme detail.

Of particular interest is the study by Patel and Chaudhuri (1963) of the tensile behavior of initially dislocation-free Si and Ge crystals within their high-temperature plastic ranges. The stress–strain curves, which otherwise strongly resemble those of the face-centered cubic metals, are characterized by large yield-point drops, with the maximum stress approximately four times the flow stress but an order of magnitude lower than the theoretical shear stress. This phenomenon was shown to arise from the dislocation microdynamics rather than from static pinning. Dislocations are first generated in the initial linear stage of deformation, at stress levels of about one-quarter of the maximum stress. The stress–strain curve rises until continuing dislocation multiplication produces a sufficient density to permit the specimen to deform at the rate imposed by the hard testing machine, at which point the maximum stress obtains. Beyond the maximum, the dislocation density continues to increase, and the stress drops down to the flow stress. The generation of dislocations at the observed low stress levels is evidently facilitated by the agglomeration of interstitial oxygen, which is ubiquitous in Si and Ge (Patel and Chaudhuri, 1962). The yield-point drop can be suppressed by the introduction of dislocations, vanishing when the initial density attains a value on the order of 10^6–10^7/cm^2. The maximum stress also decreases sharply with decreasing strain rate and with increasing temperature, which reflects the characteristically small stress dependence and large temperature dependence of the dislocation velocity in these covalent crystals (Haasen, 1962).

There is general agreement that dislocation glide in Si and Ge proceeds by thermally activated double-kink formation. Chaudhuri et al. (1962) found the stress-independent activation energy for 60° dislocations to be 2.2 eV in Si and 1.6 eV in Ge. These values were confirmed for screw dislocations in Si by Suzuki and Kojima (1966) and for 60° and screw dislocations in Ge by Schafer (1967) and Haasen (1968). Within experimental error, the observed activation energies agree with the calculated energy of double-kink formation: 2.14 eV in Si and 1.75 eV in Ge (Labusch, 1965). However, there is no consensus on the stress dependence of the dislocation velocity. The data of Kabler (1963) and of Kuczynski et al. (1972) fit the inverse exponential law of Eq. (12); the results of the studies cited previously support the power law of Eq. (11), although there is considerable variation in the reported value of the velocity–stress exponent m^*. Haasen (1968), who reviewed the literature on Ge, considered the determination of Schafer (1967) to be the most reliable and concluded that m^* is essentially equal to unity in both Si and Ge, which indicates viscous dislocation motion. On the other hand, Patel and Freeland (1970) stated that Schafer's results are incorrect because of an erroneous assignment of the Burgers vector.

It now appears that the temperature and stress dependences of the dislocation velocity are more complex than had been believed. Patel and Freeland (1971) corrected the deficiencies of the earlier studies on Ge by performing their measurements on *isolated* dislocations *proven* to be in the 60° orientation by their contrast in X-ray diffraction topographs. They found that the activation energy is a function of stress and that the velocity–stress exponent is a function of both temperature and stress. The activation energy varies from 2.9 eV at 0.6 kg/mm^2 to 1.6 eV at 2 kg/mm^2 and 1.5 eV at 8 kg/mm^2. The velocity–stress exponent goes from 5.1 at 380° C to 1.6 at 580° C within the low stress regime below 1–2 kg/mm^2; at higher stresses, the variation is much smaller, from 1.6 at 380° C to 2.1 at 500° C. These latter results are in accord with the data of Suzuki and Kojima (1966) on Si, where m^* varies from 1.85 at 600° C to 1.50 at 800° C. Significantly, m^* does not approach unity at any temperature or stress, so that the viscous dislocation motion reported previously may not, in fact, occur.

2. PREDOMINANTLY COVALENT BONDING (THE ZINC BLENDE STRUCTURE)

The zinc blende structure (cubic, space group $F \bar{4} 3 m$) is identical to the diamond cubic except that two atomic species are present, each species occupying one of the component face-centered cubic sublattices. As would be expected, the slip morphology is identical to that of Si and Ge, $\{111\}\langle \bar{1}10\rangle$, with no secondary slip. However, the cleavage is not octahedral but dodecahedral, $\{110\}$, which has been attributed by Abrahams and Ekstrom (1960) to microcrack nucleation at sessile Lomer dislocations. Zinc blende-type crystals show greater plasticity than the diamond cubic elements; the ductile-to-brittle transition usually occurs at a temperature of several hundred degrees C, and room temperature dislocation glide has been observed. The tensile behavior of initially dislocation-free GaSb and InSb crystals is similar to that of Si and Ge, with smaller yield-point drops (Patel and Chaudhuri, 1963). The nominally stress-independent activation energies are lower, e.g. for 60° dislocations, 1.3 eV in GaSb and 0.7 eV in InSb; the stress dependence of the dislocation velocity apparently conforms to the power law of Eq. (11), with a velocity–stress exponent of 2.0 for GaSb at 450° C and 1.87 for InSb at 218° C (Chaudhuri et al., 1962).

There is one fundamental crystallographic difference between the diamond cubic and the zinc blende structures: the latter lacks a center of symmetry. Consequently, the composite {111} slip planes are polar in zinc blende-type compounds, and the extra half-plane of an edge component can terminate either with a row of Group II or III atoms or with a row of Group VI or V atoms; i.e. positive and negative dislocations in the same nonscrew orientation are chemically distinct (Haasen, 1957; Holt, 1966). In InSb, for example, parallel 60° dislocations of opposite sign introduce acceptor levels of different energies (Bell et al., 1966; Bell and Willoughby, 1970). They also produce different plastic behavior; at 270° C, the lower yield stress for deformation by Sb dislocation glide is 34% greater than that for deformation by In dislocation glide (Bell and Willoughby, 1966).

Steinhardt and Schafer (1971) measured the velocities of In dislocations and Sb dislocations in the 60° orientation in initially dislocation-free InSb crystals. The In dislocations are 20–40 times more mobile over the temperature range from 330° C to 270° C, and their motion was observed down to room temperature. The activation energies are 0.78 eV for In dislocations and 1.1 eV for Sb dislocations; the stress dependence of the In type was found to follow the power law with m^* equal to unity, i.e. viscous behavior. Presumably, the difference in activation energy reflects a lower Peierls barrier opposing the glide of the In dislocations. Indirect evidence for different velocities of the two chemical types of dislocation has also been obtained in (In, Ga)P and Ga(As, P) mixed crystals (Abrahams et al., 1972).

3. PARTIALLY IONIC BONDING (THE WURTZITE STRUCTURE)

In the wurtzite structure, (hexagonal, space group $P \, 6_3 \, m \, c$), the two atomic species separately occupy sublattices of the close-packed hexagonal type which are displaced by $(c_0/8)[0001]$. It differs from the zinc blende structure only in the disposition of the second nearest neighbors, and polymorphic transformations between the two are common among the II–VI compounds. Slip occurs preferentially on $(0001)\langle 11\bar{2}0 \rangle$, with secondary $\{\bar{1}100\}\langle 11\bar{2}0 \rangle$ slip observed at moderately low temperatures. In α-CdS, which has an ionicity comparable to the predominantly covalent zinc blende-type compounds, the secondary prismatic slip system is activated at about 200° C when the Schmid factor is a maximum for $\{\bar{1}100\}$ shear (Blistanov et al., 1969); in ZnO, where the bonding is on the order of 50–60% ionic, prismatic slip becomes common near 75° C (Carlsson, 1971). The preferred cleavage plane is (0001) or $\{\bar{1}100\}$, both of which can be rationalized on the basis of either the partially ionic bonding or the coalescence of dislocations to nucleate microcracks. The wurtzite-type compounds generally exhibit dislocation glide at room temperature.

The wurtzite structure, like the zinc blende structure, is noncentrosymmetric, and the composite (0001) slip plane is polar, so that there are two chemical types for each nonscrew dislocation orientation on the basal plane. On the other hand, the secondary $\{\bar{1}100\}$ is a simple stoichiometric form, and the two atomic species alternate along the singular line of an edge component in this plane. Furthermore, the observed electric inactivity of secondary slip dislocations, which lie along the [0001] edge orientation, indicates the absence of broken bonds (Carlsson, 1971); apparently, the interlocking of the broken bonds is facilitated by the stoichiometric arrangement.

V. Suppression of Brittle Fracture

The notch sensitivity and susceptibility to brittle fracture of the crystalline nonmetals can be reduced by several obvious techniques. Significant improvements have already been achieved by the removal of impurities which raise the flow stress to undesirably high levels and by the elimination of flaws which nucleate premature fracture. An additional possibility is the use of mixed crystals in which the different slip modes of the end members operate at comparable shear stresses to provide the five independent systems required for polycrystalline ductility. Of far greater promise is the application of two well-known metallurgical phenomena to modify the intrinsic semibrittle behavior of these numerous and diverse materials: solid-solution softening, which can lower the flow stress significantly, and precipitation hardening, which can increase the fracture stress by a factor of two or more. These two effects of solute additions are discussed in this final section.

A. Solid-Solution Softening

1. RELATIONSHIP WITH THERMALLY ACTIVATED FLOW

Solid-solution softening is a low-temperature, low-concentration effect which was originally described as the reduction of the strength of an alloy below that of the pure component. The solute addition increases the athermal component of the flow stress, as predicted by the general theory of impurity hardening, but it produces an unexpected decrease of the thermal component. These opposing changes give rise to softening, in the domain of low-temperature deformation, provided that the increase of the athermal component can be overcome by the decrease of the thermal component. Thus, the essential phenomenon is the inverse dependence of the latter, effective stress on the solute concentration, which can occur without macro-scopic weakening of the alloy. For this reason, solid-solution softening has been redefined by Gamble (1969) as the reduction of the temperature depen-dence of the strength, or, equivalently, as the decrease of the effective stress. Furthermore, as emphasized by Clark and Wolfson (1971), the transition from macroscopic softening to hardening can be correlated with the vari-ance between the concentration dependences of the two stress components, with the temperature of the transition itself a strong function of solute concentration.

Referring to the shear strain rate equation,

$$\dot{\gamma} = bNv \tag{6}$$

it is clear that softening must occur either by an increase in the density of mobile dislocations N or by an increase in their mean velocity v. Most investigators believe that the rate-controlling factor is the dislocation velo-city. Sakamoto et al. (1968) and Christ et al. (1969), however, have attributed the softening in interstitial solid solutions of nitrogen in iron to an increase in the mobile dislocation density; it was estimated that this density must increase 100-fold to account for the magnitude of the effect, but there has been no metallographic corroboration of such a change. In Ge, the change was found to be in the opposite direction. Patel and Chaudhuri (1966) used their measured values of the mean dislocation velocity to calculate the mobile dislocation densities at the upper yield point, where the plastic strain rate is equal to the known strain rate imposed by the cross-head speed. The density in the hardened Ge is almost three times greater than that in the softened Ge; e.g. $2 \times 10^6/\mathrm{cm}^2$ and $7.5 \times 10^5/\mathrm{cm}^2$, respectively. Clearly, the variation in the maximum stress, both hardening and softening, is due solely to the difference in the dislocation velocity. On the other hand, a fivefold increase in the density in HF-doped ice, which exhibits considerable softening, has been reported by Jones and Gilra (1972); these investigators concluded that their results are consistent with the idea of an increased dislocation velocity, the higher density being due to the resultant greater ease of dislocation multiplication in their softened specimens. Although the possibility of solute-induced dislocation generation cannot be discounted, if only as a secondary influence (Kan, 1971), it cannot provide a general explanation.

Solid-solution softening is thus inseparable from the imperfectly under-stood temperature dependence of the flow stress, and, like the latter, it has been correlated with the thermally activated dislocation velocity. There are two general explanations for the strongly temperature-dependent strength: the double-kink model for overcoming intrinsic Peierls effects, with the sessile-to-glissile transformation mechanisms and the pseudo-Peierls

mechanisms as special cases; and the thermal activation of mobile dislocations past residual impurity atoms. Similarly, there are two general explanations for solid-solution softening: intrinsic Peierls barrier reduction, and
impurity scavenging.

If the thermal dependence of the strength is controlled by the intrinsic
resistance of the crystal to dislocation glide, then the solute atoms which
produce softening must lower the effective value of the Peierls stress. The
most obvious way for this to come about is by the reduction of the shear
modulus (Arsenault, 1969), but the modulus change is frequently too small
to account for the effect (Gamble, 1969), and solid-solution softening is even
accompanied in some systems by a modulus increase (Urakami, 1970). The
effective value of the Peierls stress can also be lowered by locally enhanced
kink nucleation due to solute-induced elastic interactions (Weertman, 1958;
Arsenault, 1967; Christ, 1969), lattice disorder (Mitchell and Raffo, 1967;
Raffo and Mitchell, 1968), and interatomic-bond modification (Klopp, 1968;
Gamble, 1969; Hildebrandt and Dickenscheid, 1971). In qualitative terms,
these tend to "smear out" the screw dislocations, whether total or extended,
resulting in a decrease in the energy for kink formation.

If, on the contrary, the low-temperature strength is due largely to hardening by residual impurity atoms, solid-solution softening can be explained by
internal scavenging. The solute atoms which produce softening can form a
compound with the residual impurity or can catalyze compound formation
within the impure solvent (Allen and Jaffee, 1963); they can also cause
segregation into large, presumably ordered complexes (Smialek et al., 1970).
The net result is the removal of the potent hardening species from solution.

Although these two models are not mutually exclusive (Ono and Sommer,
1970), most investigators have argued in favor of one or the other. The more
prevalent opinion is that thermally activated flow proceeds by double-kink
formation, with solid-solution softening arising from locally enhanced kink
nucleation. This latter interpretation has been supported by the recent work
of Sato and Meshii (1973), who proposed a single double-kink mechanism
capable of predicting both softening and hardening in the appropriate temperature ranges in the same body-centered cubic metal alloy. They considered the motion of a screw dislocation through the combined field of
Peierls potential and misfit strain centers, and they found that the latter can
promote double-kink nucleation while impeding the subsequent lateral propagation of the two half-kink segments. Whether macroscopic softening or
hardening results is controlled by the relative magnitudes of these opposing
effects, which are determined by the temperature, the concentration, the
misfit parameter, and the scale of the dispersion. This unifying treatment is
particularly satisfying since it accounts for the softening–hardening transition, which neither the scavenging mechanism nor the solute-induced dislocation multiplication mechanism can do.

On the other hand, as Sato and Meshii acknowledge, representing the point obstacles by misfit strain centers may be an oversimplification, even in the body-centered cubic metals. In the crystalline nonmetals, where the substitutional and displacement disorders produced by the solute cannot be screened by perturbations in the density of an "electron gas," misfit may be of lesser importance than the local modification of the bonding pattern. There is, in fact, a study which suggests that this is true in some of the body-centered cubic metals: the review by Klopp (1968) of the mechanical properties of the Group VIA metals Cr, Mo, and W alloyed with the metals of Groups VIIA and VIIIA. He discovered that the magnitude of the softening effect varies systematically with the position of the solute in the periodic table, and he discerned two trends. The major trend has the solute concentration for maximum softness decreasing with increasing number of s and d electrons, while the minor trend has this necessary solute concentration increasing with increasing period number; i.e. with increasing atomic radius and mass of the solute. Klopp's major trend indicates that the efficacy of the solute in producing solid-solution softening is dominated by the relative number of valence electrons. His minor trend shows that maximum softening tends to occur at lower concentrations for solutes with smaller misfit parameters. This latter observation appears to be consistent with the mechanism of Sato and Meshii, which predicts maximum softening by a relatively weak strain center; their mechanism may also explain the occurrence of large softening effects in Fe–Ni alloys, where Ni causes very little change in the lattice parameter (Jolley, 1968).

In common with the other mechanisms cited in this section, the double-kink mechanism of Sato and Meshii was developed specifically for the body-centered cubic metals, where size misfit is most important and where the glide of screw dislocations controls the low-temperature deformation. Indeed, their treatment indicates that the interaction between an edge dislocation and either a size misfit or a modulus defect is incapable of producing softening. However, in the crystalline nonmetals discussed in Sections III and IV, the electrical effects of a solute appear to predominate over misfit strain, and edge dislocations can move more slowly than either screw or mixed dislocations. The literature on softening in these crystals will now be reviewed.

2. Experience on Ionic Crystals

Unfortunately, the published data on solid-solution softening in ionic crystals are fragmentary, consisting, for the most part, of the bare observation of a softening effect. In the course of an investigation of the diffusivities of Co^{2+} and Ni^{2+} ions in MgO, Zaplatynsky (1962) noted that Co^{2+}-doped specimens have a lower microhardness than the pure compound, while

Ni^{2+}-doped specimens are harder. Parasnis (1961) found that the micro-hardness of single-crystal AgCl is lowered by the presence of 500 ppm of Cu^{+}; Cu^{2+} and Cd^{2+} increase the hardness. Wimmer *et al.* (1963), in their measurement of the room temperature critical resolved shear stress of NaCl–NaBr single crystals, observed softening between 0.1 and about 1.0 mole % NaBr, with the maximum at 0.5 mole %; interestingly, they were unable to detect this softening by microhardness, perhaps because of the high degree of work hardening produced by indentation.

A detailed study of softening in single-crystal CsI was performed by Uru-sovskaya *et al.* (1969). They measured the macroscopic yield stress in compression, the microhardness, and the dislocation velocity under constant stress, and they investigated the structure of the dislocation rosettes around the microhardness indentations. The CsI crystals were grown from melts containing up to 0.5% deliberately added impurity: Cr_2O_3, $KMnO_4$, AgCl, $CuCl_2$, or CdCl$_2$, or FeO. Cr_2O_3 and $KMnO_4$ harden the crystal, approxi-mately doubling the yield stress. AgCl and $CuCl_2$ have no perceptible effect. $CdCl_2$ and FeO soften the crystal, decreasing the yield stress by a factor of from 1.5 to 2 and increasing the dislocation velocity by a factor of roughly 3 over that in the hardest specimens. In the hardened crystals, the dislocation rosettes comprise short, broad bands suggestive of cross slip; the disloca-tions themselves are quite straight. In the softened crystals, the rosettes are made up of long, narrow bands indicative of high mobility on a single slip system, and the dislocations are extremely irregular. These investigators analyzed the compositions of their specimens by X-ray spectroscopy. They were unable to detect either of the softening cations Cd^{2+} or Fe^{2+} in the solid, but they did note that the traces of $(OH)^-$ present in all other specimens had disappeared from the softened specimens. On this basis, they concluded that Cd^{2+} and Fe^{2+} soften CsI by gettering oxygen in the melt, preventing it from forming the $(OH)^-$ ions which would otherwise enter the crystal and harden it.

The circumstantial evidence for this conclusion seems reasonable, but there is one observation that does not fit the notion that the softened crystals are even purer than the undoped material: the dislocations become markedly irregular in the softened specimens. This can be taken to indicate that the effective Peierls barrier of CsI has been reduced, presumably by the incorporation of undetected impurities. Furthermore, Urusovskaya and her associates are not justified in attributing the changes in the mechanical properties solely to the foreign cations, for the anion solutes are capable of producing both hardening and softening. Indeed, the failure of Cu^{2+} to give rapid Fleischer-type hardening might be due to a softening effect of the accompanying Cl^- anion, although divalent impurities may behave dif-ferently in compounds with the cesium chloride structure (Chin *et al.*, 1972).

The available information on softening in these ionic crystals is most unsatisfactory. Too few systems have been examined, the studies are not sufficiently detailed, and the specimens are obviously contaminated by residual impurities. It is tempting to ascribe the observed softening to impurity scavenging, and, indeed, there is considerable indirect support for such an interpretation. However, intrinsic Peierls effects cannot be ruled out, and two highly tentative correlations are possible. First, the four softening cations, and here are included Cd^{2+} and Fe^{2+} in CsI, have relatively low values of the polarizing power, defined as the ratio of the cationic charge to cationic radius; they and Br^- in NaCl all tend to form bonds which are weaker than those of the species they replace in the host crystal. Second, the hardening cations tend to have high polarizing powers and to form stronger bonds. Thus, it might be argued that softening occurs when the strength of the ionic bond is reduced locally without an increase in directionality.

Solid-solution softening is particularly pronounced in ice, where Jones and his colleagues (Jones, 1967; Jones and Glen, 1969; Nakamura and Jones, 1970) have produced very large effects with concentrations on the order of 5 ppm of HF and HCl. Two types of measurement were used: the tensile creep test and the compressive constant-strain rate test. In the tensile creep test, the crystal is deformed under constant stress, and the strain developed is measured as a function of time. Although it is impossible to deform pure ice plastically in tension at $-90°$ C without fracture, ice containing about 5 ppm of HF deforms by 0.4% in 50 hr. At $-70°$ C, where pure ice exhibits limited plasticity, the doping increases the strain rate by a factor of approximately 30. In the compressive constant-strain rate test, the stress required to deform the crystal at a constant rate is measured, the results being plotted as the resolved shear stress versus resolved shear strain on the basal (0001) slip plane. At $-70°$ C, the presence of from 3 to 12 ppm of either HF or HCl in solution reduces the stress level for constant strain rate by a factor of about 5; at $-26°$ C, it is reduced by a factor of about 3; and at $-5°$ C, by a factor of about 1.5. Similar effects were observed at various strain rates in the range from 10^{-8}/sec to 10^{-4}/sec, and at other temperatures in the interval from $-70°$ C to $-5°$ C. These results have been interpreted according to the model proposed by Glen (1968), discussed in Section III,E; on this model, the resistance of the crystal to glide arises from the need to reorient the randomly distributed hydrogen bonds, which is accomplished by the passage of Bjerrum defects. The softening effect of substitutional HF and HCl is attributed to the fact that they increase the number of vacant bonds, or L-defects, leading to a more rapid bond reorientation and, therefore, to a higher dislocation velocity. The decreased temperature dependence of the strength in the doped ice supports this explanation. The effect of the impurities becomes greater as the temperature is lowered, since the number of thermally generated Bjerrum defects is decreasing. In addition, the activation energy for creep in the doped crystals is in reasonable agreement with that for the motion of L-defects; in pure ice, it corresponds to the combined activation energies for formation plus motion.

3. EXPERIENCE ON COVALENT CRYSTALS

As noted in Section IV,A,4, the dislocation velocity in Si and Ge generally increases with n-type doping, decreases with p-type doping, and remains unaffected by doping with neutral Group IV impurities; this behavior has been widely attributed to the effect of the carrier density on the energy for double-kink formation. However, a more complex behavior has been discovered by Erofeev *et al.* (1970) in their investigation of the stress and temperature dependences of the velocity. They found that the donors As and Sb and the acceptor B all increase the dislocation velocity and decrease the activation energy in Si at 450° C, with the acceptor producing by far the smallest change. As the temperature rises, the effect of the acceptor is altered, and, by 750° C, it has been reversed; it now decreases the velocity, while the donors continue to increase the velocity. An even more complicated temperature dependence has been reported by Mil'vidskii *et al.* (1968) in Si doped with As and B. By measuring the indentation microhardness as a function of temperature in the range from 20° C to 800° C, they revealed yet another change of the doping effect. From 20° C to about 300° C, both the donor and the acceptor impurities produce a softening which decreases with increasing temperature. Between 400° C and 800° C, the acceptor hardens the crystal slightly, while the donor softens it considerably; this variance reaches its maximum near the midpoint of the range. Finally, above 800° C, doping has no effect at all. Mil'vidskii and his colleagues observed similar behavior in GaAs doped with the acceptor Zn and the donor Te: from 20° C to 100° C, both dopants produce softening; from 150° C to 600° C, the acceptor has no effect, but the donor hardens the crystal slightly; above 600° C, neither has an effect.

These results seem to fit together and to form a definite pattern. At low temperatures, where the dislocations are relatively immobile and the behavior largely brittle, softening is produced by all of the electrically active dopants; these are substitutional impurities which introduce bonding defects: an electron-deficient bond for an acceptor atom, an extra unpaired electron for a donor atom, and bond asymmetry for both. The donors, which also increase the electron density, give by far the greater softening. At intermediate temperatures, beginning with the onset of pronounced plasticity, the dopants have variable effects, but only the donors soften the crystal, and, in order to do so, they must be present in a high enough concentration to yield an electron density in excess of the intrinsic carrier density. Finally, at high temperatures, where the thermal energy is so great that rapid dislocation motion occurs, neither acceptors nor donors have any effect. Here, the thermal energy is also great enough to generate a high carrier density, equivalent to a dopant concentration on the order of a few tenths of an atomic percent, which is an order of magnitude larger than any of the concentrations investigated. Unfortunately, it is also larger than the solid solubilities of most dopants in the crystals.

Thus, solid-solution softening in these covalent crystals may involve two associated and complementary phenomena: the local disruption of the bonding pattern, which is important only at lower temperatures, and the increase in the electron density, which dominates at higher temperatures. There are two additional observations that tend to support this interpretation. First of all, the entire sequence of low-, intermediate-, and high-temperature regions is shifted to lower temperatures in going from Si to Ge to GaAs, which parallels the decrease in the bond strength and suggests a Peierls effect. Second, the electrically neutral impurities Si and Sn have no influence on the mechanical behavior of Ge in the intermediate-temperature region; these are Group IV elements that normally form tetrahedral bonds, and they introduce neither bonding defects, aside from a slight asymmetry, nor free carriers.

4. SUMMATION

In summary, the scanty information on solid-solution softening in the crystalline nonmetals indicates that electronic interactions are more important than size misfit. In ice and the covalent crystals, this appears to be due partly to the solute-induced modification of the bonding, but, in the ionic crystals, it may be merely a reflection of the importance of electrostatic effects in determining the degree of solid-solution hardening. The softening in ice and in Si, Ge, and GaAs can be ascribed with reasonable certainty to locally enhanced kink nucleation; this is in agreement with the conclusion that thermally activated flow in these materials proceeds by some double-kink mechanism. The origin of softening in the ionic crystals, however, is obscured by the presence of residual impurities. Despite the fact that these compounds are believed to have moderately high Peierls potentials, there is no direct evidence of important Peierls effects. This, in itself, may be sufficient justification to conclude that the softening is an impurity scavenging effect; by logical extension, the effective stress in ionic crystals with the rocksalt and the cesium chloride structures could then be attributed to the thermal activation of mobile dislocations past residual impurity atoms. Further clarification will undoubtedly follow improvements in the techniques of purification and impurity detection.

For those crystals in which low-temperature deformation is controlled by the Peierls barrier, the double-kink mechanism of Sato and Meshii (1973) provides a promising approach to the variable effects of solute additions on the strength. The diamond cubic elements apparently belong to this category, and a simple experiment suggests itself. Deform suitably doped, dislocation-free Si or Ge to introduce pure edge dislocations along $\langle 112 \rangle$, as was done by Patel (1958), and repeat the solid-solution softening studies already in the literature, all of which involved the motion of screw dislocations. If softening arises solely from the couple force exerted by misfit strain defects on screw dislocations, then only normal hardening will be observed.

Finally, the general occurrence of softening in the two most thoroughly investigated groups of nonmetals, the alkali halides and the simple adamantine crystals, is probably not coincidental. On the contrary, it can be argued that solid-solution softening will prove to be widespread among nonmetallic crystals (Wolfson and Clark, 1971), at least among those in which screw dislocations dominate the deformation.

B. Increase of Fracture Strength by Precipitation

1. Conditions for Brittle Fracture

In a very real sense, the low-temperature deformation of semibrittle crystals comprises not only the dislocation glide that produces plastic flow but also the brittle fracture that terminates it. Both processes are governed by the nature of the chemical bonding and the crystal structure, both are opposed by the cohesion of the ideal crystal, and both are intrinsically inhomogeneous. Indeed, the relationship between them is even more intimate, for fracture can be nucleated by glide, and fracture competes with glide in the relief of the applied stress. Semibrittle materials are extremely notch sensitive, and cleavage is usually initiated prematurely by surficial cracks and steps, internal microcracks, two-dimensional crystal defects, grain boundaries, and particles of a second phase. These flaws are not inherent, and they can be avoided by suitable material preparation, in principle at least. In their absence, cracks are nucleated by microscopically inhomogeneous slip in both the ordinary cleavage of the crystalline nonmetals (Johnston and Parker, 1963) and the low-temperature brittle fracture of metals (Low, 1963). Thus, the fracture process may be divided into three stages: crack nucleation, crack growth to some critical size beyond which spontaneous propagation can take place, and crack extension as a cleavage surface.

Nucleation occurs when a glide band encounters an extended obstacle, either another glide band or an internal surface. (External surfaces are not effective barriers, evidently, without the presence of a surface layer.) At the intersection of two glide bands, a variety of mechanisms may come into play; e.g. dislocation coalescence (Cottrell, 1959), kink formation (Argon and Orowan, 1961). At the intersection of a glide band with a surface, the mechanism depends upon the crystallographic nature of the obstacle; specific models have been proposed for crack formation at a grain boundary (Zener, 1948) and at a tilt boundary (Orowan, 1954), among others. The general energy expressions can be formulated in terms of cleavage dislocations, as was done by Meakin and Petch (1963), but such detailed considerations are not essential to the present treatment.

A crack forms because the inherent cohesive strength of the crystal has been exceeded locally by a tensile stress, and it grows because of the local plastic constraint that it exerts. The role of stress concentration in the growth of a microcrack from subcritical to critical dimensions was demonstrated by the work of Stokes and Li (1963) in NaCl. They found that even when the plastic constraint is relieved by cross slip and rotation, the relaxation is only temporary, and the growth of the crack is jerky rather than continuous. When the mutual orientation of a microcrack and the applied stress is least favorable, a period of relatively slow growth may precede cleavage, provided that the strain rate is extremely small. In general, however, the promotion of a microcrack from subcritical to critical size is very rapid.

The critical size for spontaneous propagation is reached, according to the classic Griffith criterion, when the strain energy released by the extension of the crack equals the attendant increase in its surface energy. This energy balance can be solved for the corresponding critical value of the tensile stress, the fracture stress σ_f.

$$\sigma_f = (A\gamma_f/C)^{1/2} \tag{15}$$

where A is a combination of geometrical factors and elastic constants determined by the shape and orientation of the crack, γ_f is the effective surface energy, and C is the critical dimension of the crack. The effective surface energy has been analyzed by Stokes (1971) into a modified surface energy component which is proportional to the specific surface energy γ, and a plastic work component γ_p.

$$\gamma_f = R\gamma + \gamma_p \tag{16}$$

The factor R specifies the roughness of the fracture surface, and γ_p accounts for the plastic deformation accompanying crack propagation.

The effective value of the surface energy deviates widely from the specific surface energy in semibrittle crystals. In LiF, the plastic work component is appreciable even at high strain rates and low temperatures, and its temperature dependence is sufficiently strong to account for the ductile-to-brittle transition which occurs near 300° C (Burns and Webb, 1970). The effective surface energy for {100} cleavage in NaCl has been found to range from 370 erg/cm^2 to 3000 erg/cm^2 (Wiederhorn *et al.*, 1970). Even in Si and Ge at temperatures approaching that of liquid N$_2$, where the behavior is completely brittle and the specific surface energy of the {111} cleavage plane is high, viz. 1230 erg/cm^2 in Si and 1060 erg/cm^2 in Ge, the effective surface energy can be increased significantly by surface roughness (Jaccodine, 1963).

2. Precipitation Hardening

Equations (15) and (16) describe the conditions for macroscopic brittle fracture, and they illustrate the complexity of the process. The microscopic crack parameters A, C, and γ_f are difficult to assess. This is especially true of the effective surface energy, which is a function of the microrelief of the fracture surface and the flow stress of the crystal, together with all of the implied secondary dependences; e.g. on crack velocity, on temperature. Nevertheless, the equations are useful conceptually, and they indicate how fracture can be suppressed by decreasing C and by increasing γ_f, both of which can be accomplished by precipitation hardening. If the precipitates cause the crack to change direction after minute increments of fracture, the surface develops a considerable degree of roughness, and R increases. If the precipitates promote plastic flow, e.g. by facilitating dislocation generation at the crack tip or by being themselves more ductile than the matrix, σ_p increases. Furthermore, either of these effects can blunt the tip of the crack, which reduces the stress–concentration factor and increases the critical value of C directly.

Precipitation hardening is capable of greatly increasing the resistance to brittle fracture. In their study of precipitation in NaCl–KCl mixed crystals, Wolfson *et al.* (1966b) found that the fracture stress in crystals containing 12.4 mole % KCl can be doubled by aging at 250° C for 260 hr; much larger increases are produced with higher concentrations and aging temperatures. Furthermore, since the maximum in the fracture stress coincides with the minimum in the yield stress, these strengthened crystals exhibit extensive plastic deformation. The fact that the fracture stress reaches its maximum long after the peak in the critical resolved shear stress suggests that the KCl precipitates are relatively coarse and interfere with crack propagation, a surmise supported by the observed microrelief of the fracture surface. Similar strengthening was obtained by Kruse and Fine (1972) by precipitation of MgFe$_2$O$_4$ in MgO single crystals containing up to 3.65 cation % Fe^{3+}. The fracture stress passes through a broad peak in the region of "over-aging," where the flow stress has dropped back to a low level; its value for optimum aging is approximately twice that of solution-treated specimens of the same composition, and four times that of nominally pure MgO. Here, however, the increased fracture stress is associated with increased notch sensitivity.

There have been very few other studies of the suppression of brittle fracture by precipitation hardening; see, for example, the reviews by Fine (1970, 1972). Thus, although aging is widely used to increase the flow strength of metallic alloys, its potential for increasing the fracture strength of the crystalline nonmetals has been largely ignored.

References

Abrahams, M. S., and Ekstrom, L. (1960). *Acta Met.* **8**, 654.
Abrahams, M. S., Blanc, J., and Buiocchi, C. J. (1972). *Appl. Phys. Lett.* **21**, 185.
Ackley, S. F., and Itagaki, K. (1969). *Bull. Amer. Phys. Soc.* [2] **14**, 411.
Ahlquist, C. N., and Carlsson, L. (1972). *Scr. Met.* **6**, 1129.
Alexander, H., and Haasen, P. (1972). *Annu. Rev. Mater. Sci.* **2**, 291–312.
Allen, B. C., and Jaffee, R. I. (1963). *Trans. Amer. Soc. Metals* **56**, 387.
Amelinckx, S. (1958). *Acta Met.* **6**, 34.
Argon, A. S. (1962). *Acta Met.* **10**, 574.
Argon, A. S., and Orowan, E. (1961). *Nature (London)* **192**, 447.
Arsenault, R. J. (1967). *Acta Met.* **15**, 501.
Arsenault, R. J. (1969). *Acta Met.* **17**, 1291.
Authier, A., and Lang, A. R. (1964). *J. Appl. Phys.* **35**, 1956.
Baeta, R. D., and Ashbee, K. H. G. (1969). *Amer. Mineral.* **54**, 1551.
Ball, A., and Smallman, R. E. (1966). *Acta Met.* **14**, 1517.
Barraclough, K. G., and Beevers, C. J. (1969). *J. Mater. Sci.* **4**, 518.
Bassett, G. A. (1959). *Acta Met.* **7**, 754.
Bell, R. L., and Willoughby, A. F. W. (1966). *J. Mater. Sci.* **1**, 219.
Bell, R. L., and Willoughby, A. F. W. (1970). *J. Mater. Sci.* **5**, 198.
Bell, R. L., Latkowski, R., and Willoughby, A. F. W. (1966). *J. Mater. Sci.* **1**, 66.
Bernal, J. D., and Fowler, R. H. (1933). *J. Chem. Phys.* **1**, 515.
Blistanov, A. A., Kugaenko, O. M., and Shaskol'skaya, M. P. (1969). *Sov. Phys.—Solid State* **11**, 642.
Bowen, D. K., Christian, J. W., and Taylor, G. (1967). *Can. J. Phys.* **45**, 903.
Braekhus, J., and Lothe, J. (1972). *Phys. Status Solidi B* **51**, 149.
Brar, N. S., and Tyson, W. R. (1972). *Scr. Met.* **6**, 587.
Brown, N., and Ekvell, R. A. (1962). *Acta Met.* **10**, 1101.
Budworth, D. W., and Pask, J. A. (1963). *J. Amer. Ceram. Soc.* **46**, 560.
Buerger, M. J. (1928). *Amer. Mineral.* **13**, 35.
Buerger, M. J., and Washken, E. (1947). *Amer. Mineral.* **32**, 296.
Burn, R., and Murray, G. T. (1962). *J. Amer. Ceram. Soc.* **45**, 251.
Burns, S. J., and Webb, W. W. (1970). *J. Appl. Phys.* **41**, 2078. 2086.
Carlsson, L. (1971). *J. Appl. Phys.* **42**, 676.
Carlsson, L., and Ahlquist, C. N. (1972). *J. Appl. Phys.* **43**, 2529.
Carlsson, L., and Svensson, C. (1968). *Solid State Commun.* **7**, 177.
Carlsson, L., and Svensson, C. (1970). *J. Appl. Phys.* **41**, 1652.
Celli, V., Kabler, M., Ninomiya, T., and Thomson, R. (1963). *Phys. Rev.* **131**, 58.
Chao, H. C., Thomassen, L., and Van Vlack, L. H. (1964). *Trans. Amer. Soc. Metals* **57**, 386.
Charpentier, P., Rabbe, P., and Manenc, J. (1968). *Mater. Res. Bull.* **3**, 69.

Chaudhuri, A. R., Patel, J. R., and Rubin, L. G. (1962). *J. Appl. Phys.* **33**, 2736.

Chin, G. Y. (1973). *Met. Trans.* **4**, 329.

Chin, G. Y., Van Uitert, L. G., Green, M. L., and Zydzik, G. (1972). *Scr. Met.* **6**, 475.

Christ, B. W. (1969). *Acta Met.* **17**, 1317.

Christ, B. W., and Smith, G. V. (1967). *Scr. Met.* **1**, 123.

Christ, B. W., Gamble, R. P., and Smith, G. V. (1969). *Scr. Met.* **3**, 521.

Clark, W. R., and Wolfson, R. G. (1971). Presented at the Symposium on Alloy Softening. Metallurgical Society, AIME, Detroit, Michigan.

Class, W. H., Machlin, E. S., and Murray, G. T. (1961). *Trans. AIME* **221**, 769.

Cochardt, A. W., Schoek, G., and Wiedersich, H. (1955). *Acta Met.* **3**, 533.

Conrad, H. (1965). *J. Amer. Ceram. Soc.* **48**, 195.

Conrad, H. (1970). *Mater. Sci. Eng.* **6**, 265.

Conrad, H., and Hayes, W. (1963). *Trans. Amer. Soc. Metals* **56**, 125, 249.

Cottrell, A. H. (1953). "Dislocations and Plastic Flow in Crystals." Oxford Univ. Press (Clarendon), London and New York.

Cottrell, A. H. (1959). *In* "Fracture" (B. L. Averbach *et al.*, eds.), p. 117. Wiley, New York.

Dash, W. C. (1956). *J. Appl. Phys.* **27**, 1193.

Davis, L. A., and Gordon, R. B. (1968). *J. Appl. Phys.* **39**, 3885.

Davis, L. A., and Gordon, R. B. (1969). *J. Appl. Phys.* **40**, 4507.

Day, R. B., and Stokes, R. J. (1967). *J. Amer. Ceram. Soc.* **50**, 445.

Dickinson, S. K., Jr. (1970). Physical Sciences Research Paper, No. 439. Air Force Cambridge Research Laboratories, Bedford, Massachusetts.

Dorn, J. E., and Mukherjee, A. K. (1969). *Trans. AIME* **245**, 1493.

Duesbery, M. S., and Hirsch, P. B. (1968). *In* "Dislocation Dynamics" (A. R. Rosenfield *et al.*, eds.), pp. 57–85. McGraw-Hill, New York.

Eisenstadt, M. E. (1971). "Introduction to Mechanical Properties of Materials," p. 179. Macmillan, New York.

Erofeev, V. N., Nikitenko, V. I., and Osvenskii, V. B. (1970). *Sov. Phys.—Dokl.* **14**, 1108.

Eshelby, J. D. (1949). *Phil. Mag.* [7] **40**, 903.

Eshelby, J. D., Newey, C. W. A., Pratt, P. L., and Lidiard, A. B. (1958). *Phil. Mag.* [8] **3**, 75.

Fine, M. E. (1970). *Advan. Mater. Res.* **4**, 1–53.

Fine, M. E. (1972). *Amer. Ceram. Soc., Bull.* **51**, 510.

Fleischer, R. L. (1962a). *Acta Met.* **10**, 835.

Fleischer, R. L. (1962b). *J. Appl. Phys.* **33**, 3504.

Fleischer, R. L. (1967). *Acta Met.* **15**, 1513.

Foreman, A. J. E. (1955). *Acta Met.* **3**, 322.

Foreman, A. J. E., Jaswon, M. A., and Wood, J. K. (1951). *Proc. Phys. Soc., London, Sect. A* **64**, 156.

Frank, W. (1968). *Phys. Status Solidi* **29**, 391 and 767.

Frank, W. (1970). *Mater. Sci. Eng.* **6**, 121 and 132.

Frank, W., and Sestak, B. (1970). *Scr. Met.* **4**, 451.

Frenkel, J. (1926). *Z. Phys.* **37**, 572.

Friedel, J. (1964). "Dislocations." Pergamon, Oxford.

Gamble, R. P. (1969). Ph.D. Thesis, Cornell University, Ithaca, New York.

Geach, G. A., Irving, B. A., and Phillips, R. (1957). *Research* **10**, 411.

Gillin, L. M., and Kelly, A. (1969). *J. Mater. Sci.* **4**, 439.

Gilman, J. J. (1959). *Acta Met.* **7**, 608.

Gilman, J. J. (1961). *Progr. Ceram. Sci.* **1**, 146–199.

Gilman, J. J. (1968). *In* "Microplasticity" (C. J. McMahon, Jr., ed.), pp. 17–43. Wiley (Interscience), New York.

Gilman, J. J. (1969). "Micromechanics of Flow in Solids." McGraw-Hill, New York.

Glen, J. W. (1968). *Phys. Kondens. Mater.* **7**, 43.

Gorum, A. E., Parker, E. R., and Pask, J. A. (1958). *J. Amer. Ceram. Soc.* **41**, 161.

Granzer, F., Wagner, G., and Eisenblatter, J. (1968). *Phys. Status Solidi* **30**, 587.

Gridneva, I. V., Milman, Yu. V., and Trefilov, V. I. (1969). *Phys. Status Solidi* **36**, 59.

Groves, G. W., and Fine, M. E. (1964). *J. Appl. Phys.* **35**, 3587.

Groves, G. W., and Kelly, A. (1963). *Phil. Mag.* [8] **8**, 877.

Gupta, I., and Li, J. C. M. (1970). *Mater. Sci. Eng.* **6**, 20.

Guyot, P., and Dorn, J. E. (1967). *Can. J. Phys.* **45**, 983.

Haasen, P. (1957). *Acta Met.* **5**, 598.

Haasen, P. (1962). *Z. Phys.* **167**, 461.

Haasen, P. (1968). *In* "Dislocation Dynamics" (A. R. Rosenfield *et al.*, eds.), pp. 701–718. McGraw-Hill, New York.

Hartley, N. E. W., and Wilshaw, T. R. (1973). *J. Mater. Sci.* **8**, 265.

Haworth, W. L., and Gordon, R. B. (1970). *Phys. Status Solidi A* **3**, 503.

Heine, V. (1966). *Phys. Rev.* **146**, 568.

Hesse, J., and Matucha, K. H. (1972). *Scr. Met.* **6**, 865.

Hildebrandt, U., and Dickenscheid, W. (1971). *Acta Met.* **19**, 49.

Hollox, G. E. (1968/1969). *Mater. Sci. Eng.* **3**, 121.

Hollox, G. E., Novak, D. L., Huntington, R. D., and Slattery, P. W. (1971). *J. Mater. Sci.* **6**, 171.

Holt, D. B. (1962). *J. Phys. Chem. Solids* **23**, 1353.

Holt, D. B. (1966). *J. Mater. Sci.* **1**, 280.

Honig, J. M., and Wahnsiedler, W. E. (1972). *J. Solid State Chem.* **5**, 452.

Hornstra, J. (1958). *J. Phys. Chem. Solids* **5**, 129.

Hsu, S. E., Kobes, W., and Fine, M. E. (1967). *J. Amer. Ceram. Soc.* **50**, 149.

Hulse, C. O. (1966). *Amer. Ceram. Soc., Bull.* **45**, 369.

Irving, P. E., and Beevers, C. J. (1972). *J. Mater. Sci.* **7**, 23.

Itagaki, K. (1969). *Bull. Amer. Phys. Soc.* [2] **14**, 411.

Itagaki, K. (1970). *Advan. X-Ray Anal.* **13**, 526–538.

Iyer, K. R., and Kuczynski, G. C. (1971). *J. Appl. Phys.* **42**, 486.

Iyer, K. R., Kuczynski, G. C., and Allen, C. W. (1971). *J. Appl. Phys.* **42**, 485.

Jaccodine, R. J. (1963). *J. Electrochem. Soc.* **110**, 524.

Jaccodine, R. J. (1964). *Appl. Phys. Lett.* **4**, 114.

Johnson, L. D., and Pask, J. A. (1964). *J. Amer. Ceram. Soc.* **47**, 437.

Johnson, O. W. (1965). *J. Appl. Phys.* **36**, 3247.

Johnson, O. W. (1966). *J. Appl. Phys.* **37**, 2521.

Johnston, T. L., and Parker, E. R. (1963). *In* "Fracture of Solids" (D. C. Drucker and J. J. Gilman, eds.), p. 267. Wiley (Interscience), New York.

Johnston, W. G., and Gilman, J. J. (1959). *J. Appl. Phys.* **30**, 129.

Johnston, W. G., and Gilman, J. J. (1960). *J. Appl. Phys.* **31**, 632.

Jolley, W. (1968). *Trans. AIME* **242**, 306.

Jones, S. J. (1967). *Phys. Lett. A* **25**, 366.

Jones, S. J. (1970). *J. Appl. Phys.* **41**, 2738.

Jones, S. J., and Gilra, N. K. (1972). *Appl. Phys. Lett.* **20**, 319.

Jones, S. J., and Glen, J. W. (1969). *Phil. Mag.* [8] **19**, 13.

Joshi, M. L., and Wilhelm, F. (1965). *J. Electrochem. Soc.* **112**, 185.

Kabler, M. N. (1963). *Phys. Rev.* **131**, 54.

Kamenetskaya, D. S., Piletskaya, I. B., and Shiryaev, V. I. (1970). *Sov. Phys.—Dokl.* **15**, 177.

Kan, T. (1971). *Scr. Met.* **5**, 607.

Katz, R. N., and Coble, R. L. (1970). *J. Appl. Phys.* **41**, 1871.

Keh, A. S., and Nakada, Y. (1967). *Can. J. Phys.* **45**, 1101.

Keig, G. A., and Coble, R. L. (1968). *J. Appl. Phys.* **39**, 6090.

Keith, R. E., and Gilman, J. J. (1960). *Acta Met.* **8**, 1.

Kelly, A. (1966). "Strong Solids." Oxford Univ. Press (Clarendon), London and New York.

Kelly, A., Tyson, W. R., and Cottrell, A. H. (1967). *Can. J. Phys.* **45**, 883.

Kelly, P. M. (1969). *Scr. Met.* **1**, 149.

Klopp, W. D. (1968). *NASA Tech. Note* **NASA TN D-4955**.

Krausz, A. S. (1968). *Acta Met.* **16**, 897.

Krausz, A. S. (1970). *Mater. Sci. Eng.* **6**, 260.

Kroupa, F., and Vitek, V. (1967). *Can. J. Phys.* **45**, 945.

Kroupa, F., and Vitek, V. (1969). *Phys. Status Solidi* **36**, K143.

Kruse, E. W., III, and Fine, M. E. (1972). *J. Amer. Ceram. Soc.* **55**, 32.

Kuczynski, G. C., Iyer, K. R., and Allen, C. W. (1972). *J. Appl. Phys.* **43**, 1337.

Kulkarni, S., and Williams, W. S. (1972). *Amer. Ceram. Soc., Bull.* **51**, 337.

Labusch, R. (1965). *Phys. Status Solidi* **10**, 645.

Lachenmann, R., and Schultz, H. (1972). *Scr. Met.* **6**, 731.

Lad, R. A., Stearns, C. A., and DelDuca, M. G. (1958). *Acta Met.* **6**, 610.

Lahiri, S. K., and Fine, M. E. (1970). *Met. Trans.* **1**, 1495.

Lau, S. S., and Dorn, J. E. (1968). *Scr. Met.* **2**, 335.

Lawley, A., and Gaigher, H. L. (1964). *Phil. Mag.* [8] **10**, 15.

Lee, F. F. M., Shyne, J. C., and Nix, W. D. (1969/1970). *Mater. Sci. Eng.* **5**, 179.

Leslie, W. C., and Sober, R. J. (1967). *Trans. Amer. Soc. Metals* **60**, 99.

Leslie, W. C., Sober, R. J., Babcock, S. G., and Green, S. J. (1969). *Trans. Amer. Soc. Metals* **62**, 690.

Li, J. C. M. (1968). *In* "Dislocation Dynamics" (A. R. Rosenfield *et al.*, eds.), pp. 87–116. McGraw-Hill, New York.

Liu, T. S., Stokes, R. J., and Li, C. H. (1964). *J. Amer. Ceram. Soc.* **47**, 276.

Low, J. R., Jr. (1963). *In* "Fracture of Solids" (D. C. Drucker and J. J. Gilman, eds.), p. 197. Wiley (Interscience), New York.

Luyckx, S. B. (1970). *Acta Met.* **18**, 233.

Lyall, K. D., and Paterson, M. S. (1966). *Acta Met.* **14**, 371.

Macmillan, N. H. (1972). *J. Mater. Sci.* **7**, 239.

Matthews, J. W., and Isebeck, K. (1963). *Phil. Mag.* [8] **8**, 469.

Meakin, J. D., and Petch, N. J. (1963). *In* "Fracture of Solids" (D. C. Drucker and J. J. Gilman, eds.), p. 393. Wiley (Interscience), New York.

Mehan, R. L., and Herzog, J. A. (1970). *In* "Whisker Technology" (A. P. Levitt, ed.), pp. 157–195. Wiley (Interscience), New York.

Mendelson, S. (1972). *J. Appl. Phys.* **43**, 2102.

Mil'vidskii, M. G., Osvenskii, V. B., and Stolyarov, O. G. (1968). *Sov. Phys.—Crystallogr.* **12**, 896.

Mitchell, T. E., and Raffo, P. L. (1967). *Can. J. Phys.* **45**, 1047.

Mitchell, T. E., Foxall, R. A., and Hirsch, P. B. (1963). *Phil. Mag.* [8] **8**, 1895.

Muguruma, J., and Higashi, A. (1963). *Nature (London)* **198**, 573.

Nabarro, F. R. N. (1947). *Proc. Phys. Soc., London* **59**, 256.

Nabarro, F. R. N. (1952). *Advan. Phys.* **1**, 269.

Nabarro, F. R. N., and Duncan, T. R. (1967). *Can. J. Phys.* **45**, 939.

Nakamura, T., and Jones, S. J. (1970). *Scr. Met.* **4**, 123.

Newman, P. C. (1963). *J. Phys. Chem. Solids* **24**, 45.

Ono, K., and Sommer, A. W. (1970). *Met. Trans.* **1**, 877.

Orava, R. N. (1967). *Scr. Met.* **1**, 153.

Orowan, E. (1954). *In* " Dislocations in Metals," p. 69. AIME, New York.

Palache, C., Berman, H., and Frondel, C. (1944). "The System of Mineralogy," Vol. I. Wiley, New York.

Palache, C., Berman, H., and Frondel, C. (1951). "The System of Mineralogy," Vol. II. Wiley, New York.

Parasnis, A. S. (1961). *Acta Met.* **9**, 165.

Patel, J. R. (1958). *J. Appl. Phys.* **29**, 170.

Patel, J. R., and Chaudhuri, A. R. (1962). *J. Appl. Phys.* **33**, 2223.

Patel, J. R., and Chaudhuri, A. R. (1963). *J. Appl. Phys.* **34**, 2788.

Patel, J. R., and Chaudhuri, A. R. (1966). *Phys. Rev.* **143**, 601.

Patel, J. R., and Freeland, P. E. (1967). *Phys. Rev. Lett.* **18**, 833.

Patel, J. R., and Freeland, P. E. (1970). *J. Appl. Phys.* **41**, 2814.

Patel, J. R., and Freeland, P. E. (1971). *J. Appl. Phys.* **42**, 3298.

Pauling, L. (1935). *J. Amer. Chem. Soc.* **57**, 2680.

Pauling, L. (1960). " The Nature of the Chemical Bond." Cornell Univ. Press, Ithaca, New York.

Peierls, R. E. (1940). *Proc. Phys. Soc., London* **52**, 34.

Pfann, W. G., and Lovell, L. C. (1955). *Acta Met.* **3**, 512.

Pfann, W. G., and Vogel, F. L., Jr. (1957). *Acta Met.* **5**, 377.

Phillips, W. L., Jr. (1961). *J. Amer. Ceram. Soc.* **44**, 499.

Potter, D. I. (1969/1970). *Mater. Sci. Eng.* **5**, 201.

Rachinger, W. A. (1956). *Acta Met.* **4**, 647.

Rachinger, W. A., and Cottrell, A. H. (1956). *Acta Met.* **4**, 109.

Raffo, P. L., and Mitchell, T. E. (1968). *Trans. AIME* **242**, 907.

Ramqvist, L. (1971). *J. Appl. Phys.* **42**, 2113.

Rauch, H. W., Sr., Sutton, W. H., and McCreight, L. R. (1968). "Ceramic Fibers and Fibrous Composite Materials." Academic Press, New York.

Ravi, K. V., and Gibala, R. (1969). *Scr. Met.* **3**, 547.

Ravi, K. V., and Gibala, R. (1971). *Met. Trans.* **2**, 1219.

Read, W. T., Jr. (1954). *Phil. Mag.* [7] **45**, 775.

Readey, D. W., and Kingery, W. D. (1964). *Acta Met.* **12**, 171.

Reppich, B. (1972). *Acta Met.* **20**, 557.

Rice, R. W. (1969). *J. Amer. Ceram. Soc.* **52**, 428.

Riewald, P. G., and Van Vlack, L. H. (1969). *J. Amer. Ceram. Soc.* **52**, 370.

Riewald, P. G., and Van Vlack, L. H. (1970). *J. Amer. Ceram. Soc.* **53**, 219.

Robinson, P. M., and Scott, H. G. (1967). *Acta Met.* **15**, 1230.

Rose, R. M., Ferriss, D. P., and Wulff, J. (1962). *Trans. AIME* **224**, 981.

Rowcliffe, D. J., and Hollox, G. E. (1971). *J. Mater. Sci.* **6**, 1261–1270.

Rueda, F., and Dekeyser, W. (1963). *Acta Met.* **11**, 35.

Sakamoto, K., Wada, H., and Sugeno, T. (1968). *Trans. Jap. Inst. Metals* **9**, Suppl., 720.

Sashital, S. R., and Vedam, K. (1972). *J. Appl. Phys.* **43**, 4396.

Sastry, D. H., and Vasu, K. I. (1972). *Acta Met.* **20**, 399.

Sato, A., and Meshii, M. (1973). *Acta Met.* **21**, 753.

Schafer, S. (1967). *Phys. Status Solidi* **19**, 297.

Schoeck, G. (1961). *Acta Met.* **9**, 382.

Scholl, R., Larson, D. J., Jr., and Freise, E. J. (1968). *J. Appl. Phys.* **39**, 2186.

Schwuttke, G. H., and Queisser, H. J. (1962). *J. Appl. Phys.* **33**, 1540.

Schwuttke, G. H., and Queisser, H. J. (1964). *Bull. Amer. Phys. Soc.* [2] **9**, 146.

Schwuttke, G. H., and Rupprecht, H. (1966). *J. Appl. Phys.* **37**, 167.

Seeger, A. (1956). *Phil. Mag.* [8] **1**, 651.

Seeger, A. (1957). *In* "Dislocations and Mechanical Properties of Crystals" (J. C. Fisher *et al.*, eds.), p. 243. Wiley, New York.
Seeger, A., Stein, D. F., and Hahn, G. T. (1968). *In* "Dislocation Dynamics" (A. R. Rosenfield *et al.*, eds.), pp. 747–756. McGraw-Hill, New York.
Sherwood, P. J., Guiu, F., Kim, H. C., and Pratt, P. L. (1967). *Can. J. Phys.* **45**, 1075.
Shetty, M. N., and Taylor, J. B. (1968). *J. Appl. Phys.* **39**, 5145.
Shockley, W. (1953). *Phys. Rev.* **91**, 228.
Singleton, G. A. M., and Crocker, A. G. (1971). *Phys. Status Solidi* **6**, 645.
Sleeswyk, A. W. (1971). *Scr. Met.* **5**, 217.
Smialek, R. L., Webb, G. L., and Mitchell, T. E. (1970). *Scr. Met.* **4**, 33.
Smidt, F. A., Jr. (1969). *Acta Met.* **17**, 381.
Sole, M. J., van der Walt, C. M., and Truzenberger, M. (1968). *Acta Met.* **16**, 667.
Solomon, H. D., McMahon, C. J., Jr., and Leslie, W. C. (1969). *Trans. Amer. Soc. Metals* **62**, 886.
Soo, P., and Galligan, J. M. (1969). *Scr. Met.* **3**, 153.
Soule, D. E., and Nezbeda, C. W. (1968). *J. Appl. Phys.* **39**, 5122.
Spitzig, W. A. (1970). *Acta Met.* **18**, 1275.
Srinivasan, M., and Stoebe, T. G. (1970). *J. Appl. Phys.* **41**, 3726.
Starodubtsev, S. V., Kaipnazarov, D., Khiznichenko, L. P., and Kromer, P. F. (1966). *Sov. Phys.—Solid State* **8**, 1521.
Stein, D. F. (1967). *Can. J. Phys.* **45**, 1063.
Steinhardt, H., and Schafer, S. (1971). *Acta Met.* **19**, 65.
Stevens, R. (1972). *J. Mater. Sci.* **7**, 517.
Stokes, R. J. (1971). *In* "Fracture" (H. Liebowitz, ed.), Vol. 7, pp. 157–241. Academic Press, New York.
Stokes, R. J., and Li, C. H. (1962). *Acta Met.* **10**, 535.
Stokes, R. J., and Li, C. H. (1963). *In* "Fracture of Solids" (D. C. Drucker and J. J. Gilman, eds.), pp. 289–314. Wiley (Interscience), New York.
Stokes, R. J., Johnston, T. L., and Li, C. H. (1961). *Acta Met.* **9**, 415.
Straumanis, M. E., Faunce, C. A., and James, W. J. (1967). *Acta Met.* **15**, 65.
Suzuki, T., and Kojima, H. (1966). *Acta Met.* **14**, 913.
Takeuchi, S. (1970). *Scr. Met.* **4**, 73.
Takeuchi, T. (1969). *J. Phys. Soc. Jap.* **27**, 436.
Tanaka, T., and Conrad, H. (1972). *Acta Met.* **20**, 1019.
Taylor, A., and Kagle, B. J. (1963). "Crystallographic Data on Metal and Alloy Structures." Dover, New York.
Teichler, H. (1967). *Phys. Status Solidi* **23**, 341.
Thomas, J. M., and Williams, J. O. (1971). *Progr. Solid State Chem.* **6**, 119–154.
Turner, A. P. L., and Vreeland, T., Jr. (1970). *Acta Met.* **18**, 1225.
Unwin, P. N. T., and Muguruma, J. (1971). *J. Appl. Phys.* **42**, 3640.
Urakami, A. (1970). Ph.D. Thesis, Northwestern University, Evanston, Illinois.
Urusovskaya, A. A., Dobrzhanskii, G. F., Sizova, N. L., Govorkov, V. G., and Martyshev, Yu. N. (1969). *Sov. Phys.—Crystallogr.* **13**, 899.
Van Bueren, H. G. (1961). "Imperfections in Crystals." North-Holland Publ., Amsterdam.
van der Walt, C. M., and Sole, M. J. (1967). *Acta Met.* **15**, 459.
Vitek, V., and Kroupa, F. (1966). *Phys. Status Solidi* **18**, 703.
Vogel, F. L., Jr. (1955). *Acta Met.* **3**, 245.
Vukcevich, M. R. (1972). *Phys. Status Solidi B* **50**, 545.
Washburn, J., Groves, G. W., Kelly, A., and Williamson, G. K. (1960). *Phil. Mag.* [8] **5**, 991.

Washburn, J., Thomas, G., and Queisser, H. J. (1964). *J. Appl. Phys.* **35**, 1909.
Webb, W. W., and Hayes, C. E. (1967). *Phil. Mag.* [8] **16**, 909.
Weertman, J. (1958). *J. Appl. Phys.* **29**, 1685.
Westbrook, J. H., and Jorgensen, P. J. (1965). *Trans. AIME* **233**, 425.
Westwood, A. R. C., and Goldheim, D. L. (1968). *J. Appl. Phys.* **39**, 3401.
Wiederhorn, S. M., Moses, R. L., and Bean, B. L. (1970). *J. Amer. Ceram. Soc.* **53**, 18.
Williams, J. O., Adams, I., and Thomas, J. M. (1969). *J. Mater. Sci.* **4**, 1064.
Wimmer, F. T., Kobes, W., and Fine, M. E. (1963). *J. Appl. Phys.* **34**, 1775.
Wolfson, R. G., and Clark, W. R. (1971). Presented at the Symposium on Alloy Softening. Metallurgical Society, AIME, Detroit, Michigan.
Wolfson, R. G., Juleff, E. M., and Lapierre, A. G. (1966a). *Int. J. Electron.* **21**, 37.
Wolfson, R. G., Kobes, W., and Fine, M. E. (1966b). *J. Appl. Phys.* **37**, 704.
Zaplatynsky, I. (1962). *J. Amer. Ceram. Soc.* **45**, 28.
Zener, C. (1948). *In* "Conference on Fracturing of Metals," pp. 3–31. Amer. Soc. Metals, Cleveland, Ohio.

Recovery and Recrystallization during High Temperature Deformation

H. J. McQUEEN

Department of Mechanical Engineering
Concordia University
Montreal, Canada

and

J. J. JONAS

Department of Mining and Metallurgical Engineering
McGill University
Montreal, Canada

I. Introduction

Almost all metals, excluding some castings and sintered parts, are hot-worked during processing into a final product. Most of these hot-worked materials go into service in the as-deformed state in the form of products such as steel plate and skelp and aluminum extrusions. The remainder is cold-worked in a further series of operations. Whether as an intermediate or as a final product, the properties of such hot deformed materials are markedly influenced by the characteristics of the microstructure that is produced by the working process. It is the main purpose of the present article to consider the microstructural changes occurring both during and after hot working and to describe the influence of these changes on the mechanical properties of the worked material.

A. Static and Dynamic Restoration Processes

Recovery and recrystallization have traditionally been considered as mechanisms of restoration through which *cold-worked* metal returns partially or completely to its condition prior to working. These mechanisms generally operate when a metal is annealed at a high temperature for a period of time. In this respect, the term "high temperature" is of course relative; it depends upon the individual metal being heated and is best expressed in terms of homologous temperature, that is as a fraction of the absolute temperature of melting. A high homologous temperature in relation to annealing is anything over about one-half.

Annealing is usually carried out in the absence of stress or strain, in which case the recovery or recrystallization that occurs is termed static. However, recovery and recrystallization can also take place under stress and during concurrent straining, i.e. under conditions of high temperature deformation. In such an event, they are distinguished from the respective static processes by the terms dynamic recovery and dynamic recrystallization. Schematic representations of some of the combinations of the dynamic and static softening processes occurring in hot working are shown in Fig. 1. Since dynamic recovery and dynamic recrystallization have been observed to take place

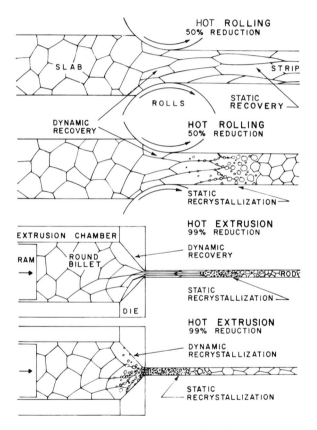

Fig. 1. In a commercial hot-working operation in which the strain is low, as in rolling, the grains soften during deformation by dynamic recovery. In metals of high stacking fault energy, static recovery takes place during cooling, whereas in metals of low stacking fault energy, static recrystallization readily occurs as well. In operations producing high strains, such as extrusion, metals of high stacking fault energy undergo dynamic recovery followed by static recovery and static recrystallization; on the other hand, metals of low stacking fault energy undergo both dynamic recovery and recrystallization followed by static recovery and recrystallization.

under both hot working and creep conditions, there often is considerable similarity between the microstructures produced by these two modes of deformation. Consequently, it will be useful to take note of some of the observations made in creep studies where these are relevant to our examination of hot-worked structures. The similarities and differences between loading in the creep and hot-working modes will therefore be considered briefly before attention is given to the microstructural processes themselves.

B. Comparison of Creep and Hot-Working Modes of Deformation

Under creep conditions, a sample of metal is deformed under a constant or nearly constant stress, and the strain rate is the dependent variable. The stresses are considerably lower than the yield stress in a standard tensile test at the temperature in question, and the strain rates accordingly lie in the range 10^{-10}–10^{-3} sec^{-1}. In practice, creep also occurs under conditions of stress relaxation; that is, in a sample of metal held at a constant value of total (elastic plus plastic) strain. Here the stress gradually diminishes as the elastic strain is replaced by plastic flow; however, the plastic deformation is limited to that required to relieve the elastic strain, and is therefore only of the order of 0.1%.

The strain rate under creep conditions is largely controlled by dynamic recovery processes. Between intervals of creep deformation, static recovery also plays an important role if the temperature is maintained. By contrast, the possibility of dynamic recrystallization, which can lead to strain rate increases of up to an order of magnitude, is generally prevented by the alloying elements present in commercial alloys. These alloy additions also prevent static recrystallization from occurring between intervals of loading.

In hot-working practice, a sample of heated metal is deformed into the required shape at a rate defined by the process machinery. This usually involves a strain rate–time profile which is unique to the process. Although it also takes place at elevated temperatures, the hot-working operation can be distinguished from the process of creep by three factors. First, in the case of hot working, the strain rate profile is an arbitrary one, and is therefore the independent variable, with the developed stress as the dependent variable (in creep, the converse applies). Furthermore, hot-working tests and operations are conducted in the much higher strain rate range of 10^{-3}–10^{3} sec^{-1}, so that the developed stresses are considerably higher than the stresses which will produce creep at the same temperature. Finally it should be noted that, whereas softening under creep conditions is usually restricted to the static and dynamic *recovery* processes, all four of the principal softening processes operate under hot-working conditions.

II. Static Recovery and Recrystallization after Cold Working

A. Static Recovery of Cold-Worked Substructures

Before turning to a detailed examination of the four principal restoration processes associated with high temperature deformation, it will be useful to review briefly some of the relevant features of static recovery and static recrystallization as they occur after low temperature deformation. When a cold-worked metal is held at any homologous temperature T_H above about 0.1, some recovery always occurs. The mechanisms of recovery operating in the region of $0.1 < T_H < 0.3$, however, are restricted to the motion of point defects and are not therefore of concern here, inasmuch as they lead to changes in physical rather than mechanical properties (Cahn, 1965). Changes in mechanical properties, on the other hand, require the motion and rearrangement of dislocations, and it is the latter process which we will be referring to when the term static recovery is used.

In worked metals of high stacking fault energy, a cell structure forms during deformation (Fig. 2) (Leslie et al., 1963; Swann, 1963). When straining is interrupted, static recovery can gradually lead to considerable softening (e.g. to a 40% decrease in residual strain hardening), and, if the strain has not exceeded the so-called critical strain for recrystallization, may even be the only restoration process that operates (Perryman, 1956). There is almost no motion of grain boundaries during recovery, so that there is little observable change in the microstructure of the elongated grains produced by working. In worked metals of low stacking fault energy, the dislocations are present in planar arrays of high energy, which constitute a strong driving force for recrystallization, so that recrystallization is likely to proceed before recovery progresses significantly (Cahn, 1965; Byrne, 1965).

The process of static recovery involves the following sequential changes in the dislocation cell structure (Fig. 2). In the rapid phase, called polygonization, the dislocations within the cells are attracted into the cell walls and the redundant dislocations in the cell walls annihilate each other, leaving neat dislocation networks separating the polygonal crystallites (Leslie et al., 1963; Vandermeer and Gordon, 1963). Eventually the subgrains grow larger as the shared subboundary between neighboring subgrains is unraveled, and the individual dislocations move away, becoming incorporated in other boundaries or annihilated (Hu, 1963; Weissmann et al., 1963). By such a mechanism of growth, a cluster of subgrains becomes a recrystallization nucleus, preparing the scene for the next stage of softening. These changes can be observed by optical techniques, such as polarized light microscopy and etch pitting, by X-ray diffraction methods involving the splitting of Laue

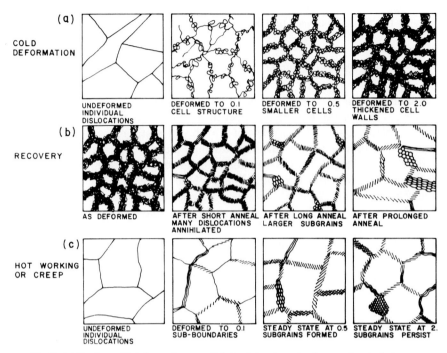

Fig. 2. (a) During cold deformation, a cellular substructure develops as the strain increases; by about 0.5 the cells reach their limiting size and thereafter the cell walls become much more dense. (b) Recovery during annealing commences with the annihilation of dislocations in the cell walls which turn into subboundaries. As recovery continues, the subgrains grow in diameter and the overall dislocation density decreases. (c) In hot working, subgrains develop in the strain hardening stage and reach an equilibrium size which is maintained during steady state flow.

arcs into discrete spots, and by transmission electron microscopy through the direct observation of dislocations.

The softening associated with recovery can be attributed to the gradual loss of the dislocations which were generated and became entangled during working. The movement of an individual dislocation occurs in response to the long range internal stress fields of the other dislocations. However, the motion of the dislocations is also opposed by various short range barriers, which are overcome by thermal activation. The thermally activated mechanisms of importance at high temperatures are climb and cross-slip, as well as the unpinning of attractive junctions (Kivilahti et al., 1974; Sastry et al., 1974). Since the overall activation energy for recovery is in many cases lower than that for recrystallization, the observation of recovery as a separate process is favored by the use of lower temperatures, for which the incubation times for recrystallization can be fairly long.

From a practical point of view, the effect of alloying on the operation of the recovery mechanisms is of considerable importance (Byrne, 1965; Cahn, 1965). Although many aspects of these mechanisms are still unclear, it appears that solute addition can hamper climb through the binding energies that tie vacancies to solute atoms. In a similar way, solute additions usually permit the dislocations to extend into partials separated by a ribbon of stacking fault. The extension of the dislocations makes climb and cross-slip more difficult and renders network nodes more resistant to unpinning, thus hindering the recovery or growth of the dislocation network. The presence of precipitates, in turn, can stabilize networks of subboundary dislocations and in this way retard recovery (Leslie *et al.*, 1963; Humphreys and Martin, 1968). Because of their industrial interest, these are matters to which we will return below, when we consider static recovery after high temperature deformation.

B. Static Recrystallization after Cold Working

After static recovery has proceeded to some degree, a further and larger decrease in strength occurs, which is indicative of static recrystallization. It is attributable to the formation and growth of new grains which have much lower dislocation densities ($\sim 10^{10}$ m^{-2}) than the worked grains they replace ($\sim 10^{16}$ m^{-2}) (Byrne, 1965; Cahn, 1965). Several mechanisms of nucleation have been proposed based on observations of different metals after various strains (Fig. 3). The subgrain coalescence mechanism has already been briefly described above. It occurs in metals of high stacking fault energy, and leads to the establishment of regions of low internal dislocation content surrounded by boundaries of high misorientation capable of migrating (Bourelier and Montuelle, 1968). In metals of low stacking fault energy, the concentration of dislocations in the tangles formed by straining is higher, and the cell walls appear to develop high misorientations, and therefore become capable of migrating directly upon annealing (Bailey, 1963).

The occurrence of the subgrain coalescence mechanism is favored by the application of relatively large prestrains. When the amount of prior straining is reduced (to less than about 0.4), an alternative mechanism, that of grain boundary bulging, takes place in many metals. This process entails the bulging out of a short segment of an existing high angle boundary to produce a roughly spherical volume, relatively free of dislocations, which then acts as a nucleus and is free to grow (Bailey, 1963).

Once nuclei are formed, they grow into the deformed material by the migration of their boundaries. The driving force for migration is provided by

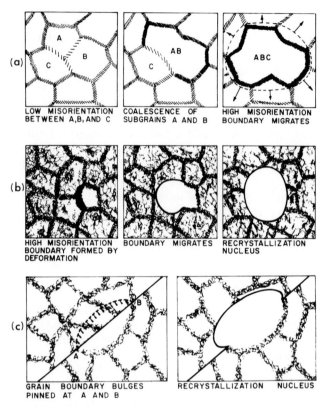

Fig. 3. The mechanism of nucleation of recrystallized grains depends upon the metal and the strain. (a) In metals of high stacking fault energy, cells such as A, B, and C, coalesce into one crystallite to form a nucleus. This occurs as a result of the dislocations between the cells moving to surrounding subboundaries and increasing their misorientation until they can migrate freely. (b) In metals of lower stacking fault energy, straining creates localized high densities of dislocations which transform into high angle boundaries upon annealing. (c) When the strain is low, so that the dislocation density varies from grain to grain, an existing boundary anchored by the substructure in one grain may bulge out into a region of higher density.

the difference in dislocation density between the interior of a nucleus and the surrounding worked metal. The driving force continuously decreases because of two factors. On the one hand, the dislocation density in the worked area diminishes with time as a result of recovery (Vandermeer and Gordon, 1963). On the other hand, the most heavily deformed regions, containing the highest densities of dislocations, tend to undergo recrystallization first. The process of recrystallization is considered complete when all of the worked regions have become absorbed, and the growing nuclei impinge upon each other. As in the case of static recovery, the rate of static recrystallization is strongly affected by the presence of impurities, either in solution or as fine particles; however, here it is the result of their effect on the rate of migration of the boundaries.

Both nucleation and growth are thermally activated processes and consequently their rates increase rapidly as the temperature is increased. Nevertheless, the grain size at the end of recrystallization depends primarily on the density of nuclei, and hence on the initial degree of deformation within the crystals. The severity of deformation depends in turn on the amount, rate, and temperature of prior straining. The rate of recrystallization also increases with the severity of straining because of the increasing density of nucleation sites and the higher driving force.

III. Dynamic Recovery at Elevated Temperatures

A. The Flow Curve at High Temperatures

When a recrystallized metal is loaded at constant nominal strain rate, the flow curve that results can be divided into three distinct stages (Luton and Jonas, 1970). The first stage is that of microstrain deformation, which is the interval during which the plastic strain rate in the sample increases from zero to the approximate strain rate of the test (Fig. 4). During this interval, the state of stress in the material rises rapidly, although not quite as steeply as it does at conventional temperatures. Typical loading slopes during initial loading range from $E/50$ at high temperatures and low strain rates to $E/5$ at relatively low temperatures and high strain rates† (Immarigeon, 1974). The loading slopes are not in general comparable with the modulus because of the plastic strain produced during the loading interval prior to macroscopic flow. This strain arises from the operation of the thermally activated mechanisms that are rate controlling in this temperature range, and therefore increases with a high temperature and a low rate of loading. When the rate of loading is low and the temperature is high, the amount of microstrain deformation is about 1%; by contrast, at higher loading rates and lower temperatures, it is only a few tenths of a percent.

The end of the microstrain interval is signified by a decrease in the slope of the loading curve of about an order of magnitude. Yield drops are not, in general, observed, and the "yield" stress is defined instead in terms of a plastic strain offset of 0.1 or 0.2%. The region of the yield stress marks the

---

† E represents the value of Young's modulus at the temperature of the test.

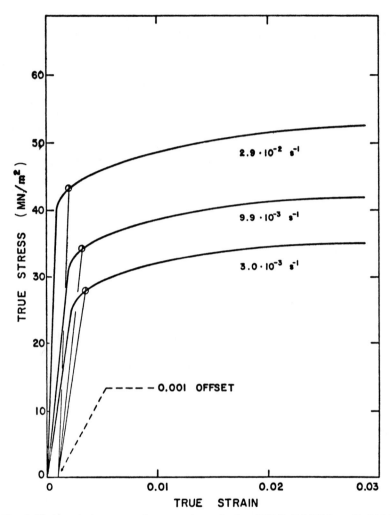

Fig. 4. Three typical stress–strain curves determined at 775° C (0.5 T_m) for a Zr–0.7% Sn alloy and plotted on an expanded strain scale. The transition from microstrain flow to generalized yielding can be seen, as well as the offset method used to establish the yield stress (Luton and Jonas, 1972a).

beginning of work hardening and, as before, the slopes are temperature and rate sensitive. Typical values range from $E/500$ at high temperatures and low strain rates to $E/100$ at low temperatures and high strain rates (Immarigeon, 1974).

During this second stage of deformation, the work hardening rate gradually decreases, until finally, in the third or steady state region, the net rate of work hardening is zero (Fig. 5). The steady state regime in hot working is

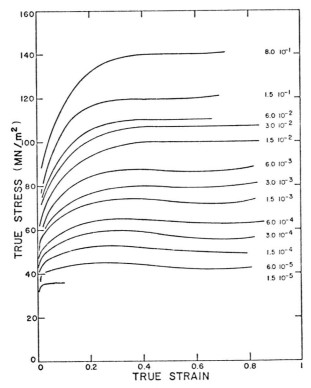

Fig. 5. Typical true stress–true strain curves for Armco iron, illustrating the strain rate dependence of the flow stress at 700° C (0.54 T_m) (Immarigeon and Jonas, 1974).

characterized, as is secondary creep, by the constancy of the three parameters: stress, temperature, and strain rate.

It should be noted that the simple shape of flow curve illustrated in Fig. 5 pertains to the case where the flow rate is limited by dynamic recovery processes alone. In practice, the flow curve may drop as a result of the operation of other mechanisms. Examples of such processes are dynamic recrystallization, which will be considered below in more detail, adiabatic heating, precipitate coarsening, modification of "hard" textures, and superplastic flow (Jonas and Luton, 1975). More rarely, the flow stress may rise as a result of precipitation or of the elimination of "soft" textures.

B. Microstructural Development during Plastic Flow

It will now be of interest to consider the microstructural changes that accompany the three stages of flow depicted in Fig. 5. Although little electron microscopy has been carried out to date on samples deformed to the vicinity of the yield strain, the limited evidence available (D. J. Abson and J. J. Jonas, unpublished work, McGill University, Montreal, Canada, 1974) suggests that some dislocation multiplication takes place during the micro-strain interval, and that the dislocation density increases from about 10^{10}–10^{11} m^{-2} in annealed, polycrystalline samples to 10^{11}–10^{12} m^{-2} at the commencement of macroscopic flow. The dislocation density continues to increase with strain after "yielding," albeit at a slower rate, and attains a value in the neighborhood of 10^{14}–10^{15} m^{-2} in the steady state region.

During the phase of positive strain hardening, the dislocations become entangled and begin to form a cellular structure. By the time the steady state regime is attained, the dislocations have arranged themselves into subgrains whose perfection, dimensions, and misorientation depend on the metal, and on the strain rate and temperature of deformation (Fig. 6) (Jonas et al., 1969). The main features of the substructure, i.e. the dislocation density between dislocation walls, the average spacing between them, and the misorientation across them, do not change during the course of steady state deformation (Jonas et al., 1968a). This condition of essentially constant dislocation density, which is necessary for the absence of strain hardening, results from the attainment of dynamic equilibrium between the rates of dislocation generation and annihilation.

The generation rate is a function of the strain rate and the associated effective stress,† but is relatively independent of strain. The rate of annihilation, on the other hand, depends on the dislocation density, and on the ease of operation of the recovery mechanisms, such as climb, cross-slip, and node unpinning. Thus the gradual elimination of strain hardening comes about through the increase in dislocation density with strain, which leads to continual increases in the annihilation rate until this matches the generation rate.

An additional interesting feature of steady state deformation is particularly worthy of note; the subgrains appear equiaxed even at large strains (Figs. 7a and b), whereas the grains deform in conformity with the outward change in shape of the object (Fig. 7c). This cannot be explained solely by migration of the subboundaries, since not all of them are capable of migration and since it has been shown that such migration contributes only 6–10% of the strain (Exell and Warrington, 1972). Another contributing process is repolygonization, that is the repeated unraveling of the subboundaries and the subsequent reformation of new subboundaries at locations which keep their average spacing and dislocation density constant

† The effective stress σ^* is the local value of the net stress acting at a dislocation. It is related to the applied or developed stress σ by the relation $\sigma^* = \sigma - \sigma_i$, where σ_i is the local value of the internal stress associated with the presence of other dislocations, or of grain boundaries or precipitates.

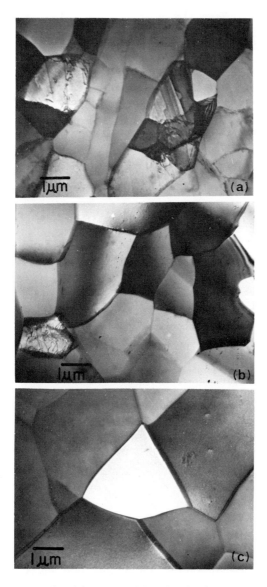

Fig. 6. Substructures produced in commercial purity aluminum by extrusion at a mean strain rate of 1 sec^{-1} and an extrusion ratio of 40 : 1 ($\varepsilon = 3.7$). (a) 250° C (0.56 T_m). (b) 350° C (0.67 T_m). (c) 450° C (0.77 T_m). × 8000. Transmission electron micrographs of longitudinal sections (Wong *et al.*, 1967).

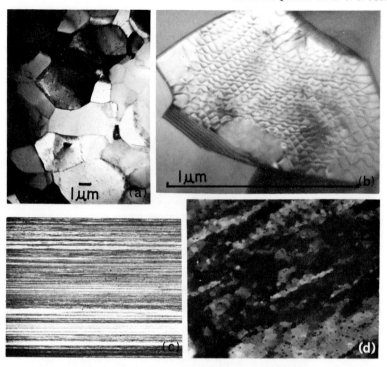

Fig. 7. (a) Substructure of commercial purity aluminum extruded at 445° C (0.76 T_m) and a strain rate of 1.2 sec^{-1}. The subgrains are equiaxed despite extension of the sample to a true strain of 3.7 (3900% elongation). × 2600 (McQueen *et al.*, 1967). (b) Some of the subboundaries produced by hot working are simple, low energy arrays like this twist boundary. Extrusion was carried out at 250° C (0.56 T_m) and a strain rate of 1.2 sec^{-1}. × 52,000 (Jonas *et al.*, 1968a). (c) The fibrous appearance of the grains in a sample of commercial purity aluminum rolled to a reduction of 96% (2400% elongation) in 15 passes in the temperature range 525–300° C (0.85–0.62 T_m). × 13 (Vaughan, 1968). (d) The formation of subgrains has resulted in an irregular or scalloped grain boundary in this sample of aluminum extruded at 420° C (0.74 T_m) and 0.1 sec^{-1} to a strain of 2.3 (900% elongation). × 130 (Demianczuk, 1967).

(McQueen *et al.*, 1967). The stable subgrain size depends in turn on the equilibrium dislocation density (Holt, 1970), which is established by the balance between generation rate and annihilation rate. Holt demonstrated that a uniform distribution of dislocations, when allowed some mobility, will form "clusters." The wavelength of clustering, which is in effect the cell wall spacing, will be proportional to the mean dislocation spacing, $l = \rho^{-1/2}$. This result is based on the condition that the stress level at a dislocation in a wall due to a dislocation in an adjacent wall is a small fixed fraction of the stress level due to a nearest-neighbor dislocation. In addition, when the subboundary dislocation density is higher, the long range stress field of an individual dislocation in the boundary is neutralized to a greater degree. As a result, the higher the subboundary density, the shorter the distance over which the stress field of the boundary as a whole has an influence, and the smaller is the equilibrium subgrain size.

In concluding this section, the reader is reminded that the process of dynamic recovery as it occurs during hot working cannot be considered as equivalent to the superposition of static recovery and cold working. The concurrence of straining and recovery prevents the high dislocation densities of cold working from ever developing. In fact, the small subgrains produced by hot working cannot be synthesized by cold working and annealing because of the intervention of recrystallization. The marked effectiveness of an applied stress in accelerating recovery at high temperatures was originally observed in a phenomenon called stress annealing (Wood and Suiter, 1951); this matter will be considered later with the deformation of metal initially containing a worked substructure.

C. Effect of Strain Rate and Temperature on the Substructure

The higher the temperature of deformation and the lower the strain rate, the larger are the subgrains that are formed during high temperature deformation (Bird *et al.*, 1969). As they increase in size, the subgrains contain fewer dislocations and have boundaries in which there are less dislocations arranged in more orderly arrays (Fig. 8). These substructural variations reflect the diminution in the equilibrium dislocation density arising from the increase in temperature and decrease in strain rate through their effects on the generation and annihilation rates. The generation rate decreases because the effective stress decreases with strain rate decrease and with temperature increase. The annihilation rate, on the other hand, does not decrease as much with a decrease in strain rate because it is not as sensitive to a stress decrease as the generation rate. Furthermore, increases in temperature tend to increase the recovery rate, thereby offsetting in part the effect of a decrease in stress.

The onset of steady state deformation is also rate sensitive and occurs at strains varying from about 0.1 at high temperatures and low strain rates to 0.5 or more at relatively low temperatures and high strain rates. This increase in strain arises because both the decrease in temperature as well as the increase in strain rate decrease the number of recovery events per unit strain. The amount of generation per unit strain, on the other hand, actually increases with strain rate and with temperature decrease.

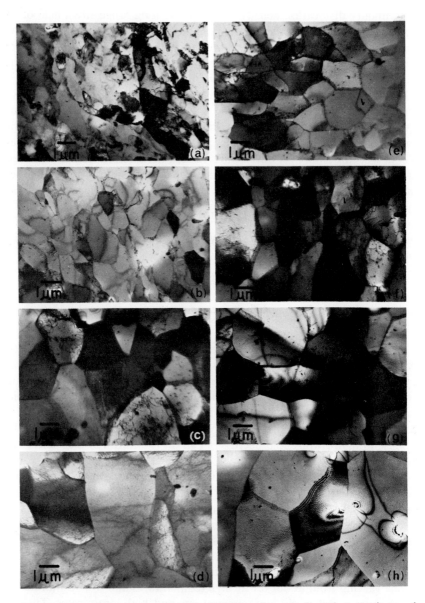

Fig. 8. The effects of temperature and strain rate are shown in these electron micrographs (all × 5000) of commercial purity aluminum which has been compressed to a strain of 0.7 (50% reduction). The first series is compressed at a strain rate of about 11 sec^{-1} at temperatures of (a) 20° C (0.36 T_m), (b) 200° C (0.50 T_m), (c) 400° C (0.72 T_m), (d) 500° C (0.82 T_m). The second series is compressed at 400° C (0.72 T_m) at rates of approximately (e) 220 sec^{-1}, (f) 12 sec^{-1}, (g) 1.4 sec^{-1}, and (h) 0.1 sec^{-1} (McQueen and Hockett, 1970).

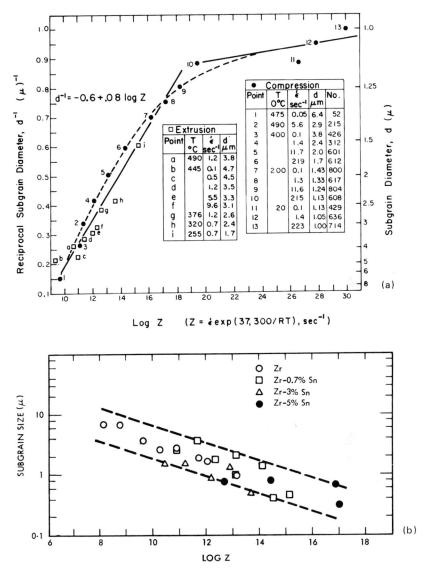

Fig. 9. The subgrain diameter increases as the temperature-compensated strain rate Z decreases (i.e., temperature increases). (a) Commercial purity aluminum extruded to a strain of 3.7 or compressed to a strain of 0.7. Points 11, 12, and 13 are not on the fitted line since they correspond to cold-worked materials (McQueen and Hockett, 1970). (b) This dependence also holds for α-zirconium and zirconium–tin alloys. The experimental activation energy used to calculate Z for each alloy increases with tin content, yet the sizes fall in a single band, indicating that there are compensating decreases in diameter associated with the increases in activation energy (Abson and Jonas, 1972).

The dependence of mean subgrain diameter d on temperature T and strain rate $\dot{\varepsilon}$ is illustrated in Fig. 9 and is described by the relation

$$d^{-1} = a + b \log Z \tag{1}$$

Here a and b are empirical constants and the temperature-corrected strain rate Z is given by

$$Z = \dot{\varepsilon} \exp (Q/RT) \tag{2}$$

where Q is the experimental activation energy associated with the temperature dependence of the flow stress and R is the universal gas constant.

D. The Flow Stress at High Temperatures

In the room temperature region, it is common to express the flow stress σ as

$$\sigma = \sigma_0 + \alpha \mu b \sqrt{\rho} \tag{3}$$

Here σ_0 is a friction stress, which may or may not be rate dependent, α is a geometrical constant, μ is the shear modulus, b the Burgers vector, and ρ the dislocation density. A similar relation appears to hold at high temperatures (Luton and Jonas, 1970), as given by

$$\sigma = \sigma^* + \sigma_i \tag{4}$$

Here σ^*, known as the effective stress, is the reversible component of the flow stress, changes in which are concurrent with changes in temperature or strain rate. The term σ_i refers to the "irreversible" or structural component of the flow stress, also known as the internal stress. Changes in σ_i do not follow immediately on changes in strain rate or temperature, but are associated with structural modifications, and therefore either with strain or with the effects of annealing.

1. THE YIELD STRESS AT HIGH TEMPERATURES

It was pointed out earlier that at yielding, the dislocation density is about three orders of magnitude below that attained during steady state flow. Setting

$$\sigma_i = \alpha \mu b \sqrt{\rho} \tag{5}$$

it can be seen that the internal stress at yielding is in the neighborhood of 3% of the internal stress present under steady state conditions. As a first approximation, we can therefore write that

$$\sigma_y \simeq \sigma^* \tag{6}$$

where σ_y is the yield stress.

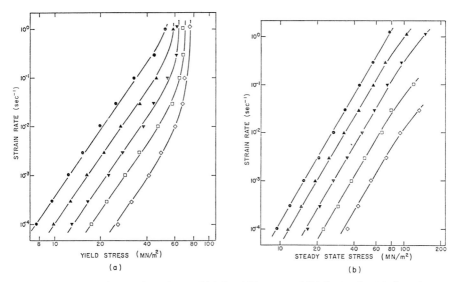

Fig. 10. The strain rate dependence of (a) the yield stress and (b) the steady state flow stress of α-zirconium plotted in the manner of creep data. The temperature range lies between 0.44 T_m and 0.63 T_m. The straight line fits at the higher temperatures and lower strain rates exhibit a power law exponent of 4.6 for the yield stress and 4.4 for the steady state stress (Luton and Jonas, 1972b). ◇, 625° C; □, 675° C; ▼, 725° C; ▲, 775° C; ●, 825° C.

The yield stress of alpha zirconium is shown as a function of strain rate and temperature in Fig. 10. It is evident that σ_y (and therefore σ^*) is strongly temperature and rate sensitive, unlike the "friction" stress commonly observed in the room temperature range. Qualitatively similar dependences on temperature and strain rate have been reported for yielding in bcc materials by Immarigeon (1974) and in fcc materials by Petkovic (1975).

The mechanism controlling yielding at high temperatures has not been established with certainty but has been suggested to be the thermally activated unpinning of dislocation nodes (Sastry *et al.*, 1974). This conclusion was supported by the following common features of high temperature yield data: (a) the relatively large experimental activation volumes, (b) the stress dependent nature of the experimental activation energies, (c) the rapid increase in certain alloys of experimental activation energy with solute concentration, and (d) the absence, at yielding, of a well-developed substructure or of subgrains, and therefore the inapplicability of the secondary creep models based on the presence of a steady state substructure.

The node unpinning model for yielding involves a zero-stress activation energy of about 0.4 μb^3 in high stacking fault energy metals. This value is somewhat higher than the activation energy for self-diffusion; it increases with decreasing stacking fault energy to as much as 0.5 to 0.7 μb^3.

2. The Steady State Stress

The steady state flow stresses σ_s observed in alpha zirconium are also shown in Fig. 10. It is clear from the diagram that the temperature and rate dependence of σ_s is closely similar to that of σ_y. When a creep power law,[†] e.g.

$$\dot{\varepsilon} = A\sigma_s^n \exp\left(-Q/RT\right) \tag{7}$$

is fitted to these data, stress exponents of 4.6 and 4.4 and activation energies of 226 and 234 kJ/mol are obtained for σ_y and σ_s, respectively. It is of interest that the ratio σ_y/σ_s for alpha zirconium, as well as for a series of Zr–Sn alloys (Luton and Jonas, 1972b), is in the range 0.7 to 0.85. For bcc alpha iron and silicon iron, σ_y/σ_s has been reported to vary from 0.5 to 0.9 (Immarigeon, 1974), and for fcc copper and gamma iron to range from 0.4 to 0.7 (Petkovic, 1975).

The broad similarity in the temperature and rate dependences of σ_y and σ_s in many materials leads to an interesting speculation. The activation parameters associated with σ_y, and therefore with σ^*, are characteristic of the rate controlling glide process, which may for example be node unpinning or the breakup of attractive junctions, as suggested above. The temperature and rate dependence of σ_s, on the other hand, will depend on the combined dependences of the glide mechanism and of the internal stress σ_i. The stable value of σ_i is, however, determined by the detailed nature of the dynamic recovery processes that are operating at high temperatures. The conclusion can therefore be drawn that the similarity in the behavior of σ_y and σ_s implies that the glide obstacles and the obstacles to recovery have similar features and may even involve the same rate controlling events.

It is of interest to note that, if the high temperature glide obstacle is the breakup of attractive junctions, and if these junctions remain stable, as expected, at intermediate temperatures, then the considerable difference between the amount of work hardening attainable at room temperature $(\sigma_i/E \propto b/\sqrt{\rho} \simeq 3 \times 10^{-2})$ and that attainable at high temperatures $(\sigma_i/E \propto b/\sqrt{\rho} \simeq 3 \times 10^{-3})$ is readily explained.

3. The Transition to Athermal Flow

The similarities in the temperature and rate dependences of σ_y and σ_s have already been noted above. When the dissimilarities indicated in Fig. 10 are also taken into account, a clearer picture emerges of the differences between ambient and elevated temperature deformation. In Fig. 11, the σ_y and σ_s values of Fig. 10 are collected and compared as functions of the

---

† Here $\dot{\varepsilon}$ is strain rate, A and n are empirical constants, Q is an experimental activation energy, and R is the gas constant.

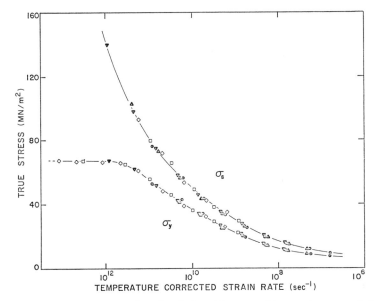

Fig. 11. The data of Fig. 10 collected and plotted in terms of temperature-corrected strain rate Z. As the temperature is reduced (leftward motion on the diagram), the yield stress σ_y stops increasing and reaches a limiting (athermal) value. By contrast, the steady state flow stress σ_s increases rapidly as the athermal flow regime is approached.

temperature-corrected strain rate. The small difference in experimental activation energy for the two types of flow is neglected for this purpose. It can be seen from the figure that as the temperature is decreased or the strain rate is increased, the yield stress becomes progressively less sensitive to temperature and strain rate, until the deformation is virtually athermal. It can also be seen from Fig. 11 that at low temperatures and high strain rates, steady state flow becomes increasingly difficult to attain, because of the progressive inhibition of the recovery processes that accompanies the transition to athermal flow. At sufficiently low temperatures and high strain rates, i.e. well within the athermal range, steady state flow cannot be attained at all.

The transition to athermal flow is considered further in Fig. 12 in which the three regimes of flow stress–temperature dependence are compared. The figure demonstrates that, in each region, it is the lower of the two flow stresses, thermal or athermal, which is operative, so that the transition under consideration is likely to involve alternative mechanisms, which do not operate at all outside their respective temperature and strain rate ranges. Thus the athermal mechanism is unlikely to be controlled by obstacles such as solute or impurity atoms, which must also be present and active at high temperatures. It is more probable instead that the athermal stress corresponds to the stress required for the dislocations to bow out between the nodes by a Frank-Read or Orowan mechanism (Héritier *et al.*, 1974).

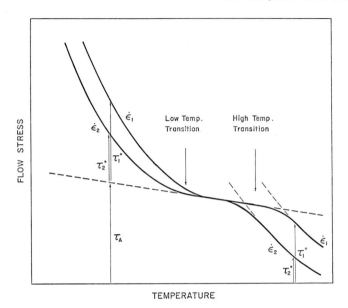

Fig. 12. Schematic representation of the athermal flow regime and its relation to the low and high temperature thermally activated regions. The transition between athermal and high temperature thermally activated flow is seen to be rate sensitive. The low yield stresses observed at high temperatures signify that the obstacles to athermal flow are bypassed or inoperative at elevated temperatures (Jonas, 1973).

E. Substructure and Flow Stress

In many investigations carried out under both creep and hot-working conditions, it has been observed that there is a unique relationship between the scale of the substructure and the flow stress once steady state flow is achieved. The relationship, which is illustrated in Fig. 13, is as follows:

$$\sigma_s = K_1 \, d^{-1} \qquad (8)$$

Here K_1 is an empirical constant. This relation can be given a physical interpretation in terms of Eq. (4) above:

$$\sigma_s = \sigma^* + \sigma_i \qquad (4)$$

First it must be recalled that, at yielding, $\sigma^* \simeq \sigma_y$. If σ^* remains approximately constant with strain (Immarigeon, 1974), and setting

$$\sigma_i = K_2 \, d^{-1} \qquad (9)$$

where K_2 is the subboundary strengthening coefficient,[†] we can write that

$$\sigma_s = \sigma_y + K_2 \, d^{-1} \tag{10}$$

But σ_y in most materials is an approximately constant fraction $f < 1$ of σ_s, i.e. $\sigma_y \simeq f\sigma_s$. Thus

$$\sigma_s \simeq K_2 \, d^{-1}/(1 - f) \tag{11}$$

and we see that the empirical constant $K_1 \simeq K_2/(1 - f)$.

In the above, it has been assumed that the internal stress σ_i is uniquely related to the subboundary density $3/d$, as suggested by Abson and Jonas (1970). This could arise through a causal relationship in which the subboundaries are the principal source of internal stress at high temperatures. Alternatively, the dislocations within the subgrains ρ_s could be directly responsible for the internal stress, as given by Eq. (5):

$$\sigma_i = \alpha\mu b\sqrt{\rho_s} \tag{12}$$

In this event, the empirical correlation (7) takes its physical base from Eq. (5) through a relation of the type introduced by Holt (1970) and discussed above. From the evidence available to date (e.g., Cuddy, 1970), it has not yet been possible to determine whether the principal source of the internal stress (i.e. of work hardening at high temperatures) is the subboundary structure directly on the one hand or the dislocation network within the subgrains on the other.

F. Substructure Formation in Pure Metals

Since dynamic recovery involves the occurrence of climb, cross-slip, and node unpinning, it takes place more readily in metals which have higher stacking fault energies. However, comparisons must be made at a common homologous temperature, so that the fluctuations which promote thermal activation will be of similar magnitude in each metal. Such a comparison is made in terms of the substructures produced by hot working in Figs. 14a, b, and c for the fcc metals. It can be seen from the figure that aluminum, in which dynamic recovery proceeds with ease, contains subgrains which are well formed, and in which the subboundaries are simple arrays (Jonas et al., 1969). Nickel and copper, on the other hand, with their much lower stacking fault energies, have very tangled substructures (Drube and Stüwe, 1967). It will be seen later that, at higher strains, these high dislocation densities lead to dynamic recrystallization.

In the bcc metals, stacking fault energies are relatively high, so that cross-glide, climb, and node unpinning are not unduly difficult, and the facility for

---

[†] The term $K_2 \, d^{-1}$ will be considered in more detail below in the section on the strengthening effect of subboundaries at room temperature.

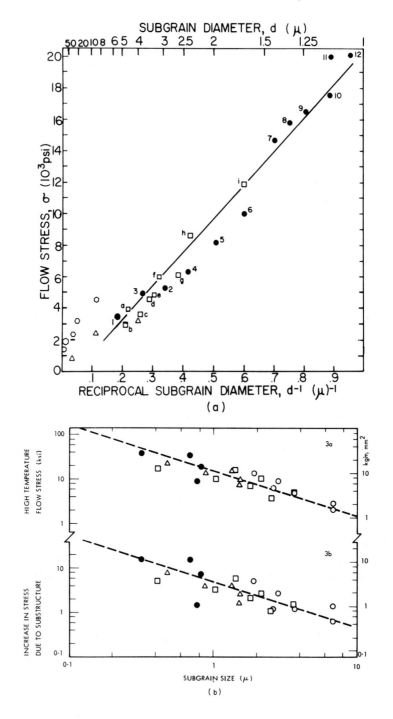

(a)

(b)

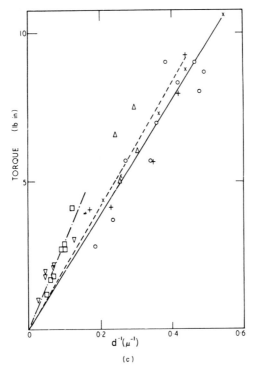

(c)

Fig. 13. The correlation between steady state flow stress and mean subgrain diameter for a variety of deformation conditions and metals. (a) Commercial purity aluminum (fcc) deformed to a variety of strains from 0.7 to 3.7 within the steady state regime (McQueen and Hockett, 1970). ●, Compression; □, extrusion (McQueen, et al., 1967); △, creep (Wood and Rachinger, 1949); ○, hot torsion (Farag, et al., 1968). (b) Subgrain size versus both flow stress and internal stress in α-zirconium and zirconium–tin alloys (hcp). The alloy data points are shifted toward decreasing subgrain size and increasing flow stress (Abson and Jonas, 1972). ○, Zr; □, Zr–0.7% Sn; △, Zr–3% Sn; ●, Zr–5% Sn. (c) The torque for steady state flow is related to the subgrain size for bcc iron and its alloys with Ni and Cr (Redfern and Sellars, 1968). × ——○, S102 iron, alloy 2, 0.62% Cr; + – – – –△, alloy 7, 2.29% Ni, alloy 9, 3.05% Ni; ▽— · —□, alloy 3, 12.9% Cr, alloy 4, 14.0% Cr.

dynamic recovery approaches that of aluminum (Rossard, 1960; Jonas et al., 1968b; Glover and Sellars, 1973). Not unexpectedly then, the deformation microstructures of the ferritic irons (Fig. 14e) and of aluminum are more closely similar than those of aluminum and of fcc metals and alloys with low stacking fault energies.

Subgrains also form in hexagonal close packed metals during high temperature deformation and have been observed in zinc (Hardwick and Tegart, 1961), beryllium (Webster et al., 1973), and zirconium. The subgrains in zirconium are similar to those in aluminum in size and perfection (Fig. 15a).

Fig. 14. Transmission electron micrographs of substructures resulting from hot working. (a) through (d) are for fcc metals rolled to 90% reduction ($\varepsilon = 2.3$) in one pass, at an average strain rate of 20 sec^{-1}, and an exit temperature of 0.65 T_m. (a) Aluminum, 340° C, × 10,000. (b) Nickel, 840° C, × 10,000. (c) Copper, 600° C, × 10,000. (d) 70/30 brass, 600° C, × 20,000 (McQueen, 1968). (e) The substructure of Armco iron deformed at 675° C (0.52 T_m) and 10^{-1} sec^{-1} closely resembles that of aluminum × 9000 (Immarigeon and Jonas, 1971).

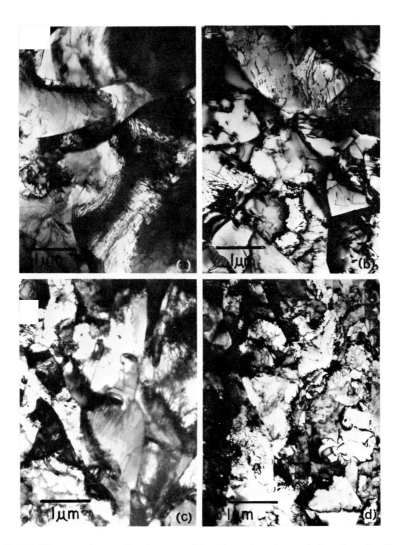

Fig. 15. Electron micrographs of commercial purity α-zirconium and zirconium–tin alloys deformed into the steady state regime at 675° C (0.44T_{m}) at a strain rate of $3 \times 10^{-2} \sec^{-1} \times 13{,}000$. (a) Zr. (b) Zr–0.7% Sn. (c) Zr–3% Sn. (d) Zr–5% Sn. The subgrains become smaller and the density of dislocations becomes greater as the tin content increases (Abson and Jonas, 1972).

G. *Effect of Solute Addition*

When solute elements are present in the range 1–20% by weight, both the
flow stress and the resistance to creep increase by up to an order of magnitude (Fig. 16). The increase in yield and steady state stress with concentration diminishes as the temperature is raised, although the increases relative
to those of the pure metal remain approximately constant. This is illustrated
in Fig. 17, from which it can be seen that the yield and steady state flow
strengths of binary solid solutions are temperature and rate sensitive, as are
those of a pure metal (Luton and Jonas, 1972b). The increase in flow stress of
solid solution alloys is sufficiently great that considerably more force and
energy is required to deform these materials in industrial hot working

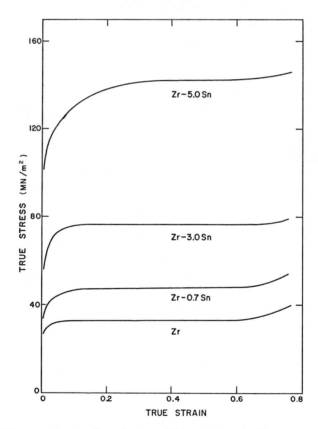

Fig. 16. The strengthening effect of solute addition is shown in this group of stress–strain
curves for Zr and three of its alloys with tin. The temperature is 775° C ($0.5T_m$) and the strain
rate 10^{-2} sec^{-1} (Luton and Jonas, 1972b).

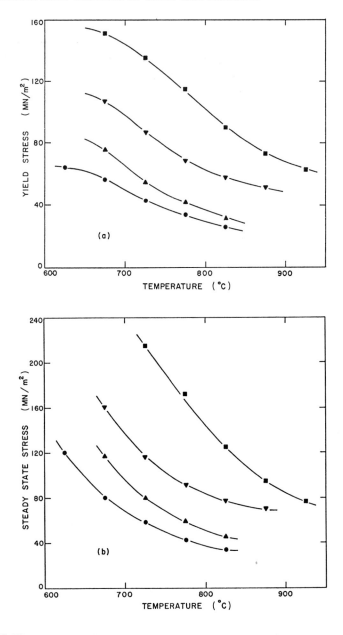

Fig. 17. The temperature dependence of the flow stress for alpha zirconium and three of its alloys with tin. For deformation at 3×10^{-2} sec^{-1}, the yield stresses appear in (a) and the steady state stress data in (b) (Luton and Jonas, 1972b) ●, Zr; ▲, Zr–0.7Sn; ▼, Zr–3.0Sn; ■, Zr–5.0Sn.

(Sabroff et al., 1968; Sellars and Tegart, 1972). For the same reason, alloys developed specifically for creep resistant service are based on solid solutions which are further strengthened by means of precipitation.

The effect of solid solution alloying on the microstructures produced by working are rather diverse and require the individual discussion of each alloy system. In general, solute addition lowers the stacking fault energy, which makes dynamic recovery more difficult. This is observed in brass, which does not form a well-defined subgrain structure under conditions of temperature and strain rate for which it is found in copper (Fig. 14d). Similarly, in Zr–Sn alloys, the addition of up to 5% tin reduces the stacking fault energy from 240 to 60 mJ/m^2 and also increases the fault width from about 2.5 to 10 b (Sastry et al., 1974). Consequently, the equilibrium subgrain size produced in these alloys at the same temperature and strain rate decreases from about 3.5 to 0.7 μm as the tin concentration is increased (Fig. 15).

In the Al–Mg system, on the other hand, while the strength increases with greater concentration of Mg, the size of the subgrains also increases for tests at the same temperature and strain rate (Fig. 18). The latter observation can be reconciled with the increased amount of work hardening in the alloys if the internal stress does not depend directly on subboundary density, as has been suggested above, but increases with $\sqrt{\rho}$ instead. A further alternative, that the internal stress contribution of the subboundaries is strongly dependent on the dislocation structure of the boundary itself, and that this changes with alloying, is also possible. In the austenitic stainless steels, a well developed substructure forms, one which is more "recovered" than that of copper under equivalent conditions, suggesting that the stacking fault energy is higher than in copper at these temperatures (Garofalo et al., 1963; Buhler et al., 1970). It is not possible to compare the stainless substructures with those formed in unalloyed austenite, since the substructures in the latter material are lost during the phase transformation produced by cooling. When the nickel-base superalloys are worked in the single phase condition, a recovered substructure is produced which is apparently as well defined and orderly as that observed in nickel (Fulop and McQueen, 1972).

In bcc alloys, additions of chromium or nickel increase the flow stress of the alloy (Redfern and Sellars, 1968). In iron–silicon alloys, the addition of silicon produces strengthening from about 600 to 800° C (Immarigeon and Jonas, 1974). Above 800° C, however, the addition of silicon leads to weakening, a phenomenon known as solute softening in the low temperature range. This has been attributed to the increase in the density of mobile dislocations associated with silicon addition, which overrides the increase in node stability. The solution of the above elements in iron does not markedly affect the scale of the substructure under given conditions of temperature

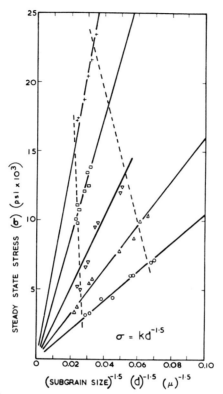

Fig. 18. The effect of magnesium concentration and subgrain size on the steady state flow stress of Al–Mg alloys. Samples were deformed in torsion between 400° C and 550° C at a strain rate of 2 sec^{-1}. Since the data points for each alloy follow the same sequence of temperatures, it can be seen that for a given temperature the subgrain size increases with Mg concentration (Cotner and Tegart, 1969). ○, S.P. Al; △, Al–$\frac{1}{3}$% Mg; ▽, Al–1% Mg; □, Al–2% Mg; +, Al–5% Mg.

and strain rate (Jonas *et al.*, 1968b). The steady state flow stress for a given subgrain size is greater for the alloys than for pure iron (Fig. 13c). As in the case of the Al–Mg alloys, these observations suggest that the subboundaries are not the primary source of internal stress in these materials. Finally, it should be noted that the addition of Si or particularly of Cr stabilizes the α phase and extends the working range to higher temperatures.

A special application of alloying and hot working is the ausforming process in which the objective is the transformation of austenite to martensite immediately after hot working so that the dislocation structure is carried into the martensite. Alloying of the austenite creates a bay in the isothermal transformation curves in which the working can be carried out and rapid quenching achieved (Schmatz *et al.*, 1963; Irani and Taylor, 1968).

H. Dynamic Recovery in Two-Phase Alloys

The simplest two-phase materials are those in which the second phase is present as roughly spherical particles which are relatively hard at the temperature of deformation. In situations where the particles are not cut by the dislocations and are relatively stable, they serve as obstacles which help initially to build up the substructure and finally to stabilize it. This effect has been observed at strains not exceeding two in many systems: e.g. Al with Al_2O_3 particles, Ni with ThO_2 particles (Clauer and Wilcox, 1967), Ni-base superalloy with γ' precipitate (Oblak and Owczarski, 1972), ferrite with carbide precipitates, and austenite with NbC particles (Le Bon et al., 1973) or other carbides (Garofalo et al., 1961). Under these conditions, the subgrain diameter may be reduced to the order of the interparticle spacing, which leads to an increase in flow stress. The presence of precipitates can prevent the formation of a subgrain structure altogether if the spacing is small enough, or if, as a result of their shearing, they induce the formation of matrix superdislocations, which cannot leave their slip planes. Furthermore, by preventing grain boundary migration, the particles can shift the initiation of dynamic recrystallization to higher strains, which can be useful in controlled rolling (Irvine et al., 1970; Tegart, 1971). It appears that the presence of NbC in austenite can completely inhibit dynamic recrystallization at moderate temperatures, although this may actually be caused by the inhibiting effect of the particles on subgrain formation, and therefore on the nucleation of new grains (Petkovic, 1975).

If the elements comprising the second phase are soluble in the matrix, even to a very limited degree, it can coalesce during deformation into fewer and larger particles. In hot working this can cause a significant decrease in flow stress over a strain increment of 1 or 2 (Fig. 19). Coalescence proceeds more rapidly at higher temperatures and higher strain rates. At higher temperatures and lower strain rates, the second phase particles are coarser when steady state flow is finally achieved. This phenomenon has been observed in pearlitic-ferritic steels (Sherby et al., 1969) and in eutectoid steel; for example, at 700° C (1292° F), a degree of spheroidization which would take 660 hr of annealing takes place in 210 sec of deformation at 0.016 sec^{-1} (Chojnowski and Tegart, 1968). The diffusional process producing the spheroidization is considered to be accelerated by the hot-work subboundaries which provide paths for pipe diffusion, which is considerably more rapid than lattice diffusion under the same conditions. Another example of this effect is found in the warm working of bainite. Here spheroidization of the acicular particles increases the ductility considerably without loss in overall strength because of the strengthening contribution of the substructure introduced. In Waspalloy, the $Ni_3(TiAl)$ particles appear to

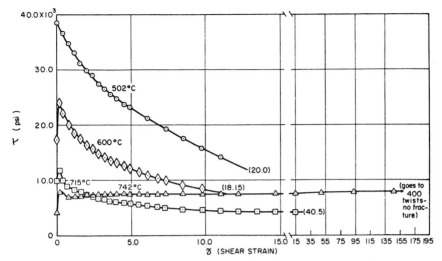

Fig. 19. Torsional stress–strain curves for an alloy of pure iron and 0.8% carbon. At 502° C (0.43T_m), 600° C (0.48T_m), and 715° C (0.56T_m), the microstructure at the beginning of the test is completely pearlitic. In the course of deformation, the carbide spheroidizes, giving rise to a continuous decrease in flow stress. The figures in parentheses indicate the number of twists to failure (Robbins *et al.*, 1967). $\dot{\gamma} = 4.10\%$ sec^{-1}.

coalesce during straining, giving rise to stress–strain curves similar to those for eutectoid steel.

If precipitation occurs *during* high temperature deformation, it causes the flow stress to increase or the creep rate to decrease. The presence of a dislocation substructure during precipitation assures a uniform distribution of precipitate in most cases. For example, when the thermomechanical working of a nickel-base superalloy is designed to permit precipitation during working, the resultant properties are better than those obtained when precipitation precedes or follows working (Oblak and Owczarski, 1972). This phenomenon is also exploited in the isoforming process, in which austenite is deformed as it is transforming to ferrite and carbide (Irani and Taylor, 1968). In this case, the precipitated carbides are nucleated on dislocations in the ferrite and the particles grow as spheres. Ultimately the carbide particles serve to stabilize and strengthen the substructure.

I. The Ease of Dynamic Recovery and Its Effect on Ductility

In hot working as well as in creep, the most common cause of fracture is the formation of grain boundary cracks (Tegart, 1968). The cracks are nucleated as a result of grain boundary sliding and are propagated by the tensile stresses generated during the working operation. In creep studies, it has been observed that as the stress and the strain rate are increased, the sites for crack initiation change from grain boundary ledges to the triple junctions. In hot working, because the flow stress is much higher than in creep, and because less time is available for grain boundary sliding and the diffusional growth of pores, the nucleation sites are almost entirely limited to grain boundary junctions.

The formation of subgrains in metals displaying a susceptibility to dynamic recovery results in the scalloping of the grain boundaries, which diminishes intergranular sliding (Fig. 7d). Furthermore, such materials have fairly low flow stresses and flow readily at triple junctions to relieve stress concentrations and thus diminish the initiation of W cracks. Both of these effects contribute to the high ductility observed in aluminum, α-iron, and α-iron–chromium alloys. In such metals, the ductility usually increases with temperature because the accommodation processes are more sensitive to temperature than those promoting crack nucleation (Gittins, 1970). However, in metals such as nickel and the nickel-base superalloys, in which the rate of recovery is relatively low, the ductility *decreases* with temperature in the lower part of the high temperature range (where dynamic recrystallization does not intervene) because of increases in the amount of grain boundary sliding (Shapiro and Dieter, 1971; White and Rossard, 1968).

A reduction in ductility with increasing temperature can also arise as a result of phase changes. For example, fcc austenitic iron above the transformation temperature is less ductile than bcc ferritic iron below the transformation temperature because of the higher yield stress of the austenite, as well as its lower capacity for recovery (Reynolds and Tegart, 1962). Similarly, at a given temperature in the range 900–1200° C, the ductility of the Fe–Cr–Ni alloys is higher for the high recovery rate δ-ferrite structure (bcc) than for the low recovery rate γ-austenite structure (fcc).

In general, solid solution alloying tends to reduce ductility (Luton and Tegart, 1969) and the effect becomes more pronounced with increasing concentration (Fig. 20). This is partly because the smaller subgrains formed in these materials are not as effective in producing scalloping of the grain boundaries, so that grain boundary sliding is not impeded to the same extent. The scalloping is also reduced by the retarding effect solutes exert on grain boundary migration. In addition, the decreased ease of dynamic recovery raises the flow stress considerably, which is instrumental in nucleating and opening up the cracks.

The presence of precipitate particles, e.g. carbides in ferritic steel, usually reduces the ductility (Fig. 21). However, the size and distribution of the particles are very important. Fine discrete precipitates uniformly distributed on the grain boundaries improve the ductility by impeding sliding (Cremisio *et al.*, 1969). Particles larger than 10 μ or semicontinuous films of precipitate

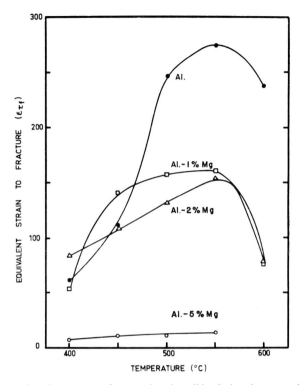

Fig. 20. Increasing the amount of magnesium in solid solution decreases the ductility of Al–Mg alloys; at 5% Mg, the ductility is severely reduced over the entire temperature range. The deformation rate employed in these hot torsion tests was 2.3 sec^{-1} (after Cotner and Tegart, 1969).

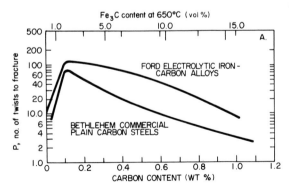

Fig. 21. The influence of carbon on the ductility of bcc iron tested in torsion at 0.03 sec^{-1} and 650° C ($0.5T_m$). The ductility initially increases as a result of the decrease in concentration of oxide inclusions. At higher carbon content, the ductility decreases as the volume fraction of carbide increases (Robbins et al., 1967).

that are harder or more brittle than the matrix are very damaging, since they help to initiate wedge cracks and prevent boundary migration (Gittins *et al.*, 1972). The overheating of certain alloys before hot working can also give rise to poor ductility because of the increase in size of the precipitate particles which form on the grain boundaries (Fig. 22).

A ductility reduction is also observed when the second phase is a metallic one and is present in much larger volume fractions. Examples are γ in α-iron (Robbins *et al.*, 1967) or in δ-iron and β-phase in Ti alloys (Reynolds, 1968). The magnitude of this effect depends upon the volume fractions of the two phases, on their relative ductilities, and on the softening mechanisms operating within the phases. The effect is typified by the Fe–Ni–Cr alloys, in which the fcc γ and bcc δ-phases coexist at high temperatures. If the δ phase is taken as the reference matrix, then, as the proportion of the harder γ-phase is gradually increased, the ductility understandably decreases (Fig. 23). The effect is more marked, however, if the proportion of δ (which softens by dynamic recovery) is increased in a γ matrix (which softens by dynamic

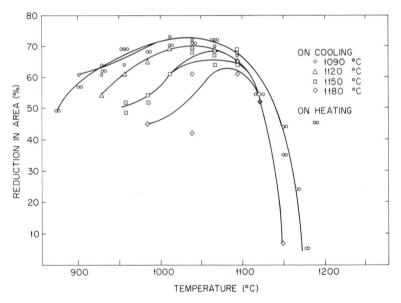

Fig. 22. The ductility of a nickel-base superalloy (Inconel 718) increases with temperature on heating from room temperature as a result of enhanced dynamic recrystallization. The ductility drops rapidly at high temperatures as the solvus temperature is approached. The ductility decrease resulting from cooling from a high preheating temperature, such as 1150 or 1180° C, is due to the increase in size of precipitate particles. The yield stress diminishes with temperature and is unaffected by the preheating temperature. The tests were carried out in tension and a crosshead speed of 5 in./sec was used (Bailey, 1969).

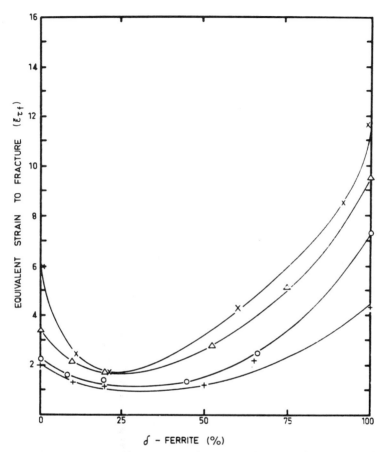

Fig. 23. The effect on ductility of the relative volume fractions of bcc δ-ferrite (which undergoes dynamic recovery) and of fcc austenite (which softens by dynamic recrystallization). The torsion tests were conducted at a strain rate of 0.02 $\sec^{-1}$ (Müller, 1967). $\times$, 1200° C; $\triangle$, 1100° C; $\bigcirc$, 1000° C; $+$, 900° C.

recrystallization). In this case the addition of a ductile phase to a less ductile one *decreases* the ductility! A possible explanation of this phenomenon is that the presence of the second phase inhibits dynamic recrystallization in the γ by restricting grain boundary mobility. It is interesting to note that, despite the presence of a ductility minimum, the strength follows approximately the law of mixtures, with the alloy gradually softening as the volume of the softer phase is increased.

Similarly, in Cu–Ni–Zn–Mn alloys at 800° C, the duplex alloy of 72% bcc β-phase is more ductile than that of 71% fcc α-phase (Ward and Helliwell, 1970). One can therefore conclude that the ductile bcc phases (which soften by dynamic recovery, and in which grain boundary migration is not essential) have their ductility reduced less than the harder fcc phases, which soften by dynamic recrystallization. An exception to this generalization occurs when the two phases are uniformly distributed as fine grains of diameter 3 μ or less. In this case, the material becomes superplastic, but only at strain rates that span the division between creep and hot working, as discussed in more detail in the article by T. H. Alden, in this volume.

In commercial alloys which contain inclusions, the fracture mechanism involves the linking up of voids nucleated at the inclusions. In such cases the ductility decreases with inclusion content, and this effect completely overshadows the alloying effects already described (Sellars and Tegart, 1972).

J. *Effect of Prior Substructure on Subgrain Formation*

In the previous sections, it was assumed that the starting material for a particular cycle of working was fully annealed. However, in many hot-working operations, such as rolling or forging, successive operations are performed on materials containing substructures inherited from previous cycles of deformation. The only common exceptions are extrusion and planetary hot rolling. It will therefore be of interest to consider the effect on the processes of dynamic recovery, and in particular on that of subgrain formation, of the presence of the substructures produced in earlier treatments.

This subject is, unfortunately, rather complex because of (1) the variety of substructures from which a given test may start, (2) differences in behavior associated with the relative strength of the initial substructure and the final equilibrium substructure at steady state, and (3) the diversity of experimental procedures employed in such tests. For simplicity, consideration will first be given to cases where deformation is initiated with a softer substructure than the stable one. In a later section, we will consider commencement with a stronger substructure. One phenomenon is common to the two types of experiment; the final equilibrium substructure is dependent only on the deformation conditions, regardless of the initial substructure, provided recrystallization or failure does not intervene (Bird *et al.*, 1969; Jonas and McQueen, 1973). However, when the starting structure is a denser one than the final equilibrium substructure, the probability of recrystallization intervening during deformation under given conditions is much greater than for a nondeformed starting material.

1. Initial Substructure Weaker than the Stable Substructure

In this case, the starting material contains relatively large subgrains with low wall densities. As straining proceeds, the wall density increases and intermediate subboundaries begin to form. Since the final subgrains must be equiaxed and an integral number is unlikely to fit into an initial subgrain, considerable adjustment is required to reach the final structure. In constant strain rate tests, the initial yield stress is higher than that of a recrystallized specimen, and in constant stress creep, the creep rate is initially less than that of recrystallized material. Considerable strain is required to produce the equilibrium substructure and the flow stress (or strain rate) may reach its steady state value before the gross features of the substructure do. The strain required for such evolution may be greater than that for recrystallized metal to attain steady state flow for the given conditions of deformation. The following results for α-iron may be taken as an example: when the strain rate is increased by a factor of 100, the initial flow stress is less than that of annealed material deformed to the same total strain entirely at the higher strain rate (Fig. 24). A strain of about 0.2 to 0.3 is required at the new strain

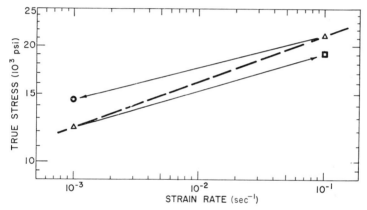

Fig. 24. Probable range of flow stresses developed in industrial processes in which marked strain rate variations occur. The broken line represents the steady state flow stresses measured in tests on Armco iron at 675° C ($0.54T_m$). The upper arrow indicates the decrease in flow stress from 10^{-1} to 10^{-3} sec^{-1}. The lower arrow indicates the increase in flow stress when the strain rate is rapidly increased from 10^{-3} to 10^{-1} sec^{-1}. The final flow stress is less than that in a continuous test at the higher strain rate. The actual increase is only 63% of the increase expected on the basis of continuous tests at constant strain rate (Immarigeon and Jonas, 1971).

rate for the flow stress and the substructure to reach their normal steady state values (Fig. 25).

Changes of the above nature can arise in creep service when the stress is raised either as a singular event or as a cyclic variation. Rapid changes in temperature or strain rate are also used in experiments to determine the activation energy or activation volume at constant structure. In industrial hot working, inherited substructures can play a role in a variety of processing schedules, some examples of which will now be considered.

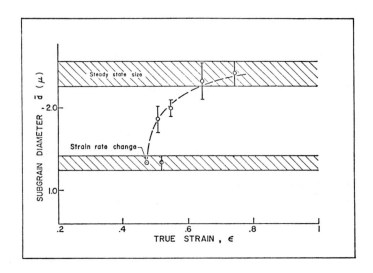

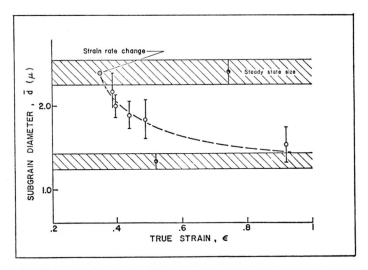

Fig. 25. Variation in subgrain size during transient deformation in Armco iron at 675° C ($0.54T_m$). The substructural changes were initiated by a rapid change in strain rate at the strains indicated. Upper figure: strain rate decrease from 10^{-1} to 10^{-3} sec^{-1}. Lower figure: Strain rate increase from 10^{-3} to 10^{-1} sec^{-1}. It can be seen that the substructural modifications associated with hardening take place over greater strains than those associated with softening between the same limits (Immarigeon and Jonas, 1971).

a. Multipass Deformation at Constant Temperature and Strain Rate. Here the metal is held at a constant temperature for some interval between passes in the same mill. During the interval, the metal can either be held in a furnace or can simply be cooling back to the temperature through the loss of heat of the previous partially adiabatic deformation. During the interval, the metal undergoes static recovery, so that its yield stress is lower than at the end of the previous pass. If the interval is not too long, there is little change in subgrain size, although the density of dislocations in the subboundaries has decreased. If the strain of a particular pass is insufficient for the attainment of steady state flow, then the cumulative strain of several passes may produce the equilibrium structure.

b. Interrupted Deformation at Falling Temperatures or Increasing Strain Rates. Here the metal is cooler at each stage of deformation in either rolling or forging. Alternatively, each pass is at a higher strain rate, as in a continuous mill, where the sheet or shape is speeded up by the deformation it undergoes in each stand. In this case, neglecting the effect of static recovery, each cycle of deformation is performed on metal with a softer (more highly recovered) substructure than would normally develop under the conditions of that working operation. If the strain in each pass is less than the strain required to transform the initial structure into the steady state one, the final flow stress in that pass may be considerably less than that of initially recrystallized metal (Vaughan, 1968). In aluminum, for example, a controlled series of 14 deformations was carried out, each of $\varepsilon = 0.8$, separated by 30 sec intervals, with temperatures decreasing from 600 to 400° C (Farag *et al.*, 1968). This treatment resulted in a final flow stress that was 25% lower than that pertaining to metal deformed continuously at 400° C. Increasing the number of stages and reducing the strain per pass had little effect. However, doubling the time interval, which allowed more static recovery to take place, gave a final flow stress only 15% less than that for isothermal deformation at 400° C. A similar but simpler schedule is shown in Fig. 26.

c. Continuous Deformation at Continuously Falling Temperatures or Rising Strain Rates. This case arises in certain forming operations, such as extrusion or rolling, in which the strain rate rises rapidly as the metal passes into and partially through the deformation zone, and then rapidly falls to zero as the metal moves out of the zone (Jonas and McQueen, 1973). In forging, the strain rates vary in a more complex manner as the hammer is decelerated and as different portions of the billet come into contact with the die. Although the temperature can increase due to adiabatic heating, in many forming operations it decreases as heat is lost to the tooling and to the atmosphere. When the temperature is continuously dropping or the strain

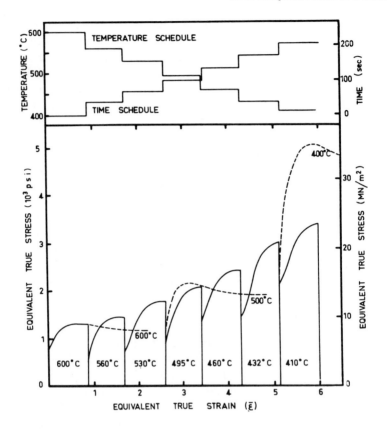

Fig. 26. The upper portion of the diagram shows the temperature–time schedule of a series of torsion tests on superpurity aluminum conducted at a strain rate of 2.3 sec^{-1}. On the temperature curve, a horizontal segment indicates the temperature of a stage of deformation and a vertical segment the cooling between stages. On the time graph, an almost horizontal segment indicates the short time elapsed during a stage of deformation, and a vertical segment indicates the delay between stages. In the lower graph, the solid lines indicate the flow stresses recorded, whereas the broken lines are for continuous isothermal tests at the temperatures noted. The slight decrease in steady state flow stress results from adiabatic heating (Farag *et al.*, 1968).

rate rising, the flow curve is lower than that calculated from isothermal isostrain rate data (Fig. 27). This is because, at each strain, the metal contains a softer substructure, inherited from the prior strain increment, than the stable one pertaining to the current temperature and strain rate. For example, in iron (Fig. 28), deformation under continuous cooling required a stress up to 30% less than that expected from isothermal testing. Similarly, a

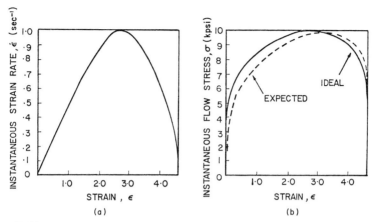

Fig. 27. (a) Strain rate profile as a function of strain for a 100 : 1 extrusion die and a ram speed of 6×10^{-3} in/sec. (b) Flow stress profiles for material flowing through the strain rate profile of (a). The full line is the ideal profile based on constant strain rate steady state data for α-Zr at 850° C. The broken line is the expected profile after the deformation history is taken into account. The area under the broken curve is the actual work done and is greater or less than the ideal area depending on whether the peak strain rate is attained early or late along the profile (Jonas and McQueen, 1973).

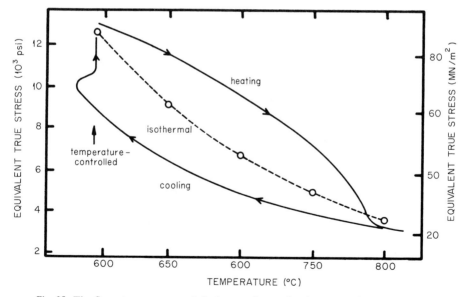

Fig. 28. The flow stresses measured during continuous heating or continuous cooling are compared to the flow stresses measured in a series of isothermal tests. The material was vacuum melted bcc iron, the strain rate in the torsion tests was 1.5×10^{-3} sec^{-1}, and the rate of heating or cooling was about 50° C/sec. (Glover, 1969).

50% upset at a constant velocity, which involves a doubling of the true strain rate, required an average flow stress 6% lower than that for reduction at the mean of the true strain rate† (Loizou and Sims, 1953).

2. INITIAL STRUCTURE STRONGER THAN STABLE SUBSTRUCTURE

In this case, the initial structure has a finer cell size with more tangled boundaries than the equilibrium substructure. Thus as deformation progresses, the dislocation density diminishes, some subboundaries disappear, and equiaxed subgrains of the stable size and wall density form. During the transition, there is an increase in creep rate in a constant stress test and a decrease in flow stress in a constant strain rate test. The latter is often referred to as work softening. Substructural softening in α-iron has been observed to require a greater strain, 0.45 compared to 0.25, than the substructural hardening described earlier (Fig. 25).

Industrially, the conversion of fine to coarser substructures occurs when multipass deformations are carried out at increasing temperatures or decreasing strain rates. This situation can arise when, with rapid forming, adiabatic deformation heats up the metal. This gives rise to higher flow curves if the strains applied are small (less than 30%), so that the stable configuration is not reached. Such a result is in opposition to the lower flow curves observed when the initial structure is softer than the stable one. When straining is continuous, with increasing temperature or decreasing strain rate, an augmented flow stress is again observed, which is in contrast to the effect of continuously dropping the temperature or increasing the strain rate (Fig. 27). Since most forming processes have a strain–time profile with a rising rate followed by a falling rate, such operations can be expected to lead to finer substructures than indicated by the final strain rate. Similar effects arise when adiabatic heating raises the finishing temperature above the entrance temperature.

The introduction of a fine substructure has been tried as a means of increasing creep resistance. Cold work, while giving a great initial improvement, almost always results in dynamic or static recrystallization, with a resulting diminution of creep resistance. Success has been attained by using substructures produced by static recovery after cold straining to the order of 10%, or by prior creep at a higher stress. Such treatments are mainly of scientific interest since they are difficult to carry out industrially and do not offer as much benefit as raising the creep resistance through alloying. A more practical alternative is the production of a suitable substructure by hot

---

† When the strain rate during a hot-working operation varies by a substantial factor, the appropriate mean strain rate to use for flow stress correlations is $[\bar{\dot{\varepsilon}}^m]^{1/m}$, where m is the rate sensitivity of the flow stress (Jonas and Chandra, 1971).

working and, when possible, its stabilization by precipitation. This method, which will be described later, is a means of further increasing the rupture life of creep-resistant alloys.

Stress Recovery. When a cold-worked metal is annealed under stress, the rate of recovery increases considerably, and recrystallization can be delayed or even prevented (Auld *et al.*, 1957; Thornton and Cahn, 1961). This can arise in industrial processes such as continuous annealing, in which the traveling strip is under tension because of its weight and for purposes of materials handling. The application of external stress, which produces a form of creep, assists the dislocations to move through the crystal and increases the opportunity for annihilation to occur. It is most effective in metals such as aluminum which normally recover, but has been observed to occur in copper as well.

K. High Temperature Dynamic Recovery and Stage III Deformation

Before concluding this section, we consider it advisable to distinguish between the sense of dynamic recovery as we have discussed it above and the more common meaning of the term associated with ambient temperature deformation. The latter is concerned with the declining rate of strain hardening in the stage III deformation of fcc single crystals at homologous temperatures below one-half. In both cases, the softening process involves the rearrangement of dislocations in the course of deformation. Furthermore, in both types, the facility of different metals for recovery increases with their stacking fault energy. On the other hand, true steady state flow, with an equilibrium substructure of constant dislocation density, does not occur in stage III deformation. Instead, although the cell size usually remains constant, the wall dislocation density increases, leading to an increase in flow stress (Kuhlman-Wilsdorf, 1966). The reason for this difference applies in the limitation of the stage III recovery mechanisms to cross-slip. By contrast, at elevated temperatures, other mechanisms are activated, such as dislocation climb and node unpinning. As a result of the availability of these additional processes at high temperatures, the accumulation of dislocations to the high concentrations associated with cold working does not occur. The dislocation density levels out instead at much reduced values, leading to the relatively low flow stresses normally associated with high temperature deformation.

As the temperature of working is raised, stage III dynamic recovery merges into the type of dynamic recovery found in hot working. Concurrently, at a given strain, the subgrain size and perfection gradually increase and the flow stress decreases. This continuous decline in flow stress and increase in ductility is employed in the warm working of steels at temperatures between $100°$ C and the lower end of the hot-working range. In this range, the surface does not become oxidized, and lubricated polished dies can be used (Reynolds and Tegart, 1962; Keegan, 1967).

IV. Dynamic Recrystallization

A. The Flow Curve

The flow curve under conditions of dynamic recovery has already been described. In that case, all the softening processes involved single dislocations, which are annihilated in individual events. We now turn to the contrasting situation where dislocations are annihilated in large numbers through the migration of a high angle boundary. By this means, the lattice in which they reside is destroyed and replaced by a new and substantially perfect one in one integral operation.

The occurrence of dynamic recrystallization modifies the appearance of the flow curves produced at constant strain rate so that they no longer have the simple shape shown in Fig. 5 above. At high strain rates in the hot-working range, the flow stress rises to a maximum at the peak strain ε_p. Thereupon, as a result of dynamic recrystallization, it diminishes to a value intermediate between the yield stress and the maximum or peak stress (Fig.29a). At lower strain rates in the hot-working range, the softening produced by dynamic recrystallization is followed by renewed hardening and a cyclic flow curve of approximately constant period but declining amplitude is traced out instead (Fig. 29a) (Sellars and Tegart, 1966a). The effect of decreasing the strain rate on the shape of torsional flow curves is shown in more detail in Fig. 30a. Here it can be seen that, at strain rates of more than 10^{-1} sec^{-1} in γ-iron, softening by recrystallization, once initiated, is continuous. By contrast, when strain rates of less than 10^{-1} sec^{-1} are used, the softening is periodic, with intervening cycles of hardening.

In a constant stress or creep test, the advent of dynamic recrystallization appears as a rapid increase in strain rate, which terminates the otherwise normal stage of primary creep, usually before steady state deformation has been attained (Fig. 29b). After the acceleration, a new stage of primary creep commences, which is also likely to terminate in dynamic recrystallization. This phenomenological cycle may be repeated several times (Fig. 30b). In general, it occurs only at high stresses in the creep range, when these are present at temperatures above 0.7 T_m (Richardson et al., 1966).

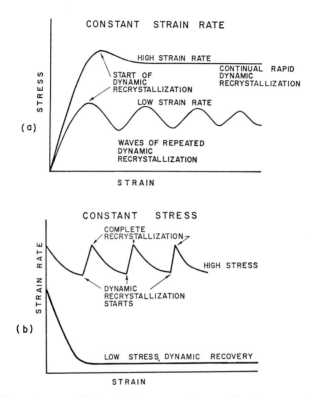

Fig. 29. Dynamic recrystallization during creep and hot working is common in metals such as lead and copper in which the rate of dynamic recovery is limited. (a) Hot working at a slow rate leads to a cyclic stress–strain curve caused by waves of recrystallization. At high strain rates, recrystallization starts at higher strains, causing the drop in stress after the peak. However, before recrystallization is complete, the oldest of the new grains deform enough to start recrystallizing again. In this way, new nuclei are continually forming throughout the material. Since these nuclei are soft, they keep the average flow stress low. Up to the peak, softening occurs by means of dynamic recovery processes. Nevertheless, after the initiation of dynamic recrystallization, dynamic recovery continues to take place in each new grain until it once again recrystallizes. (b) At low creep stresses, the amount of dynamic recovery is sufficient to keep the dislocation density below critical levels. At high stresses, the substructure becomes so dense that recrystallization nuclei form and grow rapidly to engulf the whole specimen. The specimen then begins to creep at high rates as it did initially.

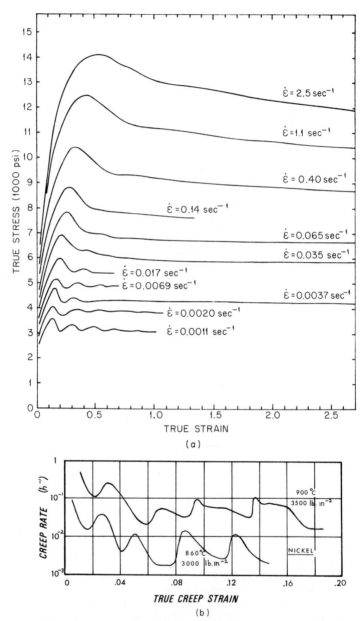

Fig. 30. (a) Flow curves of a plain 0.25% C steel in the austenitic state (fcc) at 1100° C (0.76T_m), illustrating the strong influence of strain rate. At the higher strain rates, they exhibit the typical shape for dynamic recrystallization: strain hardening to a peak stress followed by work softening to a steady state level. The work softening is caused by the start of dynamic recrystallization and the stable flow results from continuous dynamic recrystallization. At low strain rates, the flow curve is periodic due to recurrent cycles of recrystallization (Rossard, 1960). (b) Creep curves for nickel under constant stress in compression at 860° C (0.65T_m) and 900° C (0.72T_m). The sudden increases in strain rate mark the initiation of cycles of dynamic recrystallization (Hardwick *et al.*, 1961).

B. *Microstructural Changes Associated with Dynamic Recrystallization*

In the metals which undergo dynamic recrystallization, a dislocation sub-structure develops in the initial stage of deformation which is rather poorly recovered; i.e. the cells have relatively smaller size and more tangled walls compared to the metals described previously, which undergo a high degree of recovery. At lower strain rates, the evidence indicates that nucleation occurs by bulging of an existing grain boundary (Luton and Sellars, 1969). Although scalloped boundaries are also observed in the high dynamic re-covery metals, nucleation does not occur. The structural features peculiar to low recovery metals that account for the nucleation of dynamic recrystalliza-tion are: (1) the greater effectiveness with which a grain boundary is pinned by the highly tangled cell walls, and (2) the greater difference in strain energy between the region within the bulge and that in front of it. It should be noted that dynamic recrystallization has also been observed in lead single crystals undergoing creep, specimens to which the grain boundary bulge mechanism of nucleation clearly does not apply (Hirst, 1940). In this case, the nuclei were observed to form in slip bands, presumably by a nucleation mechanism similar to that described below for high strain rates.

At high strain rates, on the other hand, a fine tangled cellular structure is developed throughout all the grains, which does not leave grain boundary segments long enough to bulge. There is evidence that, with increasing strain, some tangles build up to high misorientations, giving rise to nuclei throughout each grain (McQueen and Bergerson, 1972). The density of nuclei is probably higher near grain boundaries because the strain is higher there as a result of accommodation to plastic anisotropy. By the time steady state flow is attained, new small equiaxed grains have replaced the original grains (Fig. 31). Throughout the steady state region, the grains remain con-stant in size and equiaxed in shape (Rossard, 1960).

Since straining of the material continues during nucleation and growth, the new grains are deformed as they grow. When the strain rate is low, the gradient of strain energy from the center of the recrystallized grain to the rapidly advancing boundary is low, and the region behind the boundary is relatively free of dislocations. As a result, the driving force due to the differ-ence in dislocation density on opposite sides of the boundary, and hence the rate of migration, is not reduced appreciably by the continuing deformation. In this way, recrystallization proceeds to completion, with the center of each grain in a deformed state, so that the material is not as soft as statically recrystallized metal of the same grain size. Once recrystallization is complete, the dislocation density builds up again, raising the flow stress (or decreasing the strain rate in creep) until recrystallization is again nucleated.

In the case of high strain rate deformation, the gradient of deformation

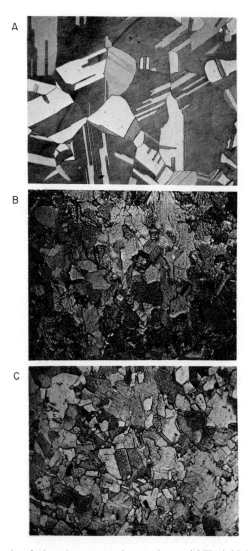

Fig. 31. Micrographs of selected copper torsion specimens: (a) The initial large grains. × 20. (b) The very small dynamically recrystallized grains present after deformation into the steady state flow region at 800° C (0.8 T_m) and 5.6 sec^{-1}, rapidly quenched. × 260. (c) The small statically recrystallized grains in metal deformed as in (b) but cooled less rapidly, × 80 (McQueen and Bergerson, 1972).

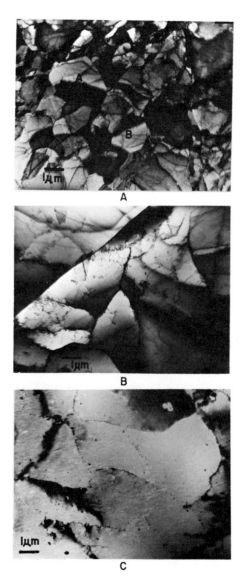

Fig. 32. Substructures in specimens of copper that have dynamically recrystallized during high temperature deformation to a strain of about 30: (a) Two recrystallization nuclei A and B, surrounded by a heavily deformed region, 800° C (0.8 T_m), 11.6 sec^{-1}. (b) Region of high dislocation density. (c) Region of low dislocation density in material deformed at 950° C (0.9 T_m) and 3.7 sec^{-1}. × 5000 (McQueen and Bergerson, 1972).

from center to boundary of a recrystallized grain is high, and so there is a considerable dislocation density just behind the advancing boundary. By diminishing the driving force, the rate of migration is reduced relative to that in static recrystallization. Before recrystallization is complete, the dislocation densities at the centers of the recrystallizing grains have increased sufficiently that another cycle of nucleation occurs and new grains begin to grow again (Fig. 32a). Thus, at any instant, there is a distribution of regions with different degrees of deformation, from zero to a value just higher than the peak strain (Fig. 32b and c). It is this distribution of dislocation substructures which maintains the average flow stress at a value intermediate between the yield stress of statically annealed material and the peak stress.

C. Critical Strains for Nucleation and Completion of Recrystallization

As already illustrated above, the start of recrystallization is quite marked on stress–strain curves, as it is on strain rate–time curves. The critical strain ε_c for the start of recrystallization is actually slightly less than the peak strain ε_p because, while the first nuclei are softening the material locally, the remaining material continues to get stronger. The difference between the two strains is even greater at higher strain rates. According to Rossard (1973), who has carried out extensive studies of this phenomenon, ε_c is given by the approximate relation:

$$\varepsilon_c \simeq 5\varepsilon_p/6 \simeq 0.83\varepsilon_p \tag{13}$$

It should be noted that the critical strain for dynamic recrystallization may not in fact be attained for a variety of reasons. For example, the material could continue to strain harden, developing higher and higher flow stresses, until failure occurs. In such an instance it can be concluded that true hot-working conditions do not apply. If, on the other hand, the operation of dynamic recovery processes alone leads to a high enough dislocation annihilation rate, the nucleation of recrystallization can be avoided entirely. This is the description that applies to bcc iron at low temperatures and high strain rates (Glover and Sellars, 1973), to aluminum over the entire hot-working range, and to most alloys in creep service.

By means of metallographic examination, it has been established that the progress of recrystallization follows the normal S-curve of volume recrystallized-versus-log time which, in constant strain rate tests, is the same as volume recrystallized-versus-log strain. For explaining some features of dynamic recrystallization, ε_x has been defined as the strain from the point of nucleation to that where the volume percent of material recrystallized is x. For convenience, the strain for "complete" recrystallization, ε_r, has been defined as the strain for 95% recrystallization.

D. Effect of Strain Rate and Temperature on ε_c and ε_r

In general, the critical strain ε_c for recrystallization increases with increasing strain rate and decreasing temperature (Luton and Sellars, 1969). [This does not, however, apply at very low strain rates, where the critical strain seems to increase with decreasing strain rate (Fig. 33)]. The increase in ε_c with temperature-corrected strain rate Z (Fig. 34a) signifies that, at higher Z, the subboundary density is higher before recrystallization is nucleated. The increase in ε_c with decrease in temperature is attributable to the decreased rate of recovery, and therefore of nucleation, at lower temperatures. When the strain rate is increased at constant temperature, there is actually a

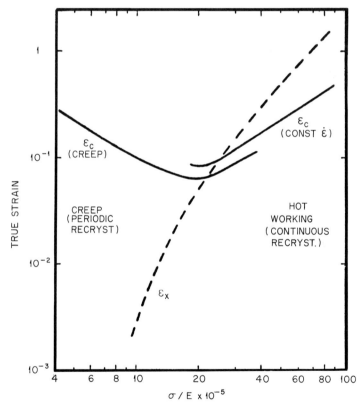

Fig. 33. Stress dependence of ε_c, the critical strain for the onset of recrystallization, and of ε_x, the strain occurring during the time for a large fraction of the volume to recrystallize. The strain scale is that for pure Ni at 935° C (0.7 T_m). At low stresses, where ε_c is greater than ε_x, the flow curve is periodic. However, at high stresses, where ε_c is less than ε_x, dynamic recrystallization is continuous (Sellars, 1971).

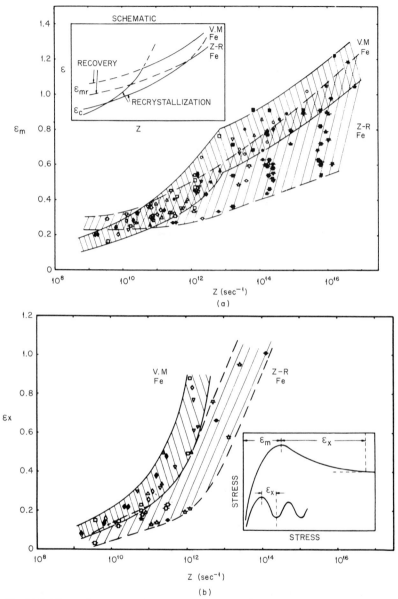

Fig. 34. (a) Dependence on Z of ε_m, the strain to the initial peak in flow stress or to the onset of steady state flow when no initial peak is observed. The insert shows a schematic interpretation of the data in terms of ε_{mr}, the strain to the onset of steady state flow if recovery is the only softening process, and of ε_c, the critical strain for the onset of dynamic recrystallization. It is seen that dynamic recrystallization is favored by low Z, i.e. low $\dot{\varepsilon}$ and high T. The materials here and in (b) are vacuum melted and zone refined iron for which Q is 281 kJ/mol. (b) Dependence on Z of the strain ε_x for the completion of recrystallization. The manner of measuring ε_x is indicated in the insert (Glover and Sellars, 1973).

decrease in the time required for recrystallization to begin, although the decrease is insufficient to offset the increase in strain rate, leading to the net increase in critical strain (Sah *et al.*, 1969).

The strain for completion of recrystallization ε_r also increases with temperature-corrected strain rate Z for reasons similar to those described for ε_c above (Fig. 34b). In addition, there is a reduction in driving force with increased strain rate resulting from deformation behind the advancing boundary. This reduction is more marked at higher strain rates than at low, so that the rate of increase of ε_r with strain rate is much greater than that of ε_c.

PERIODIC AND CONTINUOUS RECRYSTALLIZATION

In the first detailed paper on dynamic recrystallization, Luton and Sellars (1969) showed that the different dependences of ε_c and ε_r on strain rate and temperature described above lead to the periodic cycles of recrystallization at low strain rates, and to the overlapping and continuous cycles of recrystallization at high strain rates illustrated in Figs. 29a and 30a.

The periodicity arises because, at low strain rates, ε_r is much less than ε_c (Fig. 35). Thus, once recrystallization starts, it is completed long before the regions which first recrystallized can work harden and nucleate a second

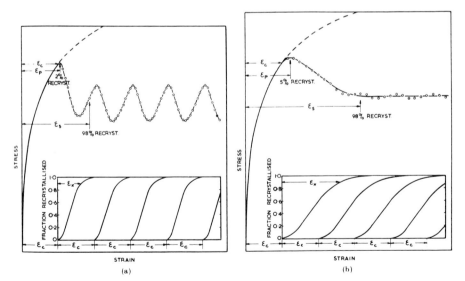

Fig. 35. Predicted stress–strain curves for dynamic recrystallization. (a) A cyclic stress–strain curve when the critical strain to initiate recrystallization $\varepsilon_c > \varepsilon_x$, the strain occurring in the time for a large fraction of recrystallization. (b) A steady state curve for the condition when $\varepsilon_c < \varepsilon_x$. The strain for completion of the first cycle ε_s equals the sum, $\varepsilon_c + \varepsilon_x$ (Luton and Sellars, 1969).

time. At high strain rates, ε_r is much greater than ε_c, so that before recrystallization is complete, the regions which first recrystallized reach the critical strain for a second nucleation (Fig. 35). In this way, more than one cycle of recrystallization may be taking place in the metal at the same time, each cycle of which is at a different stage of the recrystallization process. The result is an equilibrium distribution of regions having different strains between zero and ε_c, giving rise to a constant average flow stress.

The size of the dynamically recrystallized grains decreases as the temperature-corrected strain rate Z increases, because the associated increase in subgrain density gives rise to a higher density of nuclei (Fig. 36).

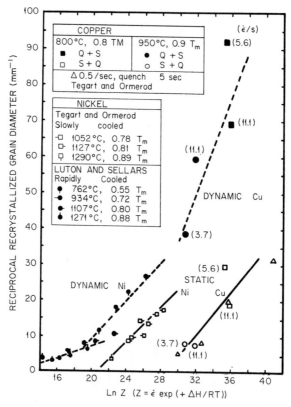

Fig. 36. Recrystallized grain sizes for Cu and Ni are larger for higher temperatures and lower rates of deformation (McQueen and Bergerson, 1972; W. J. McG. Tegart and H. Ormerod, unpublished work, University of Sheffield, England, 1971; Luton and Sellars, 1969). The statically recrystallized grains, which form during cooling, are larger than the dynamically recrystallized grains, even though cooling occurs within seconds. In the calculation of Z, the values of activation energy Q used for Cu and Ni were 302 and 234 kJ/mole.

There are also more cycles of recrystallization present simultaneously at higher values of Z. It follows from this that there is also a relationship of grain size to steady state flow stress which will be presented in the next section.

E. Relationship of Flow Stress to Structure

The grain structure of dynamically recrystallized material affects the high temperature flow stress in two distinct ways. In the first place, the *volume fraction* that has recently recrystallized determines the degree to which the flow stress approaches that of statically recrystallized material on the one hand or that of fully work-hardened material on the other. Secondly, the grain *size* itself can influence the high temperature flow stress when the grain boundaries act so as to impede the propagation of slip throughout the material.

The relationship between volume fraction recrystallized per increment of strain and the flow stress was first analyzed in detail by Luton and Sellars (1969). They showed that the volume recrystallized during each strain increment after the critical can be calculated from a knowledge of ε_c and ε_r (Fig. 35). Thus at any strain, the distribution of progressively deformed volumes and hence of progressively increased strengths can be summed to give the average flow stress. This assumes of course that all regions are undergoing the same strain, which is not unreasonable since they are quite small and are uniformly dispersed.

By this technique of calculation, it is possible to simulate, for low strain rates, cyclic stress–strain curves with the correct period, amplitude, and shape. At high strain rates, by taking into consideration the presence of more than one cycle of recrystallization, it is possible to derive a stress–strain curve with the same work-softening characteristics and approximately the same steady state flow stress as observed experimentally. Furthermore, according to the analysis, the calculated curves change from the cyclic to the steady state shape as ε_r becomes greater than ε_c at approximately the value of Z at which the experimental curves change character (Sah *et al.*, 1969; Glover and Sellars, 1973).

It should be noted that the steady state flow stress which results from dynamic recrystallization is considerably less than that which would result if only dynamic recovery operated in these metals. This reduction in flow stress can only be exploited industrially if the strain exceeds the peak strain for the conditions of deformation in effect; this can be fairly high at industrial forming rates. Nevertheless, dynamic recrystallization generally takes place in the low recovery metals, as long as the deformation is continuous, as in extrusion, or if it accumulates, as in a hot strip mill when there is no static recrystallization between stands.

  With regard to the effect of grain size on the high temperature flow stress, the situation is not quite as clear. As was stated earlier, the dynamically recrystallized grain size does not vary with strain during steady state deformation. Under most conditions, a relationship between steady state flow stress σ_s and grain size d_g of the following form is observed (Fig. 37).

$$\sigma_s = M d_g^{-q} \tag{14}$$

Here M is an empirical constant and q has been observed to be 0.71 in zone refined α-iron (Glover and Sellars, 1973), 0.75 in Ni and Ni–Fe alloys (Luton

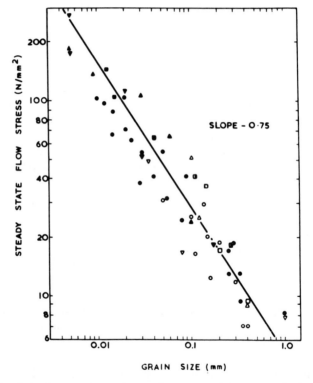

Fig. 37. Correlation between the steady state flow stress and the recrystallized grain size for the copper–aluminum alloys of Fig. 40. The data for all alloys fall on the same curve. This indicates that the pure metal and the alloy have the same grain size when the steady state flow stresses are the same; for the latter to be in effect, the alloys must be deformed at a higher temperature or a lower strain rate. In a similar way, when the pure metal and the alloy are deformed at the same temperature and rate, the alloy has a higher stress and a finer grain size (Bromley and Sellars, 1973).

and Sellars, 1969), and 0.75, 1.0, and 1.0 in Cu–Al alloys, Cu, and Ni, respectively (Tegart, 1968).

Despite the form of Eq. (14), however, the flow stress σ_s is unlikely to be causally related to the grain size d_g; rather d_g is probably linked to the magnitude of the induced stress σ_s. This arises because the subgrain size d formed by the dynamic recovery processes that are taking place (see Sections III,C and III,E) is causally related to the strain rate and temperature, and therefore to the flow stress σ_s and the amount of work hardening σ_i by Eqs. (8) and (9):

$$\sigma_s = K_1 d^{-1} \tag{8}$$

$$\sigma_i = K_2 d^{-1} \tag{9}$$

Now the dynamically recrystallized grain size will in general be related to the nucleus density and therefore to the volume density of subgrains present within the material. Under these conditions, the recrystallized grain size d_g is approximately proportional to the subgrain size d (Luton and Sellars, 1969), so that σ_s can be correlated with d_g as well as with d; i.e.

$$\sigma_s \simeq K_3 d_g^{-1} \tag{15}$$

$$\sigma_i \simeq K_4 d_g^{-1} \tag{16}$$

It should be noted that Eqs. (15) and (16) are *correlations* rather than causal relationships in view of the additional hardening contributions of both the subboundaries and the dislocations contained within the subgrains. This point can be seen more clearly if the steady state flow stress is written as

$$\sigma_s = \sigma^* + \alpha\mu b\sqrt{\rho_s} + K_2 d^{-1} + K_5 d_g^{-1/2} \tag{17}$$

Here σ^* is the effective stress, Eqs. (4) and (6), the term $\alpha\mu b\sqrt{\rho_s}$ represents the hardening contribution of the dislocations, Eq. (12), $K_2 d^{-1}$ that of the subboundaries, Eq. (9), and $K_5 d_g^{-1/2}$ the Hall–Petch strengthening contribution of the grain boundaries.† Bearing in mind that $\sqrt{\rho_s} \propto 1/d$ (Section III,E), we can write

$$\sigma_s = \sigma^* + K_6 d^{-1} + K_5 d_g^{-1/2} \tag{18}$$

and, given the dependence referred to above that $d_g \propto d$, σ_s is given by

$$\sigma_s \simeq \sigma^* + K_7 d_g^{-1} + K_5 d_g^{-1/2} \tag{19}$$

† The validity of the Hall–Petch relation at high temperatures has not been conclusively established. However, aside from the fine grain sizes associated with superplastic behavior and with the appreciable weakening attributable to grain boundary sliding, there is clear evidence that, for grain sizes above about 20 μ, the high temperature yield strength increases with decreasing grain size (Fink *et al.*, 1955; P. J. Wray, private communication, 1974). Although the exact form of the dependence was not determined in these investigations, in copper at 600° C, the strengthening component has been reported to follow the Hall–Petch relation (Fig. 38), and this is the view that is adopted here.

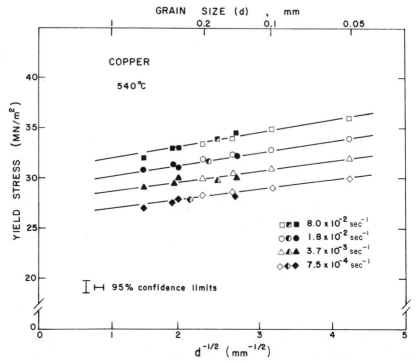

Fig. 38. Effect of grain size on the high temperature yield stress of tough pitch copper. Testing was done in compression and the different grain sizes were produced by prestraining to a constant strain of 0.4 at various strain rates and temperatures. The yield stress can be seen to be rate sensitive; the slopes of the fitted lines increase perceptibly with strain rate, indicating a slight dependence of the Hall–Petch boundary strength coefficient on strain rate (Petkovic, 1975).

The experimental values of q in Eq. (14), which are closer to 1 than to $\frac{1}{2}$, thus indicate that the component of substructure strengthening ($K_7 d_g^{-1}$) is generally greater than the Hall–Petch contribution of the grain boundaries ($K_5 d_g^{-1/2}$), and that Eq. (14) is therefore more likely to reflect the link between flow stress and nucleus density than that between grain size and resistance to flow.

F. Characteristics of Different Metals and Alloys

The metals which exhibit dynamic recrystallization are generally those that undergo a relatively low degree of dynamic recovery during high temperature deformation. This is because separation of the partial dislocations make climb, crossglide, and node detachment difficult. The low rate of dynamic recovery gives rise in turn to a dense dislocation substructure which promotes nucleation. This effect is found in fcc metals of low and moderate stacking fault energy, notably nickel (Shapiro and Dieter, 1971), cobalt (Jacquerie and Habraken, 1968), copper, and austenitic Fe (Keane *et al.*, 1968).

The capacity for dynamic recrystallization also depends on the ease of migration of the grain boundaries, which increases as the purity of the metal is improved. Thus dynamic recrystallization has been observed in vacuum melted and zone refined α-iron, but not in iron of standard purity or in steels (Glover and Sellars, 1973). Even in the former materials, dynamic recrystallization occurs only when the temperature-corrected strain rate remains below a certain value, which is higher for the more purified metal.

Effect of Alloying

In solid solution alloys, the addition of solute tends to diminish the ability of the metal to recover, thus increasing the tendency to dynamic recrystallization; however, the solute may also hinder the migration of boundaries, slowing the rate of dynamic recrystallization. Some typical flow curves in copper alloys, which soften by dynamic recrystallization, are shown in Fig.39. It can be seen that the rate of work hardening diminishes with strain, through the operation of dynamic recovery processes, until strains of 0.15 to 0.20 are reached. In relatively pure copper, when recrystallization is initiated, the flow stress drops rapidly and then levels out after a further strain of about 0.05. In copper containing 0.05% oxygen, numerous particles of Cu_2O are present, which impede the motion of grain boundaries, with the result that the softening associated with dynamic recrystallization is retarded, and an increased strain of 0.3 is required to attain stable flow. An even greater difference can be discerned in the flow curve for the Cu–Ni alloy. Here the addition of 9.5% Ni has considerably diminished the rate of dynamic recovery, and the rate of work hardening is higher than in the other two materials. The addition of Ni has also affected the processes of nucleus formation and of grain boundary migration so that, by the time the end of the test and a strain of 0.7 have been attained, softening by dynamic recrystallization has still not become fully developed.

Dynamic recrystallization has been observed in brass (Sunter and Burman, 1972), monel (Rhines and Wray, 1961), nickel-base superalloys (Fulop and McQueen, 1972), and austenites of plain carbon, low alloy, stainless, and tool steels (Rossard, 1960). In these metals, the general effect of alloying is an augmentation of strength and a retardation of dynamic recrystallization so that, at temperatures low in the hot-working range, it does not occur, giving rise to a severe ductility minimum. On the other hand, in alloys of copper in which the aluminum content varies between 1 and 8%, there is

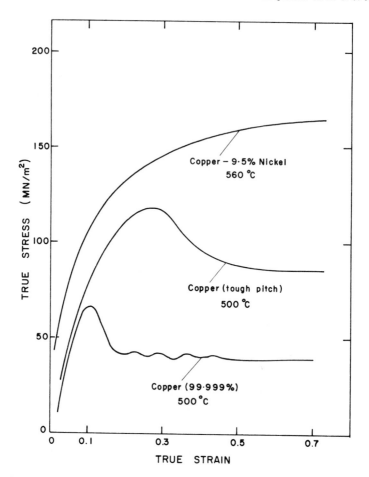

Fig. 39. Compression flow curves in three cuprous materials. Strain rate 1.8×10^{-2} sec^{-1}. In pure copper at 500° C, dynamic recrystallization begins after a strain of about 0.1 and is complete in strains of about 0.05 or less. In tough pitch copper (0.05% O), the peak strain doubles to about 0.2 and the recrystallization strain increases even more sharply to about 0.3. This can be attributed to the effect of a dispersion of Cu_2O particles. In the Cu–Ni alloy, the net rate of work hardening is increased appreciably by the addition of Ni, and the critical strain for recrystallization is greater than the strain limit imposed by the testing method (Petkovic, 1975).

little reduction in ductility in the range 0.55–0.9 T_m. While the steady state flow stress does increase with aluminum content up to 4% Al, it is almost unchanged if the concentration is increased further to 8% Al (Fig. 40). In these alloys, the presence of solute appears to have little effect on the relationship between the high temperature flow stress and the dynamically recrystallized grain size (Fig. 37). However, information is lacking in regard to

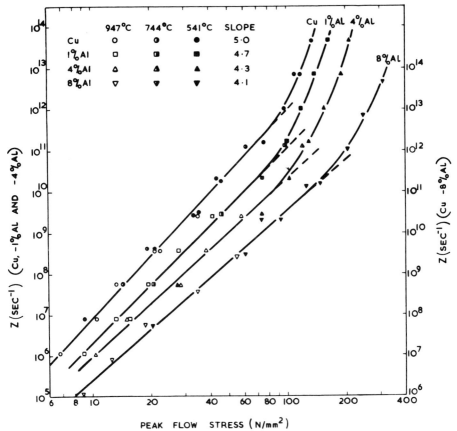

Fig. 40. The peak stress in Cu alloys increases with increase in strain rate and decrease in temperature. The peak stress also increases with the concentration of aluminum, which is known to lower the stacking fault energy at room temperature. However, the peak stresses for the 4% and 8% Al alloys are almost the same (note difference in vertical scales). The same activation energy ($Q = 200$ kJ/mole) was used for all the alloys (Bromley and Sellars, 1973).

differences in substructure in the pure metal and the alloy at the same grain size.

In Ni–Fe alloys, as the iron concentration is increased to 20%, the stresses at the peak and in the steady state region gradually increase (Fig. 41). Although the absolute magnitude of the solute strengthening component decreases with increasing temperature, the degree of strengthening relative to unalloyed material remains approximately constant. The activation energy of deformation increases with iron content from 234 kJ/mole for pure Ni to 298 for 5% Fe, 330 for 10% Fe, and 393 for 20% Fe. In Eq. (14), the constant M decreases from 1.45 to 0.82 ksi/mm$^{3/4}$ for 20% Fe, indicating

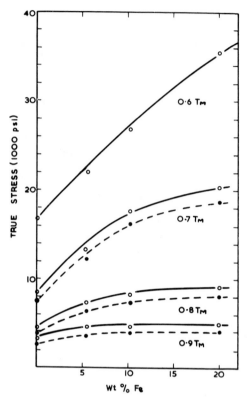

Fig. 41. The influence of the addition of Fe to Ni on the peak flow stress (open points) and steady state flow stress (filled points) at constant strain rate, $\dot{\varepsilon} = 7.7 \times 10^{-2}$ sec^{-1} (Luton and Sellars, 1969).

that, for a given flow stress, the grain size of the alloy is lower as the Fe concentration is increased, because the dissolved iron increases the probability of nucleation and decreases the rate of boundary migration. The lack of increase in flow stress with decrease in grain size supports the view described earlier that the internal stress component of the flow stress is associated mainly with the dislocation substructure.

In Fe–C alloys, an increase in carbon concentration and associated alloying elements raises the activation energy from 322 to 460 kJ/mole (Rossard, 1960). The flow stress, however, can increase or decrease somewhat with the concentration of additional elements, perhaps for reasons similar to those mentioned above (Section III,G) with respect to the effect of silicon addition to α-iron. The flow stress for Fe–18% Cr–8% Ni has been reported to be 50% higher than for a plain carbon steel and again the activation energy is higher in the alloyed material (405 kJ/mole). The addition of Cr, Mo, or W increases the flow stress still further (Gueussier and Castro, 1958).

The presence of dispersed second phase particles tends to stabilize the substructure and prevent or restrain grain boundary migration. Such inhibition or retardation of dynamic recrystallization has been observed in austenite containing NbC (Le Bon *et al.*, 1973) and in Udimet 700 with γ' precipitate (Oblak and Owczarski, 1972). The coalescence of second phases during high temperature deformation has already been described in Section III,H above; at present there is insufficient evidence to say whether or not dynamic recrystallization occurs in alloys which exhibit this behavior and in which the matrix recovers poorly.

G. Dynamic Recrystallization and Ductility

In constant strain rate tests, the metals which undergo limited dynamic recovery have a minimum in ductility in a temperature zone near the lower limit of the hot-working range. This effect has been observed principally in

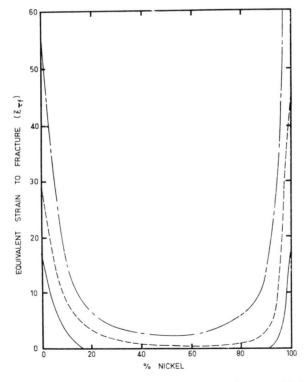

Fig. 42. The effect of solid solution alloying on the ductility of copper–nickel alloys. It can be seen that increasing the temperature greatly improves the ductility in dilute alloys, but only marginally improves the concentrated alloys, in which dynamic recrystallization does not take place before fracture (Sellars and Tegart, 1966b). ——, 0.6 T_m; – – – –, 0.7 T_m; — – —, 0.8 T_m.

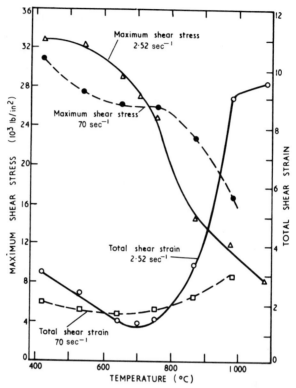

Fig. 43. The effect of temperature and strain rate on the ductility and peak stress for Inconel 600. The minimum in ductility results from the widespread occurrence of grain boundary cracking; the rapid rise in ductility at higher temperatures for a strain rate of 2.52 sec^{-1} is the result of dynamic recrystallization. At a strain rate of 70 sec^{-1}, dynamic recrystallization is delayed to greater critical strains and thus does not have an opportunity to improve the ductility in the temperature range of the tests (Dieter *et al.*, 1968).

nickel and its alloys (Figs. 42 and 43). The cause of the phenomenon has been clarified from an examination of the mode of fracture.

At temperatures below the decline, the fracture mode is similar to that at room temperature and the ductility is somewhat higher. As the test temperature is increased to that of the minimum, more and more grain boundary cracking is observed and the fracture becomes entirely intercrystalline (White and Rossard, 1968). This arises because metals which undergo little dynamic recovery have high yield stresses and high work-hardening rates. The resultant high flow stresses at finite strains prevent the accommodation, by lattice deformation, of the stress concentrations developed at triple points by grain boundary sliding. As the temperature is increased beyond that of the minimum, more and more dynamic recrystallization is observed (Bailey,

1969). The introduction of new grains isolates the cracks already formed from the grain boundaries, thus inhibiting their propagation (Fig. 44). Thus, in hot working, the initiation of dynamic recrystallization leads to a considerable increase in ductility. It is also valuable to point out that in multipass operations, static recrystallization between passes has a similar effect in increasing ductility (White and Rossard, 1968).

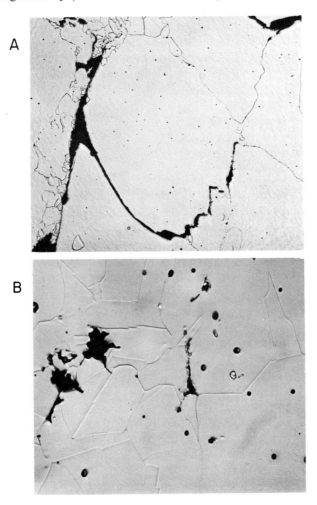

Fig. 44. Grain boundary cracking in Ni–5% Fe deformed at a strain rate of 2×10^{-3} sec^{-1}. (a) At 760° C (0.6 T_m), which is near the ductility minimum, the cracks have propagated extensively by the time a strain of 0.35 is reached. × 100. (b) At 934° C (0.7 T_m), dynamic recrystallization (as evidenced by the finer grain size) has prevented propagation even at a strain of 1.6. × 500 (Luton and Tegart, 1969).

As the strain rate is increased, the minimum in ductility tends to move to higher deformation temperatures (Rhines and Wray, 1961). However, the increase in strain rate also reduces the fraction of the strain attributable to grain boundary sliding relative to the overall lattice strain, so that the minimum becomes less noticeable (Fig. 43).

The metals which exhibit a ductility minimum in hot working usually exhibit severe grain boundary cracking in creep. Dynamic recrystallization, if it occurred, would increase their ductility, but the improvement would not be as great as at hot-working strain rates. In part this is because, at creep strain rates, where dynamic recrystallization tends to occur in distinct cycles, there is considerable opportunity for the cracks to grow in the part of the cycle in which recrystallization is not taking place. To some extent, the improvement in ductility that can be brought about by the operation of dynamic recrystallization is only of academic interest in creep service because the desirable increase in ductility would be outweighed by the undesirable increase in the average creep rate.

H. Effect of Interruption or of Continuous Variation in the Conditions of Deformation

As discussed in Section III,J, many industrial operations consists of a series of passes with intervening delays. Furthermore, in the course of any single forming operation, there is usually a continuous change in strain rate and temperature (Figs. 26, 27, and 28). While Section III,J dealt with the structures produced by prior dynamic or static recovery, this section will discuss the effect arising from prior dynamic or static recrystallization. To be of interest, the latter must not be complete, for it is only in the partially recrystallized condition that the effect of further deformation differs from that described so far in this section. A more complete description of static recovery and recrystallization following dynamic recrystallization will be presented in Section V.

Since a dynamically recrystallizing material contains a substructure, interruption of a test or operation generally leads to partial static recrystallization. On interruption, the most severely strained regions recrystallize first, so that the flow stress in a subsequent limited strain cycle is less than for steady state flow (Bromley and Sellars, 1973). As some worked grains are retained, which recrystallize with little strain upon resumption of deformation, the strain to the peak stress, as well as the height of the peak, is reduced. If complete recrystallization takes place during an interruption, then the flow curve in the next strain cycle resembles that of the previous cycle.

In tests at continuously increasing strain rates or decreasing temperatures, the flow stress is lower than that observed in isothermal, isostrain-rate tests.

This is similar to the situation described earlier under conditions of dynamic recovery, and for essentially the same reasons. The dynamically recrystallizing material contains a substructure characteristic of the temperature and strain rate of deformation. If the former is augmented and the latter reduced, further deformation is imposed on an inherited substructure which is softer than the equilibrium structure for the new conditions of deformation (Fig. 45).

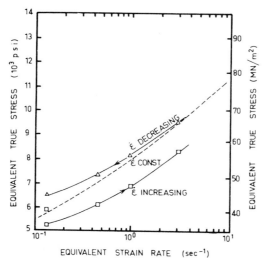

Fig. 45. The flow stresses measured during two series of deformations, one with decreasing strain rate and the other with increasing strain rate, are compared to the flow stresses measured in individual isostrain rate tests. The points plotted represent the flow stress attained after a strain of 0.9 in each stage. Since the specimens are maintained at 750° C throughout all experiments, the delays of 10 sec between stages permitted some static recovery and recrystallization to occur. The flow stress upon inheritance of a softer microstructure is less than the normal steady state stress, which in turn is less than that upon inheritance of a stronger structure (Bromley, 1969).

V. Static Recovery and Recrystallization after Hot Working

A. Static Recovery

In Sections III and IV above, we were primarily concerned with the softening processes occurring *during* high temperature deformation, i.e. with dynamic recovery and recrystallization. We turn our attention now to the static softening processes, i.e. those taking place *after* deformation is complete or between intervals of hot working. As before, these processes can be conveniently divided into two distinct categories: (a) the recovery processes, which involve the annihilation of dislocations in individual events, and (b) the recrystallization processes, in which dislocations are simultaneously eliminated in large numbers as a result of the motion of high angle boundaries. Each of these topics can be further subdivided according to whether the prior dynamic softening occurred solely by recovery or whether dynamic recrystallization was involved as well. This further distinction is particularly relevant to our discussion of recrystallization below, which will therefore be separated into two parts.

1. THE CRITICAL STRAIN FOR STATIC RECRYSTALLIZATION

The critical strain for static recrystallization is generally of the order of 10%. If the critical strain is exceeded, an incubation period must elapse before recrystallization proceeds. The length of this incubation period depends on temperature and on the amount and nature of the prior strain, as will be described in more detail below. Of interest in the present context is that, if the critical strain is exceeded, softening by static recovery can only take place during the incubation period prior to recrystallization. Furthermore, if the critical strain for recrystallization is not reached, the recovery process does not lead to full softening, but saturates at a softening level well below 100%; i.e. *full softening cannot be produced by recovery processes alone.*

The existence of a critical strain for full softening, and of a saturation limit to the effect of recovery, are readily revealed by the technique of interrupted compression testing. According to this method, individual samples are deformed at a constant strain rate, and then, at a selected strain, they are unloaded and held for increasing delay times before they are reloaded at the original strain rate. Examples of such interrupted tests are given in Fig. 46. In this type of work, the fractional softening that takes place can be evaluated in terms of the quotient:

$$\frac{\text{flow stress on interruption—reloading yield stress}}{\text{flow stress on interruption—original yield stress}}$$

Four such tests, executed on samples of low carbon steel, are shown in Fig.47. Here it can be seen that at 746° C, which is in the $\alpha + \gamma$ region, strains of 2.5% and 9% do not lead to recrystallization within 10^4 sec (Fig. 47b). When accumulated strains of 15% (10% + 5%) and 30% (15% + 15%) are imposed at a temperature of 677° C, on the other hand, recrystallization begins in less than 100 sec and is virtually complete at the end of the test (Fig. 47a).

The existence of a critical strain is shown more clearly in Fig. 48. In these experiments, conducted on a 0.68% C plain carbon steel, a prestrain of 5.5%

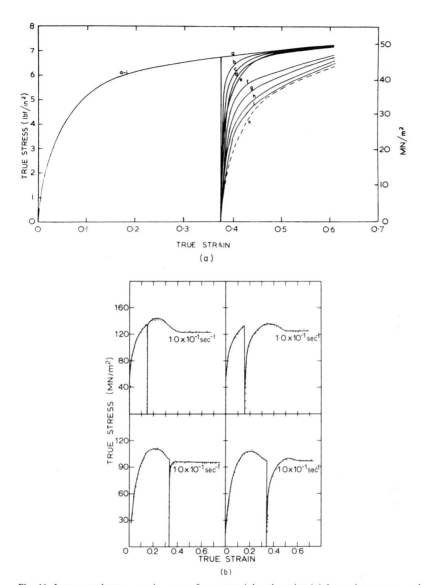

Fig. 46. Interrupted stress–strain curves for a material undergoing (a) dynamic recovery and (b) dynamic recrystallization. (a) Commercial purity aluminum tested at 410° C (0.74 T_m) and 2.7 sec^{-1}. The delay time on interruption ranged from 1.8 to 4100 sec. For comparison, the broken curve k is a trace of the initial loading curve (Evans and Dunstan, 1971). Curve and delay time (sec): b, 1.8; c, 2.5; d, 3.95; e, 8.5; f, 405; g, 490; h, 1314; i, 4100. a, Single deformation test; k, initial portion of curve a. (b) The upper curves, which are for a 0.42% C steel, have an interruption strain of 0.14 in the dynamic recovery region. Both tests were conducted at 870° C (0.65 T_m) and the delay time was 0.5 sec for the left-hand curve and 17 sec for the right-hand curve. The lower curves, for a 0.68% C steel deformed at 915° C (0.78 T_m), were strained to 0.33 in the dynamic recrystallization region. The test on the left was for a delay time of 0.5 sec, and that on the right for a delay time of 10.5 sec. All samples were austenitized at 960° C (Djaic and Jonas, 1972).

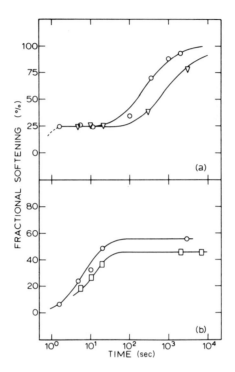

Fig. 47. Effect of temperature and interruption strain on the rate of softening of a 0.14% C steel. Tests in (a) were conducted at 677° C (0.54 T_m) on α-phase material. $\bigcirc$, $\varepsilon = 0.15 + 0.15$, $\dot{\varepsilon} = 0.1$ sec^{-1}; ∇, $\varepsilon = 0.10 + 0.05$, $\dot{\varepsilon} = 0.05$ sec^{-1}. (b) Tests were conducted at 746° C (0.57 T_m) on a mixture of $\alpha + \gamma$ phases. $\square$, $\varepsilon = 0.025$, $\dot{\varepsilon} = 0.03$ sec^{-1}; $\bigcirc$, $\varepsilon = 0.09$, $\dot{\varepsilon} = 0.075$ sec^{-1}. The upper curves exhibit recrystallization after an incubation period, whereas the lower curves are for strains which have not exceeded the critical (Djaic and Jonas, 1972).

(curve a) was followed by a recovery cycle lasting some 100 sec. As softening did not go to completion, it can be concluded that the critical strain exceeded 5.5%. From curve b it can be seen that, after an interruption strain of 9.8%, softening by recovery was complete after about 100 sec; after a further delay of about 6000 sec, recrystallization began and proceeded to completion in a further 10^6 sec. The critical strain for this material, at 780° C, is thus between 5.5 and 9.8%.

The softening curves of Fig. 48 reveal that, for the particular steel chosen, the maximum relative softening that can be produced by static recovery alone is about 40%. This compares with softening levels of about 40%, 50%, and 50% produced by recovery alone in aluminum (Evans and Dunstan, 1971), tough pitch copper (Petkovic, 1975), and in niobium-modified steels (Petkovic et al., 1975).

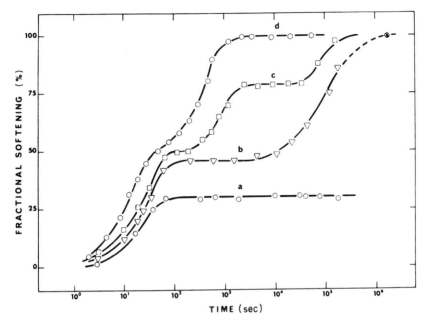

Fig. 48. Effect of strain on the static softening of a 0.68% C steel at 780° C (0.58 T_m). $\dot{\varepsilon} = 1.3 \times 10^{-3}$ sec^{-1}. The following interruption strains were used. Curve a, 0.055; curve b, 0.098; curve c, 0.24; curve d, 0.41. The final point in curve b was obtained by heating the sample to 915° C during the interruption. The first plateau in each curve indicates the completion of static recovery. The second plateau in c indicates the termination of metadynamic recrystallization (Djaic and Jonas, 1973).

2. THE EFFECT OF TEMPERATURE, STRAIN RATE, AND STRAIN ON THE RATE OF RECOVERY

The main experimental variables affecting the recovery rate are the temperature, strain, strain rate, and composition. The influence of temperature is illustrated in Fig. 49a for a medium-carbon steel. The recovery curves all represent softening cycles occurring after a prestrain of 14%; i.e. following interruption strains below the critical. The effect of temperature on the recovery rate is evident, but is not particularly marked, in part because the amount of stored energy driving the recovery process decreases as the deformation temperature is increased. It is less than the effect of temperature on the recrystallization rate, as will be shown in more detail below. It is also apparent that the rate of recovery decreases with time, indicating that the stored energy or driving force is progressively reduced by the operation of the recovery process. Because the driving force for recovery is generally different at different deformation temperatures, it is difficult to determine

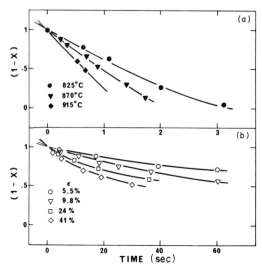

Fig. 49. (a) Effect of temperature (0.61–0.67 T_m) on the rate of static recovery in a 0.42% C plain carbon steel. The results were obtained by interrupted compression testing after an interruption strain of 0.14. $\dot{\varepsilon} = 1 \times 10^{-1}$ sec^{-1}. X represents the fractional softening, i.e. relative reduction in yield stress with time, and $1 - X$ is the residual strain hardening. (b) Effect of interruption strain on the rate of static recovery in a 0.68% C plain carbon steel at $780°$ C (0.58 T_m). $\dot{\varepsilon} = 1.3 \times 10^{-3}$ sec^{-1}. The increases in softening rate at strains of 0.24 and 0.41 include a component of concurrent softening by metadynamic recrystallization (Petkovic et al., 1975).

the activation energy associated with static recovery after high temperature deformation.

In general, increases in strain lead to increases in the recovery rate until steady state flow is achieved. This can be attributed to the attendant increase in dislocation density, and therefore driving force, with strain until equilibrium is reached. The effect upon the recovery rate of increasing the driving force is shown in Fig. 49b. These curves were constructed from the recovery or initial restoration portions of the softening curves of Fig. 48. It can be seen that the recovery rate increases continuously with the amount of interruption strain, up to a prestrain of 41%. These prestrains were selected to be: curve a—below the critical strain; curve b—above the critical strain for static recrystallization, but below that for dynamic recrystallization; curve e—just beyond the peak of the flow curve, and therefore just beyond the critical strain for dynamic recrystallization; curve d—close to the steady state region, i.e. that of continuous dynamic recrystallization. It appears that when a strain cycle is interrupted in the dynamic recrystallization range, the grain boundaries continue to move, i.e. recrystallization continues, without requiring an incubation period. This special kind of softening has been

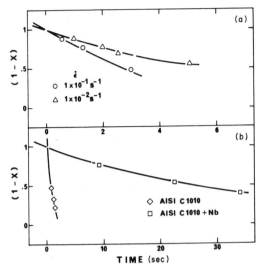

Fig. 50. (a) Effect of strain rate on the rate of static recovery in a 0.42% C plain carbon steel at 825° C (0.61 T_m), $\varepsilon = 14\%$. The increased rate of recovery that follows the higher testing strain rate is attributable to the higher dislocation density and therefore the higher stored energy that is produced under these conditions. $1 - X$ represents the residual strain hardening as determined by yield stress measurements made by interrupted compression testing. This steel and the samples referred to in Fig. 49 were austenitized at 960° C. (b) Effect of Nb addition on the rate of static recovery in a low carbon steel at 930° C (0.68 T_m), $\dot{\varepsilon} = 8 \times 10^{-2}$ sec^{-1}, $\varepsilon = 25\%$. The plain carbon steel was austenitized at 960° C and the Nb-modified steel at 1150° C. The very large decrease in recovery rate that follows on the addition of Nb appears to play a large part in the retardation of recrystallization in such steels. It is not clear whether the reduction in rate is due to the effect of Nb in solution or as a precipitate (Petkovic *et al.*, 1975).

termed metadynamic or post-dynamic recrystallization (Djaic and Jonas, 1972, 1973) and will be discussed again later. For the moment, it is sufficient to note that part of the increase in apparent recovery rate from curve c to curve d can be attributed to the simultaneous occurrence of metadynamic recrystallization.

A further effect of driving force is shown in Fig. 50a. Here two different strain rates, differing by an order of magnitude, were employed to produce the prestrain. As in the previous case, it can be seen that, at a fixed temperature, the recovery rate increases with stored energy, i.e. with the prior deformation rate.

3. EFFECT OF ALLOYING

It is evident from the discussion above that after high temperature deformation, static recovery begins as soon as the force is removed. That is, no delay period is required for softening to be initiated. In general, the factors that increase the rate of static recovery (e.g. increases in temperature, pre-strain, and strain rate), decrease the incubation time, i.e. the time available for recovery before recrystallization begins. However, when comparing different metals, an increase in stacking fault energy increases the rate of static recovery and can also increase the incubation time. The latter can be attributed to the decrease in worked dislocation density that is associated with stacking fault energy increase, or to changes in the nucleation mechanism (McQueen, 1968).

The addition of some solutes, as has already been noted above, decreases the stacking fault energy and therefore decreases the rate of dynamic recovery and increases the net work hardening rate. In a similar way, solute addition can decrease the rate of static recovery and concurrently increase the incubation time. From a practical point of view, because solute addition makes dynamic recovery more difficult, and therefore increases the flow stress at a fixed temperature, higher temperatures are frequently used to work alloys (and low stacking fault energy metals). It is the higher working temperature and the attendant diminished incubation time that makes the observation of static recovery rare or difficult in these materials.

An example of the effect of alloy addition is illustrated in Fig. 50b. Here we see that the presence of 0.07% Nb has an enormous effect on the rate of static recovery in plain carbon steel. It is not yet clear whether the retardation of recovery is due to a solute effect (i.e. to large decreases in stacking fault energy) or to the effect of the precipitation of NbC particles on the dislocation network. Whatever the cause, it is clear that the retardation of static *recrystallization* in carbon steels is attributable, in a large measure, to the retardation of *recovery* caused by niobium addition.

B. Static Recrystallization

The occurrence of static recovery, which certainly always takes place, is difficult to observe and to study because its progress is not readily followed by means of optical metallography. More dramatic, from the microstructural point of view, is the occurrence of static recrystallization, and as a result, it has been reported and described more frequently in the literature (Fig.51). A good example of the results obtained by metallographic means is displayed in Fig. 52. From Fig. 52a, the effect of temperature on the incubation time can clearly be seen. In these experiments, the interruption flow stress, and therefore the driving force, was maintained approximately constant by suitably varying the strain rate of each series of tests. Had the strain rate been held constant, the softening curves would have been closer together in time. The effect of temperature on the *rate* of recrystallization

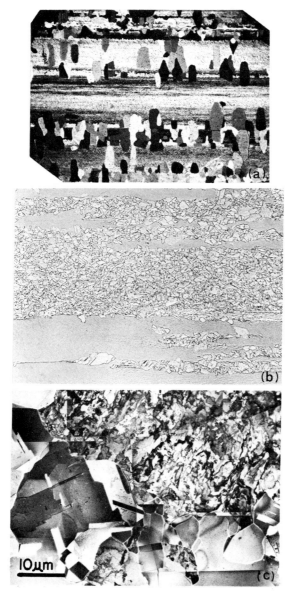

Fig. 51. Static recrystallization is usually clearly followed using optical microscopy by measuring the volume fraction of equiaxed grains and elongated deformed grains. (a) Aluminum extruded to a strain of 3.7 at 450° C (0.77 T_m) and slowly cooled. × 25 (Jonas *et al.*, 1968a). (b) Copper rolled to a strain of 2.3 at 600° C (0.65 T_m) and quenched in 3 sec. × 35. (c) Copper rolled as in (b). × 1300. Electron microscopy reveals that the statically recrystallized grains contain no dislocation substructure. This microstructure should be compared with that of the dynamically recrystallized metal in Fig. 32 (McQueen, 1968).

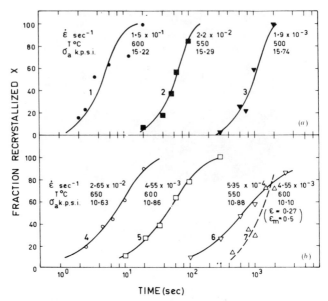

Fig. 52. Effect of temperature of deformation on the isothermal static recrystallization of zone refined iron. (a) The fraction recrystallized after dynamic recovery for specimens deformed to the same flow stress at three different combinations of temperature and strain rate. (b) The fraction metadynamically recrystallized after deformation to the same steady state flow stress at three different combinations of temperature and strain rate. The dashed curve is for static recrystallization after deformation to same stress level before the peak; it should be compared with the middle solid curve. The homologous temperatures of deformation are 0.51 (650° C), 0.49 (600° C), 0.46 (550° C), and 0.43 (500° C) (Glover and Sellars, 1972).

cannot be seen as readily because of the use of the log scale for time; it is, however, apparent that an increase in temperature of 50° C leads to a decrease in recrystallization time of about an order of magnitude and therefore to an approximate order of magnitude increase in recrystallization rate.

The effect of temperature at constant strain rate is evident from the results reproduced in Fig. 53. Here the interruption flow stress was not held constant and so the softening curves are closer together. In addition, it should be noted that Fig. 53 was determined on fcc γ-iron and that the samples were deformed in compression. By contrast, Fig. 52 represents results on bcc α-iron which were produced by torsion testing, a deformation mode leading to a different flow pattern, and a less homogeneous one than compression testing.

The accelerating effect on the recrystallization rate of raising the temperature while holding the strain rate constant has been reported by numerous workers (Sellars and Tegart, 1966b; Buhler et al., 1970) (see Fig. 36). In a

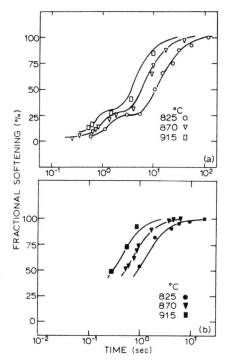

Fig. 53. The effect of temperature of deformation on the isothermal recrystallization rate for specimens deformed at the same strain rate. (a) The strain of interruption, 0.14, lies within the dynamic recovery region. (b) The strain of interruption, 0.35, lies within the dynamic recrystallization region (Djaic and Jonas, 1972).

modification of this type of test, Schey (1957) heated or cooled a series of specimens to a single holding temperature after deformation. He observed that the time for complete recrystallization *increased* with the deformation temperature (i.e. the recrystallization rate *decreased* with increasing temperature). The recrystallized grain size also increased with deformation temperature, confirming the strong influence of subgrain size and driving force on the processes of nucleation and growth.

1. EFFECT OF STRAIN RATE AND STRAIN ON RECRYSTALLIZATION

Increasing the strain rate of deformation decreases the incubation time and increases the rate of subsequent recrystallization for the reasons described above. An example is given in Fig. 54, for the case of a niobium-modified steel. It should be noted, however, that in materials like the niobium steels, in which particle precipitation takes place, decreasing the strain rate can permit more precipitation to take place during deformation. Thus

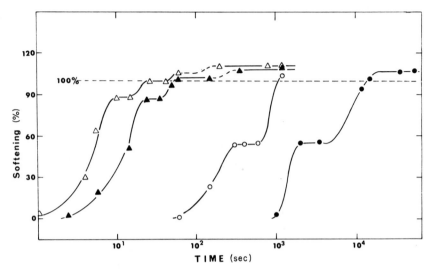

Fig. 54. The effect of strain rate and interruption strain on the isothermal recrystallization of a C1010 steel containing 0.07% Nb. The low strain interruption was in the dynamic recovery region, whereas the high strain interruption was in the dynamic recrystallization region. The softening exceeded 100% as a result of changes in the nature of the NbC precipitate. The samples were austenitized at 1150° C and tested at 1040° C (0.73 T_m) (Petkovic, 1975). Strain and strain rate (sec^{-1}): $\triangle$, 0.25, 8×10^{-2}; $\blacktriangle$, 0.25, 8×10^{-3}; $\bigcirc$, 0.1, 8×10^{-2}; $\bullet$, 0.1, 8×10^{-3}.

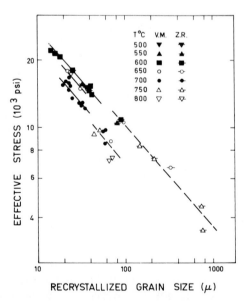

Fig. 55. Correlation between steady state flow stress and statically recrystallized grain size in vacuum melted and zone refined iron (Glover and Sellars, 1972). Solid lines, dynamic recovery; dashed lines, dynamic recrystallization.

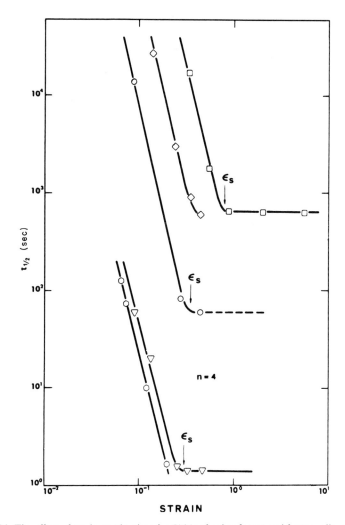

STRAIN

Fig. 56. The effect of strain on the time for 50% softening for several ferrous alloys in both the ferritic and austenitic conditions. The driving force for recrystallization increases with strain up to the steady state region in which the substructure remains dynamically stable. The symbols and references from top to bottom are: □, Vacuum melted iron at 650° C (Glover and Sellars, 1972); ◇, Fe–4% Si at 812° C (English and Backofen, 1964); ○, 0.68% C steel at 780° C (Djaic and Jonas, 1973); ▽, 0.68% C steel at 915° C (Djaic and Jonas, 1973); ○, 0.055% C, 0.46% Mn steel at 900° C (Morrison, 1972).

part of the retardation associated with the strain rate decrease may be attributed to the retarding effect of the increased particle density on the softening processes. Finer grains are produced by deformation at higher strain rates (and lower temperatures) because of the reduced subgrain sizes formed by dynamic recovery at lower values of temperature-corrected strain rate, Z. An example of this effect is given in Fig. 55, in which the correlation is plotted against stress instead of Z.

The influence of the amount of prior strain on the rate of recrystallization and on the recrystallized grain size has also been noted by numerous investigators. The work of English and Backofen (1964), for example, was carried out on bcc silicon iron, whereas that of Wusatowski (1966) and Buhler et al. (1970) was conducted on fcc austenitic stainless steel. Similar results were also obtained on plain carbon steels in the austenite range by Gueussier and Castro (1958) and by Weiss et al. (1972).

Some of the data reported recently by four groups of workers are collected and illustrated in Fig. 56. Here it can be seen that the time for 50% softening by recrystallization decreases by as much as two orders of magnitude with increasing strain, finally leveling out when the steady state regime of flow is attained. For deformations between the critical strains for static and dynamic recrystallization, the increase in rate is attributable directly to the increased driving force. At larger strains, however, part of the increase in rate is attributable to the displacement of classical static recrystallization, which requires an incubation time, by metadynamic or post-dynamic recrystallization, which does not (Djaic and Jonas, 1972).

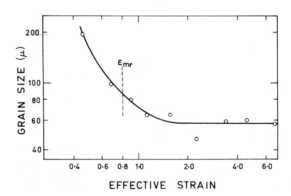

Fig. 57. Effect of strain upon the statically recrystallized grain size in vacuum melted iron after deformation at 650° C (0.53 T_m) and a strain rate of 3.8×10^{-2} sec^{-1}. Once the equilibrium dynamic recovery structure for steady state flow, ε_{mr}, is well established, the recrystallized grain size becomes independent of strain (Glover and Sellars, 1972).

The increase in recrystallization rate with strain is accompanied by a decrease in recrystallized grain size. Once again the grain size becomes independent of strain when steady state flow has been reached (Fig. 57). The preferred sites for nucleation of static recrystallization are heavily deformed regions near grain and twin boundaries.

2. EFFECT OF ALLOYING ELEMENTS ON RECRYSTALLIZATION

In many metals, the rate of recrystallization can be reduced by the addition of suitable solutes (e.g. Fe in Ni; Luton and Sellars, 1969). An increase in solute concentration can also lead to finer recrystallized grain sizes. There are several reasons why this trend is observed. One is the drag effect of impurity atoms on moving grain boundaries; another is the increase in the density of possible nucleation sites caused by impurity addition. An increase in solute concentration can also affect recrystallization indirectly through its influence on the preceding recovery processes.

The latter effect is illustrated in Fig. 58, in which two low carbon steels

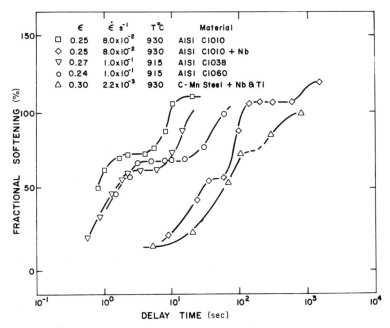

Fig. 58. Effect of Nb addition on the recovery and recrystallization rates of plain carbon steels at about 0.68 T_m. These results were obtained by interrupted compression testing and show that the addition of Nb decreases the softening rate of such steels by more than an order of magnitude in time (Petkovic, et al., 1975). The data for the final alloy (0.17% C, 1.25% Mn) in the diagram are from Weiss et al. (1972).

containing Nb are compared with three unmodified plain carbon steels. It can be seen that substantial softening by recrystallization takes at least an order of magnitude longer in the niobium steels than in the unmodified steels. However, it should be noted that the presence of niobium leads to a similar retardation in the rate of static recovery (see also Fig. 54), suggesting that the effect of solute addition on the recovery rate, and particularly on the nucleation rate or incubation time, is the rate limiting one. In solid solution alloys such as stainless steel (Sokolkov and Surkov, 1963; Irvine et al., 1969), the rate of static recrystallization is retarded to such a degree that it is possible to retain the hot-worked substructures by rapid cooling.

C. Metadynamic Recrystallization

It is not generally recognized that there are three distinct softening processes taking place after high temperature deformation. The two that we have already discussed are closely related to their counterparts occurring after room temperature deformation, and are therefore familiar processes. We turn now to a type of static recrystallization which does not have a parallel in the annealing of cold-worked materials, i.e. to metadynamic or post-dynamic recrystallization.

When high temperature deformation is halted during dynamic recrystallization, many nuclei are already present within the material, and some of the grain boundaries are migrating and sweeping regions free of dislocations. These boundaries can continue to migrate and the nuclei to grow without any classical incubation period being required. The softening process which results from the continued growth of these nuclei is known as metadynamic recrystallization. As this type of recrystallization does not require a nucleation interval, it proceeds very rapidly upon the termination of deformation. Nevertheless, nucleation for static recrystallization can still take place in regions which do not contain dynamic nuclei.

The interrelation between the three static softening processes depends on the prior or interruption strain and is depicted schematically in Fig. 59. Here we see that for deformations below the critical strain for static recrystallization, the only possible restoration process is static recovery. For strains which are larger, but still below the peak strain, ε_p, static recovery is followed, after a suitable interval, by static recrystallization. For strains greater than the peak but less than the steady state strain, there is a period of concurrent static recovery and metadynamic recrystallization, followed by a diminished cycle of static recrystallization similar to that prior to the peak, but with a shorter completion time (Fig. 48). Finally, once within the steady state range, static recrystallization is superceded entirely, and only metadynamic recrystallization takes place, with some concurrent static recovery.

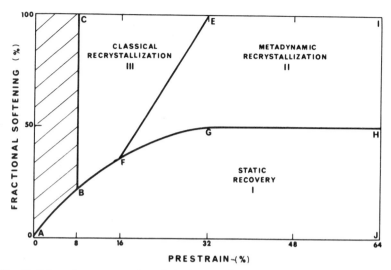

Fig. 59. Schematic representation of the interrelation between the three static softening mechanisms and of their dependence on strain for a material which dynamically recrystallizes. The proportions of softening attributable to each mechanism are also shown. The sequence of static restoration processes at each strain is read vertically upwards. B represents the critical strain for static recrystallization, F the peak strain, and G the strain for the onset of steady state flow (Djaic and Jonas, 1973).

Metadynamically recrystallized grains are shown in Fig. 31, where they may be compared with the dynamically recrystallized grains from which they formed during cooling. At these relatively high temperatures of deformation, the dynamic grains are smaller (Fig. 36) because the continuing deformation generates further nuclei. At relatively low temperatures, on the other hand, the metadynamic grains are finer because potential nuclei are rendered ineffective by continued deformation (Sandstrom and Lagneborg, 1975).

The data in Fig. 55 indicate that the relationship between the statically recrystallized grain size and the high temperature flow stress is little affected by the dynamic softening process that precedes recrystallization. This is somewhat surprising since the dynamically recovered substructure is uniform, whereas the substructure in the dynamically recrystallized material is highly nonuniform, containing regions of both high and low dislocation density. The explanation may lie in the opposing effects of the low density regions on the one hand, which should promote a coarse grain size, and the small size of the dynamically recrystallized grains on the other, which would tend to favor a fine grain size. Furthermore, as has been shown for the case of dynamic recovery (Fig. 56), metals which undergo dynamic recrystallization exhibit a decrease in the statically recrystallized grain size with increase in strain, but only until steady state flow is reached (Petkovic, 1975).

The practical importance of metadynamic recrystallization arises from the absence of any prior incubation time, with the result that it is about an order of magnitude more rapid than classical recrystallization. Allowances must therefore be made for the structural and mechanical changes associated with metadynamic recrystallization if dynamic recrystallization is taking place within a material prior to the cessation of deformation.

VI. The Effect of High Temperature Deformation on the Room Temperature Mechanical Properties

A. The Effect of Deformation Substructure on Mechanical Properties

We have been considering above the two main categories of dynamic softening process and the three types of static softening process. We turn now to the substructures produced by dynamic recovery (with or without concurrent dynamic recrystallization) and to the effect these can have on the mechanical properties of the materials in which they are found. In order to have an effect, the dynamic recovery substructures must of course be retained after deformation, which requires a suitable rate of cooling. The rate of cooling, in turn, must be high enough to prevent the occurrence of both classical and metadynamic recrystallization. From our earlier discussions of these mechanisms, it is therefore clear that the prospects for substructure strengthening are greater in the dynamic recovery than in the dynamic recrystallization metals.

It has been observed by numerous workers that, in the presence of a hot-work substructure, the room temperature yield strength, σ_{RT}, is given by a relationship of the form:

$$\sigma_{RT} = \sigma_A + N \, d^{-p} \tag{20}$$

Here σ_A is the yield strength of coarse grained material in the absence of subboundaries, N is a constant, and d is the mean subgrain size. The exponent p has been reported to be about 1 in Al (Hockett and McQueen, 1970), Armco iron, Fe–3% Si, Zr, and Zr–Sn alloys (Fig. 60). An equation of similar form, but with different values of p, has been reported for strength correlations carried out by means of hardness testing (e.g., Redfern and Sellars, 1968). It should be noted, however, that the conversion factor for hardness to yield strength is not a linear one (Abson and Jonas, 1970), so that it is more difficult to give a physical interpretation to hardness/subgrain size correlations. The relation between some of the subgrain sizes in Fig. 60 and the conditions of working required to produce them is illustrated in Fig. 61.

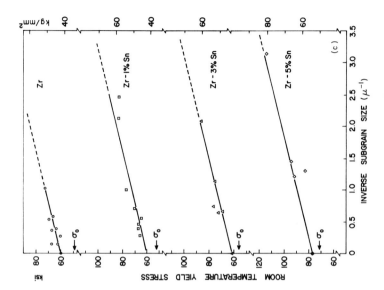

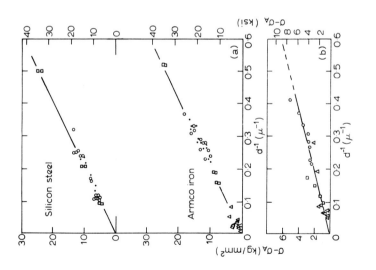

Fig. 60. The increase in yield stress attributable to the presence of a deformation substructure plotted against inverse subgrain size or subboundary density. The annealed yield strength of substructure free material is represented by σ_A. It can be seen that subgrains in the size range 1–10 μ lead to strength increases of about (a) 30,000 psi in Armco iron and silicon steel ($\bigcirc$, Uvira, 1967; $\triangle$, Warrington, 1963; $\square$, Kosik, 1970), (b) 7000 psi in commercial purity aluminum ($\square$, Cotner and Tegart, 1969; $\bigcirc$, McQueen et al., 1967; $\triangle$, after Ball, 1957) (Kosik et al., 1971), and (c) 20,000 psi in a series of Zr–Sn alloys (Abson and Jonas, 1972).

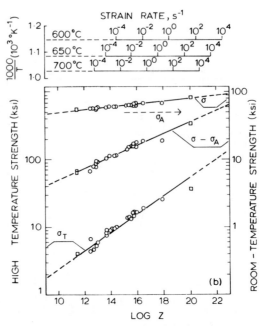

Fig. 61. The room temperature strength σ and strength increment $\sigma - \sigma_A$ and the high temperature steady state flow stress σ_T plotted against the logarithm of the temperature-compensated strain rate Z ($= \dot{\varepsilon} \exp Q/RT$) for the silicon steel of Fig. 60a. An activation energy of $Q = 335$ kJ/mole was used to prepare the diagram, which also indicates the combinations of working temperature and strain rate required to produce the substructure strengthening component $\sigma - \sigma_A$, $\bigcirc$, Uvira, 1967; $\square$, Kosik, 1970 (Kosik et al., 1971).

B. A Modified Hall–Petch Relation for Substructure Strengthening

Equation (20) above differs from the conventional Hall–Petch relation for grain boundaries:

$$\sigma_{RT} = \sigma_0 + k\, d_g^{-1/2} \qquad (21)$$

principally in that p is equal to 1 instead of $\frac{1}{2}$. Here σ_0 is the yield strength of a single crystal of similar purity and k is the grain boundary strength coefficient. The difference between the two equations is largely attributable to the decrease in subboundary strength and increase in subgrain size that are both associated with deformation at lower values of temperature-corrected strain rate.

The effect of the variation in subboundary strength with the conditions of deformation is shown in Fig. 62. For reference purposes, Eq. (21) for grains is plotted on the top left-hand side of Fig. 62b. Below this line are plotted

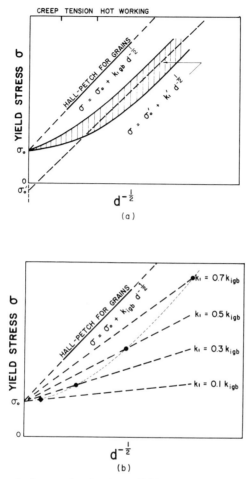

Fig. 62. (a) Schematic diagram showing the available room temperature data for Al and Fe containing subgrain boundaries. A negative intercept can be seen to result from the fitting of a straight line to part of the scatter band; the similarity in the values of k_1, the grain boundary strength coefficient, and k'_1, the apparent subboundary strength coefficient, can also be seen. (b) The data in (a) are represented as the locus of points with a varying true Hall–Petch slope k_1 and a constant friction stress σ_0. (After Abson and Jonas, 1970).

similar relations for subgrains having boundary strengths 10, 30, 50, and 70% as high as the grain boundary strength. Recalling now that subboundary thickness and strength increase as subgrain size decreases (Section III,C), it can be seen that the scatter band of strength data can be expected to pass from the region of low subboundary strength $k_{sb} \simeq 0.1\ k$, at large subgrain sizes, to that of high subboundary strength, $k_{sb} \simeq 0.7\ k$, at small subgrain sizes (Fig. 62a). The resultant curvature in the σ_{RT} vs $d^{-1/2}$ plot is equivalent to a straight line fit on a σ_{RT} vs d^{-1} plot.

The variation in subboundary strength coefficient k_{sb} with subgrain size d takes the form (Abson and Jonas, 1970);

$$k_{sb} = N\ d^{-m} \tag{22}$$

where $m \simeq \frac{1}{2}$, so that the Hall–Petch relation for subgrains can be expressed as

$$\sigma_{RT} = \sigma_0 + k_{sb}\ d^{-1/2} \tag{23}$$

$$= \sigma_0 + N\ d^{-(m+1/2)}$$

$$\simeq \sigma_A + N\ d^{-p} \tag{20}$$

C. *Industrial Significance of Substructure Strengthening*

The strengthening due to a hot-work substructure is of industrial importance in the production of nonheat-treatable aluminum and Al–Mg alloy sheet and extruded shapes. In the press heat treatment of alloys 6061, 6063, 7005, and 7039, the preheating and working serve as the solution treatment, and the quenched-in dislocations lead to substructure strengthening on the one hand, and to increases in precipitation strengthening on the other (Ashton, 1969). In steels, the partially recovered substructures produced during warm working close to the lower limit of the hot-working range are responsible for increases in the final strength of such forgings. This process is applied particularly to low carbon steels, and also to austenitic stainless steels and to superalloys (Reynolds and Tegart, 1962; Keegan, 1967). In ausforming, the substructure developed during the deformation of metastable austenite is retained during the martensite transformation to strengthen the tempered martensite by improving the dispersion of the temper carbides (Johari and Thomas, 1965; Schmatz et al., 1963).

Retained hot-work substructures can also impart improved creep resistance. For example, by quenching in the substructure, creep life can be extended by a factor of 5 in simple metals, stainless steels, and nickel-base superalloys. This must be followed by a low temperature anneal to permit precipitation on the dislocation network and its subsequent stabilization (Ivanova and Gordienko, 1968). After this treatment, the service temperature must be limited so as to prevent recrystallization, which causes the loss of the improved properties. In a similar way, Udimet 700 can be given a thermomechanical treatment by means of which working is carried out in steps, with the precipitation of γ' occurring both between and during the deformation.

This results in a strong stable substructure, which leads to much greater creep resistance at lower temperatures. If the temperature is raised to 760° C, however, the creep resistance is reduced to that of the normal heat treated material (Kear et al., 1972).

D. Dynamic Recrystallization and Substructure Strengthening

If a hot-worked material which is recrystallizing dynamically is rapidly quenched, it is possible to retain the dislocation substructure on cooling to room temperature. Furthermore, by preventing metadynamic recrystallization, the dynamic grain size is also preserved. This can be coarser or finer depending on the deformation and holding conditions. The room temperature hardness increases with increase in the temperature compensated strain

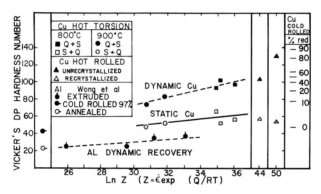

Fig. 63. The effect of temperature compensated strain rate during hot torsion on the room temperature properties of dynamically and metadynamically recrystallized copper. For comparison the hardnesses pertaining to the following treatments are included: dynamically recovered aluminum (Wong et al., 1967), hot rolled copper, both dynamically recovered and statically recrystallized, and cold rolled copper (McQueen and Bergerson, 1972).

rate of hot forming (Fig. 63) since that yields a finer substructure and grain size. Moreover, it can be seen that the hardness of dynamically recrystallized metal is intermediate between the hardness of hot-rolled dynamically recovered metal (points identified as hot rolled and unrecrystallized in Fig. 63) and that of statically recrystallized metal. Furthermore, it can be seen that the hot-worked material has a hardness similar to that found in material cold-worked 10 to 50%. Structures such as these are, unfortunately, unstable at elevated service temperatures.

In copper–aluminum alloys, the room temperature hardness has been

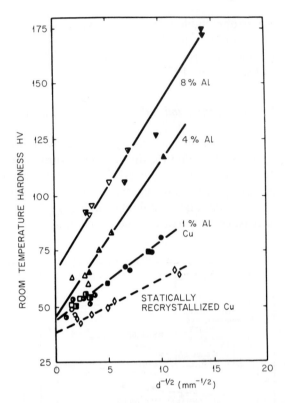

Fig. 64. The dependence of the room temperature hardness on the dynamically recrystallized grain size follows a Hall–Petch type of relationship. The increase in substructure density with decrease in grain size is responsible for the slope for the dynamically recrystallized copper being higher than that for statically recrystallized copper (Bromley and Sellars, 1973).

observed to vary inversely with the square root of the dynamically recrystallized grain size (Fig. 64). The broken line in the diagram shows the Hall–Petch relation applicable to substructure-free copper. Comparison of the first full plot, for Cu and Cu–1% Al, with that for the statically recrystallized material, shows that, for the same grain size, the dynamically recrystallized samples are stronger. This is to be expected from the presence of the subgrains in the latter material. It is also of interest to note that the slope or boundary strength coefficient is about $1\frac{1}{2}$ times higher for the dynamically than for the statically recrystallized samples. This can be attributed to the higher exponent applicable to subgrains [Eq. (20)] than to grains, and to the consequent more rapid increase in subboundary density than grain boundary density that accompanies a decrease in grain size.

E. *Static Recrystallization and Controlled Rolling*

In many hot-working operations, recrystallization takes place before the product can be cooled to room temperature. Such recrystallization eliminates the substructure and removes that source of strengthening. While this would not be beneficial if the product were to be put into service without further processing, it is useful if the metal is to undergo cold deformation. Thus, even if the substructure is not to be retained, the cooling rate following working is important in controlling grain size for both strength and uniformity of cold deformation.

The effects of hot working and of the rate of cooling on the characteristics of recrystallization are utilized in the process known as controlled rolling, which is applied mainly to steels (Irvine *et al.*, 1970; Tegart, 1971). The objective is a fine, uniform ferrite grain size which has a high yield strength and a low transition temperature, and which is produced by rapid transformation from fine grained austenite (Fig. 65). The latter is produced by static recrystallization from austenite which has been deformed up to 50% at a

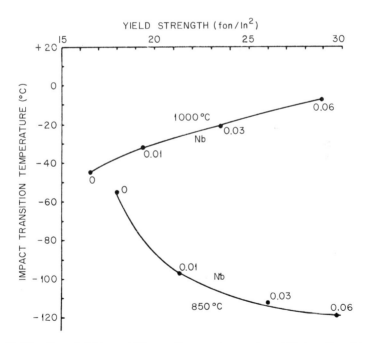

Fig. 65. The effect of niobium addition on the room temperature yield strength and impact transition temperature of a 0.04% C, 6.5% Mn steel. The improvement resulting from a lower rolling temperature is also shown (Jones and Rothwell, 1968).

finishing temperature low enough so that there is very little grain growth during run-out from the rolling mill. If the total strain in the relevant passes is much greater than 50%, there is a strong possibility that the austenite will recrystallize dynamically during deformation. However, there is also the possibility that the slab or billet will recrystallize statically while cooling to the proper finishing temperature before undergoing the final passes. Such a softening cycle can coarsen the final grain size or lead to undesirable mixed grain sizes. To slow down the recrystallization of the austenite so that the process is easier to control, small amounts of Nb are added to the steel. This has the further advantage that higher strains can be attained without dynamic recrystallization, thus increasing the number of nuclei (Jones and Rothwell, 1968). In the case where NbC inhibits static recrystallization, the elongated austenite grains give rise to ferrite grains approximately equal to their shortest dimension since nucleation is mainly at the boundaries.

A detailed review of the principles of controlled rolling is clearly beyond the limits of the present chapter. It should, however, be apparent that an understanding of the factors affecting controlled rolling requires some insight into the dynamic and static softening processes which have been described above.

VII. Summary

When metals are deformed under both hot working and creep conditions, strain hardening is counterbalanced by the concurrent softening processes of dynamic recovery and recrystallization. During dynamic recovery, mechanisms such as cross-slip, climb, and node unpinning permit the dislocations to unravel from hardening networks and annihilate each other. During dynamic recrystallization, new grains nucleate and grow; they deform as they grow, however, with the result that recrystallization takes place again and again. During dynamic recrystallization, dynamic recovery is occurring in both the old and the new grains. The characteristics of the two dynamic softening mechanisms are summarized and compared below.

A. The capacity of a metal for dynamic recovery diminishes with decrease in stacking fault energy as the dislocations become more extended. Dynamic recrystallization occurs in a metal if the degree of dynamic recovery is insufficient to maintain the substructure density below the critical level for nucleation. The characteristics of different metals and alloys are summarized in Table I.

TABLE I

RESTORATION PROCESSES IN DEFORMED METALS

Metals	Cold-worked structure	High temperature softening process		
		During annealing	During deformation	During holding
		Static	Dynamic	Static
Al α-Fe Ferritic alloys Bcc refractory metals Zr alloys Hcp metals	Well developed cell structure	Recovery followed by recrystallization (nucleation by coalescence at high strains)	Recovery even to high strains	Recovery; Recrystallization relatively slow, easily prevented by cooling after working
Ni Ni-base superalloys γ-Fe Austenitic alloys Cu Brass	Cell structure alters gradually to uniform distribution of dislocations with decrease in stacking fault energy	Limited recovery followed by recrystallization (nucleation by g.b. bulging or from high local concentration of dislocations)	Hot working at small strains Creep at normal strain rates	
			Recovery:	Recovery; followed by rapid static recrystallization
			Hot working at large strains	
			Recovery and Recrystallization:	Recovery and Metadynamic recrystallization followed by static recrystallization

B. Both dynamic processes lead to flow curves with steady state regions. With dynamic recrystallization, however, the steady state flow stress is reached only after passing through a maximum; recrystallization is initiated at this point, producing softening with further strain. At low values of temperature-corrected strain rate, dynamic recrystallization leads to periodic flow curves.

C. When dynamic recovery is taking place, the grains become elongated in the direction of flow; meanwhile, the substructure undergoes a process called repolygonization, which keeps it dimensionally stable and equiaxed. Under conditions of dynamic recrystallization, new grains continually replace the most work-hardened ones, maintaining a fine equiaxed grain structure, with a distribution of degrees of deformation.

D. Increases in the temperature and decreases in the strain rate of deformation result in larger, more perfect, subgrains or grains during dynamic recovery or recrystallization, respectively.

E. Decreases in temperature and increases in strain rate increase the strains for the nucleation and for the completion of recrystallization. As the latter becomes greater than the former, the periodic cycles of softening begin to overlap, and recrystallization becomes continuous.

F. The yield stress under hot-working conditions is highly sensitive to strain rate and temperature. In this respect, it differs from the essentially athermal yields observed at intermediate temperatures. The strain rate and temperature dependence of the steady state stress is similar to that of the yield stress, indicating that similar mechanisms are rate controlling.

G. The steady state stresses observed at high temperatures are about twice the yield stresses. The increase in flow stress after yielding can be attributed to an increase in internal stress, which is in turn proportional to both the subboundary density and the square root of the dislocation density.

H. The subgrain size developed under conditions of both dynamic recovery and recrystallization is inversely related to the steady state stress. The similar correlation between dynamically recrystallized grain size and flow stress arises from the dependence of nucleus density on subgrain density.

I. The ductility is increased many times over that at room temperature by the occurrence of either high temperature dynamic recovery or recrystallization. A high level of dynamic recovery reduces intergranular cracking through the formation of bulges in the grain boundaries which restrict sliding. The lower flow stresses brought about by enhanced dynamic recovery also permit easier accommodation at triple points, reducing crack nucleation. When dynamic recrystallization occurs, it prevents the propagation of intergranular cracks by continually displacing the grain boundaries from the ends of the cracks.

J. Solute additions usually retard dynamic recovery, especially if they cause further extension of the dislocations. This in turn tends to increase the stored energy, and therefore to promote dynamic recrystallization. However, solute atoms also retard nucleation and slow down grain boundary migration; as a result, higher deformation temperatures are required for the attainment of steady state flow. In many alloys, the occurrence of grain boundary sliding without concurrent dynamic recrystallization, at intermediate temperatures, leads to unimpeded grain boundary cracking with consequent reduction in ductility.

K. The presence of fine second phase particles usually stabilizes the substructure, thus raising the steady state flow stress and delaying dynamic recrystallization. Larger second phase particles frequently cause a considerable decrease in ductility in metals that recrystallize dynamically in so far as they pin the grain boundaries, permitting the propagation of intergranular cracks.

L. The presence of a dynamically recovered or dynamically recrystallized structure at the commencement of a further stage of deformation can have a considerable effect on further flow. The inherited structure may lead to initially higher or lower flow stresses depending on whether it is harder or softer than the equilibrium structure for the new stage of deformation. Nevertheless, at high strains, the equilibrium structure and flow stress are independent of the initial structure.

M. As soon as deformation is completed or interrupted, static recovery and recrystallization take place, following the dynamic softening processes. A high degree of dynamic recovery delays and may prevent static recrystallization. Dynamic recrystallization provides nuclei for metadynamic (postdynamic) recrystallization, which as a result does not have an incubation period and takes place concurrently with static recovery. The rates of all three static softening processes increase with the temperature, strain rate, and strain of the prior deformation.

N. The substructures produced by dynamic recovery and recrystallization, when retained to room temperature, lead to yield stresses higher than those of annealed materials. The yield strength of worked metals is proportional to the inverse subgrain size and to the inverse square root of the grain size.

While a general description of the dynamic softening processes has been presented in this chapter, for brevity, many details have been omitted. For those who wish to examine the experimental evidence in greater depth, the following reviews are recommended: Stüwe and Drube (1967), Tegart (1968), Jonas et al. (1969), Sellars and Tegart (1972), Fulop and McQueen (1972) (Ni-base superalloys), McElroy and Szkopiak (1972), Jonas and McQueen (1973), and Rossard (1973).

Those interested in experimental techniques should refer to the following publications: Rossard and Blain (1962), Henning and Boulger (1964), Nicholson (1964), Rossard (1968), Young and Sherby (1969), Weiss *et al.* (1970), and McQueen and Jonas (1971).

ACKNOWLEDGMENTS

The authors wish to thank Atomic Energy of Canada Limited, the National Research Council of Canada, and the Defence Research Board of Canada for the financial support of many of the investigations described in this chapter.

References

Abson, D. J., and Jonas, J. J. (1970). *Metal Sci. J.* **4**, 24–28.
Abson, D. J., and Jonas, J. J. (1972). *J. Nucl. Mater.* **42**, 73–85.
Ashton, R. F. (1969). *J. Metals* **22**, No. 9, 35–38.
Auld, J. H., Garrod, R. I., and Thomson, T. R. (1957). *Acta Met.* **5**, 741–746.
Bailey, J. E. (1963). *In* "Electron Microscopy and Strength of Crystals" (G. Thomas and J. Washburn, eds.), pp. 535–574. Wiley (Interscience), New York.
Bailey, R. E. (1969). "Microstructural Examination of Waspaloy and Al-718 Gleeble Hot Workability Test Specimens," Rep. SP-69-9. Allegheny Ludlum Steel Research Center, Brackenridge, Pennsylvania.
Ball, C. J. (1957). *Phil. Mag.* **2**, 1011–1017.
Bird, J. E., Mukherjee, A. K., and Dorn, J. E. (1969). *In* "Quantitative Relation Between Properties and Microstructure" (D. G. Brandon and A. Rosen, eds.), pp. 255–342. Isr. Program Sci. Transl., Jerusalem.
Bourelier, F., and Montuelle, J. (1968). *Mem. Sci. Rev. Met.* **65**, 65–76.
Bromley, R. (1969). Ph.D. Thesis, Sheffield University, Sheffield, England.
Bromley, R., and Sellars, C. M. (1973). *Proc. Int. Conf. Strength Metals Alloys, 3rd, 1973*, Vol. 1, pp. 380–385.
Buhler, H., Bobbert, D., and Rose, A. (1970). *Stahl Eisen* **90**, 21–28.
Byrne, J. G. (1965). "Recovery, Recrystallization and Grain Growth." Macmillan, New York.
Cahn, R. W. (1965). *In* "Physical Metallurgy" (R. W. Cahn, ed.), pp. 925–987. Wiley, New York.
Chojnowski, E. A., and Tegart, W. J. McG. (1968). *Metal Sci. J.* **2**, 14–18.
Clauer, A. H., and Wilcox, B. A. (1967). *Metal Sci. J.* **1**, 86–90.
Cotner, J. R., and Tegart, W. J. McG. (1969). *J. Inst. Metals* **97**, 73–79.
Cremisio, R. C., Butler, H. M., and Radavich, J. F. (1969). *J. Metals* **21**, No. 11, 55–61.
Cuddy, L. J. (1970). *Met. Trans.* **1**, 395–401.
Demianczuk, D. W. (1967). M.Eng. Thesis, McGill University, Montreal, Canada.
Dieter, G. E., Mullin, J. V., and Shapiro, E. (1968). "Deformation Under Hot Working Conditions," pp. 7–13. Iron Steel Inst., London.
Djaic, R. A. P., and Jonas, J. J. (1972). *J. Iron Steel Inst.* **210**, 256–261.
Djaic, R. A. P., and Jonas, J. J. (1973). *Met. Trans.* **4**, 621–624.
Drube, B., and Stüwe, H. P. (1967). *Z. Metallk.* **58**, 799–804.
English, A. T., and Backofen, W. A. (1964). *Trans. AIME* **230**, 396–407.
Evans, R. W., and Dunstan, G. R. (1971). *J. Inst. Metals* **99**, 4–14.
Exell, S. F., and Warrington, D. H. (1972). *Phil. Mag.* [8] **26**, 1121–1136.

Farag, M. M., Sellars, C. M., and Tegart, W. J. McG. (1968). "Deformation Under Hot Working Conditions," pp. 60–67. Iron Steel Inst., London.

Fink, K., Lueg, W., and Burger, G. (1955). *Arch. Eisenhuettenw.* **26**, 665–688.

Fulop, S., and McQueen, H. J. (1972). "Superalloys: Processing," pp. H1–H21. Metals and Ceramics Information Center, Columbus, Ohio.

Garofalo, F., von Gemmingen, F., and Domis, W. F. (1961). *ASM (Amer. Soc. Metals) Trans. Quart.* **54**, 430–443.

Garofalo, F., Richmond, C., Domis, W. F., and von Gemmingen, F. (1963). *Joint Int. Conf. Creep, 1963* pp. I31–I39.

Gittins, A. (1970). "Fracture of Metals Under Hot Working Conditions," MRL 29/1. BHP Melbourne Res. Lab., Australia.

Gittins, A., Hinton, L. G., and Tegart, W. J. McG. (1972). "Ductility of Steels in Hot Working," MRL S3/10. BHP Melbourne Res. Lab., Australia.

Glover, G. (1969). Ph.D. Thesis, Sheffield University, Sheffield, England.

Glover, G., and Sellars, C. M. (1972). *Met. Trans.* **3**, 2271–2280.

Glover, G., and Sellars, C. M. (1973). *Met. Trans.* **4**, 765–775.

Gueussier, A., and Castro, R. (1958). *Rev. Met. (Paris)* **55**, 1023–1040.

Hardwick, D., and Tegart, W. J. McG. (1961). *J. Inst. Metals* **90**, 17–20.

Hardwick, D., Sellars, C. M., and Tegart, W. J. McG. (1961). *J. Inst. Metals* **90**, 21–22.

Henning, H. J., and Boulger, F. W. (1964). "Mechanical Working of Steel," Vol. 1, pp. 107–125. Gordon & Breach, New York.

Héritier, B., Luton, M. J., and Jonas, J. J. (1974). *Metal Sci.* **8**, 41–48.

Hirst, H. (1940). *Proc. Australas. Inst. Min. Met.* **120**, 777–806.

Hockett, J. E., and McQueen, H. J. (1970). *Proc. Int. Conf. Strength Metals Alloys, 2nd, 1970* pp. 991–995.

Holt, D. L. (1970). *J. Appl. Phys.* **41**, 3197–3201.

Hu, H. (1963). *In* "Recovery and Recrystallization of Metals" (L. Himmel, ed.), pp. 311–378. Wiley (Interscience), New York.

Humphreys, F. J., and Martin, J. W. (1968). *Phil. Mag.* [8] **17**, 365–380.

Immarigeon, J.-P. A. (1974). Ph.D. Thesis, McGill University, Montreal, Canada.

Immarigeon, J.-P. A., and Jonas, J. J. (1971). *Acta Met.* **19**, 1053–1061.

Immarigeon, J.-P. A., and Jonas, J. J. (1974). *Acta Met.* **22**, 1235–1247.

Irani, J. J., and Taylor, P. R. (1968). "Deformation Under Hot Working Conditions," pp. 83–96. Iron Steel Inst., London.

Irvine, K. J., Gladman, T., and Pickering, F. B. (1969). *J. Iron Steel Inst.* **207**, 1017–1028.

Irvine, K. J., Gladman, T., Orr, J., and Pickering, F. B. (1970). *J. Iron Steel Inst.* **208**, 717–726.

Ivanova, V. S., and Gordienko, L. K. (1968). "New Ways of Increasing the Strength of Metals," pp. 13–35. Iron Steel Inst., London.

Jacquerie, J. M., and Habraken, L. (1968). *Cobalt* **38**, 13–19.

Johari, O., and Thomas, G. (1965). *ASM (Amer. Soc. Metals) Trans. Quart.* **58**, 563–578.

Jonas, J. J. (1973). *J. Sci. Ind. Res.* **32**, 528–535.

Jonas, J. J., and Chandra, T. (1971). "Metal Forming: Interrelation Between Theory and Practice," pp. 115–129. Plenum, New York.

Jonas, J. J., and Luton, M. J. (1975). "Advances in Deformation Processing" Sagamore Conference 21st, 1974, Syracuse University Press (in press).

Jonas, J. J., and McQueen, H. J. (1973). *Trans. Soc. Mfr. Eng.* **2**, 209–233.

Jonas, J. J., McQueen, H. J., and Wong, W. A. (1968a). "Deformation Under Hot Working Conditions," pp. 49–59. Iron Steel Inst., London.

Jonas, J. J., Axelrad, D. R., and Uvira, J. L. (1968b). *Trans. Jap. Inst. Metals* **9**, Suppl., 257–267.

Jonas, J. J., Sellars, C. M., and Tegart, W. J. McG. (1969). *Met. Rev.* **14**, 1–24.

Jones, J. D., and Rothwell, A. B. (1968). "Deformation Under Hot Working Conditions," pp. 78–82. Iron Steel Inst., London.

Keane, D. M., Sellars, C. M., and Tegart, W. J. McG. (1968). "Deformation Under Hot Working Conditions," pp. 21–28. Iron Steel Inst., London.

Kear, B. H., Oblak, J. M., and Owczarski, W. A. (1972). *J. Metals* **24**, No. 6, 25–32.

Keegan, J. W. (1967). *Metal Progr.* **92**, No. 2, 31–33.

Kivilahti, J. K., Lindroos, V. K., and Lehtinen, B. (1974). *In* "High Voltage Electron Microscopy" (P. R. Swann *et al.*, eds.), p. 195. Academic Press, New York.

Kosik, O. (1970). M.Eng. Thesis, McGill University, Montreal, Canada.

Kosik, O., Abson, D. J., and Jonas, J. J. (1971). *J. Iron Steel Inst.* **209**, 624–671.

Kuhlman-Wilsdorf, D. (1966). *In* "Work Hardening" (J. P. Hirth and J. Weertman, eds.), pp. 97–139. Gordon & Breach, New York.

Le Bon, A., Rofes-Vernis, J., and Rossard, C. (1973). *Mem. Sci. Rev. Met.* **70**, 577–588.

Leslie, W. C., Michalak, J. T., and Aul, F. W. (1963). *In* "Iron and its Dilute Solid Solutions" (C. W. Spencer and F. E. Werner, eds.), pp. 119–216. Wiley (Interscience), New York.

Loizou, N., and Sims, R. B. (1953). *J. Mech. Phys. Solids* **1**, 234–243.

Luton, M. J., and Jonas, J. J. (1970). *Proc. Int. Conf. Strength Metals Alloys, 2nd, 1970* pp. 1100–1105.

Luton, M. J., and Jonas, J. J. (1972a). *Can. Met. Quart.* **11**, 79–90.

Luton, M. J., and Jonas, J. J. (1972b). *Proc. Inter-Amer. Conf. Mater. Technol., 3rd, 1972* pp. 910–917.

Luton, M. J., and Sellars, C. M. (1969). *Acta Met.* **17**, 1033–1043.

Luton, M. J., and Tegart, W. J. McG. (1969). *Metal Sci. J.* **3**, 142–146.

McElroy, R. J., and Szkopiak, Z. C. (1972). *Int. Met. Rev.* **17**, 175–202.

McQueen, H. J. (1968). *Trans. Jap. Inst. Metals* **9**, Suppl., 170–177.

McQueen, H. J., and Bergerson, S. (1972). *Metal Sci. J.* **6**, 25–29.

McQueen, H. J., and Hockett, J. E. (1970). *Met. Trans.* **1**, 2997–3004.

McQueen, H. J., and Jonas, J. J. (1971). "Metal Forming: Interrelation Between Theory and Practice," pp. 393–428. Plenum, New York.

McQueen, H. J., Wong, W. A., and Jonas, J. J. (1967). *Can. J. Phys.* **45**, 1225–1235.

Morrison, W. B. (1972). *J. Iron Steel Inst.* **210**, 618–623.

Müller, T. L. F. (1967). Ph.D. Thesis, Sheffield University, Sheffield, England.

Nicholson, A. (1964). *Iron Steel* (*London*) **37**, 290–294 and 363–367.

Oblak, J. M., and Owczarski, W. A. (1972). *Met. Trans.* **3**, 617–626.

Perryman, E. C. W. (1956). "Creep and Recovery," pp. 111–145. Amer. Soc. Metals, Metals Park, Ohio.

Petkovic, R. A. (1975). Ph.D. Thesis, McGill University, Montreal, Canada.

Petkovic, R. A., Luton, M. J., and Jonas, J. J. (1975). *Can. Met. Quart.* **14** (in press).

Redfern, G. A., and Sellars, C. M. (1968). "Deformation Under Hot Working Conditions," pp. 29–37. Iron Steel Inst., London.

Reynolds, K. A. (1968). "Deformation Under Hot Working Conditions," pp. 107–116. Iron Steel Inst., London.

Reynolds, R. A., and Tegart, W. J. McG. (1962). *J. Iron Steel Inst.* **200**, 1044–1059.

Rhines, F. N., and Wray, P. J. (1961). *ASM* (*Amer. Soc. Metals*) *Trans. Quart.* **54**, 117–128.

Richardson, G. J., Sellars, C. M., and Tegart, W. J. McG. (1966). *Acta Met.* **14**, 1225–1236.

Robbins, J. L., Shepard, O. G., and Sherby, O. D. (1967). *ASM* (*Amer. Soc. Metals*) *Trans. Quart.* **60**, 205–216.

Rossard, C. (1960). *Metaux* **35**, 102–115, 140–153, and 190–205.

Rossard, C. (1968). *Rev. Met.* **65**, 181–193.

Rossard, C. (1973). *Proc. Int. Conf. Strength Metals Alloys, 3rd, 1973* Vol. II, pp. 175–203.

Rossard, C., and Blain, P. (1962). "Flat Rolled Products," Vol. III, pp. 3–28. Gordon & Breach, New York.

Sabroff, A. M., Boulger, F. W., and Henning, H. J. (1968). "Forging Materials and Practice." Van Nostrand-Reinhold, Princeton, New Jersey.

Sah, J. P., Richardson, G. J., and Sellars, C. M. (1969). *J. Aust. Inst. Metals* **15**, 292–297.

Sandstrom, R., and Lagneborg, R. (1975). *Scripta Met.* **9** (in press).

Sastry, D. H., Luton, M. J., and Jonas, J. J. (1974). *Phil. Mag.* [8] **30**, 115–127.

Schey, J. (1957). *Acta Tech. (Budapest)* **16**, 131–152.

Schmatz, D. J., Schaller, F. W., and Zackay, V. F. (1963). "The Relation Between the Structure and Mechanical Properties of Metals," pp. 613–645. HM Stationery Office, London.

Sellars, C. M. (1969). *J. Sheffield Univ. Met. Soc.* **8**, 43–47.

Sellars, C. M., and Tegart, W. J. McG. (1966a). *Acta Met.* **14**, 1136–1138.

Sellars, C. M., and Tegart, W. J. McG. (1966b). *Mem. Sci. Rev. Met.* **63**, 731–746.

Sellars, C. M., and Tegart, W. J. McG. (1972). *Int. Met. Rev.* **17**, 1–24.

Shapiro, E., and Dieter, G. E. (1971). *Met. Trans.* **2**, 1385–1391.

Sherby, O. D., Harrigan, M. J., Chamagne, L., and Sauve, C. (1969). *ASM (Amer. Soc. Metals) Trans. Quart.* **62**, 575–580.

Sokolkov, Ye. N., and Surkov, Yu. P. (1963). *Phys. Metals Metallogr. (USSR)* **16**, No. 6, 107–109.

Stüwe, H. P., and Drube, B. (1967). *Z. Metallk.* **58**, 499–506.

Sunter, B. J., and Burman, N. M. (1972). *J. Aust. Inst. Metals* **17**, No. 2, 91–100.

Swann, P. R. (1963). *In* "Electron Microscopy and Strength of Crystals" (G. Thomas and J. Washburn, eds.), pp. 131–182. Wiley (Interscience), New York.

Tegart, W. J. McG. (1968). "Ductility," pp. 133–177. Amer. Soc. Metals, Metals Park, Ohio.

Tegart, W. J. McG. (1971). *Metals Aust.* **3**, No. 1, 3–8.

Thornton, P. H., and Cahn, R. W. (1961). *J. Inst. Metals* **89**, 455–463.

Uvira, J. L. (1967). Ph.D. Thesis, McGill University, Montreal, Canada.

Vandermeer, R. A., and Gordon, P. (1963). *In* "Recovery and Recrystallization of Metals" (L. Himmel, ed.), pp. 211–240. Wiley (Interscience), New York.

Vaughan, T. B. (1968). "Deformation Under Hot Working Conditions," pp. 68–78. Iron Steel Inst., London.

Ward, D. M., and Helliwell, B. J. (1970). *J. Inst. Metals* **98**, 199–203.

Warrington, D. H. (1963). *J. Iron Steel Inst.* **201**, 610–613.

Webster, D., Crooks, D. D., and Vidoz, A. E. (1973). *Met. Trans.* **4**, 2841–2847.

Weiss, B., Grotke, G. E., and Stickler, R. (1970). *Weld. Res., Suppl.* **35**, 471s–487s.

Weiss, H., Gittins, A., Brown, G. G., and Tegart, W. J. McG. (1972). "Recrystallization of Steel in the Austenite Range," MRL 41/1. BHP Melbourne Research Lab., Australia.

Weissmann, S., Imura, T., and Hosokawa, N. (1963). *In* "Recovery and Recrystallization of Metals" (L. Himmel, ed.), pp. 241–267. Wiley (Interscience), New York.

White, F. E., and Rossard, C. (1968). "Deformation Under Hot Working Conditions," pp. 14–20. Iron Steel Inst., London.

Wong, W. A., McQueen, H. J., and Jonas, J. J. (1967). *J. Inst. Metals* **95**, 129–137.

Wood, W. A., and Rachinger, W. A. (1949). *J. Inst. Metals* **76**, 237–253.

Wood, W. A., and Suiter, J. W. (1951). *J. Inst. Metals* **80**, 501–506.

Wusatowski, R. (1966). *J. Iron Steel Inst.* **204**, 727–736.

Young, C. M., and Sherby, O. D. (1969). "Simulation of Hot Forming Operations by Means of Hot Torsion," AFML-TR-69-294. Stanford University, Palo Alto, California.

Subject Index